THE ENGINEER'S
CAREER GUIDE

THE ENGINEER'S CAREER GUIDE

JOHN A. HOSCHETTE

WILEY

A JOHN WILEY & SONS, INC., PUBLICATION

For general information on our other products and services or technical support, please contact our Customer Care Department within the United States at (800) 762-2974, outside the United States at (317) 572-3993 or fax (317) 572-4002.

Wiley also publishes its books in a variety of electronic formats. Some content that appears in print may not be available in electronic formats. For more information about Wiley products, visit our web site at www.wiley.com.

Library of Congress Cataloging-in-Publication Data

Hoschette, John A., 1952-
    The engineer's career guide / John A. Hoschette.
        p. cm.
    ISBN 978-0-470-50350-8 (pbk.)
    1. Engineering–Vocational guidance. 2. Career development. I. Title.
    TA157.H6183 2009
    620.0023–dc22
                                                            2009014023

Printed in the United States of America

10 9 8 7 6 5 4 3 2 1

CONTENTS

PART 4 IMPROVING YOUR PERFORMANCE ON THE JOB AND STANDING OUT FROM YOUR PEERS

PART 10 BE A CONSULTING ENGINEER

PREFACE

The most often asked question from attendees of my career workshops is, "When are you going to write another book and include this new career material?" This feedback has provided the inspiration to write this second book, knowing the material is time tested, it works, and people are using the guidance to accelerate and enhance their careers.

In this book, I have improved and added significantly more career material over my previous book *Career Advancement and Survival for Engineers* published in 1994. The career tips and guidance in this book now span the entire life of an engineer. Helpful career advice is provided for a recently graduated engineer, a mid-life engineer, a senior engineer, and even for those engineers soon retiring.

This book makes an excellent career reference book for engineers to drawupon when they need advice on career planning, encounter career challenges, need to recover from a setback or job loss, want a promotion or raise, or want to know how to deal with difficult work situations, coworkers, and managers.

The book has been divided into 10 separate parts each covering a different aspect of an engineer's career. Part 1 of the book deals with successful career planning. How take control and develop a career plan that leads to success as well as career strategies, making a job change, having a career discussion with your boss or mentors, how to get out of dead-end jobs, careers that have reached a plateau, what new graduates should be doing to successfully transition to industry, and finally, retirement planning.

Part 2 addresses company structures, organizations, and barriers impacting your career. Here I discuss the possible career paths open to engineers when working for small companies or in large corporations and the special challenges engineers face in order to receive a promotion. In Part 3, determining the criteria by which you are measured and what it takes to get better performance ratings is presented.

Part 4 addresses improving your performance on the job and standing out from your peers. Here we explore the engineering process and provide tips for better designs, team skills, generating career advancing ideas, and key aspects for obtaining a raise or promotion. Part 5 addresses dealing with difficult people at work including your boss. Part 6 presents strategies and tips for finding a new job including resume and cover letter writing as well as interviewing tips.

Part 7 offers tips and guidelines for returning for further education as well as how to keep your skills updated through lifelong learning. Part 8 covers the basics of financial planning for engineers and Part 9 is about fast tracking. Finally, in Part 10, tips and advice are given for those who are considering becoming a consulting engineer.

I have also received very positive feedback on the material in this book from international engineering organizations. The material allows foreign engineers to better understand how engineers in the United States think and how US companies operate.

In addition, the response from the Human Resources departments has been very positive of my material since it helps them understand how engineers are thinking. It provides an excellent book for them to refer to when they need to help or provide career guidance for engineers.

My engineering career has spanned over 35 years working for world class organizations such as Northrop, Honeywell, Loral, Alliant Tech Systems, and Lockheed Martin. I have worked on the East coast, in the Midwest, and on the West coast. My experiences have been similar at all these companies and locations when it comes to career advancement for engineers. The material presented in this book is generic in nature and can be applied at any company or in any region of the United States as well as other professions. I have had a very rewarding and successful career both technically and in my relationships.

On the technical side of my career, my helmet-mounted sight and display designs are flying in our nation's helicopters. They allow our troops to protect this nation and have also been highlighted in Hollywood movies. I have worked on the ring laser gyroscopes and inertial navigators that are guiding the jet aircraft we travel on as well as weapons that protect our nation. I have worked with the nation's most brilliant scientists at MIT, Lincoln Labs, and Stanford, developing and transferring technology to industry. I have held the position of project lead for infrared (heat) camera systems that are currently helping our troops, Homeland Security, policemen, and firemen protect and save lives. I have worked on the THAAD missile system that is the only missile capable of flying both inside and outside the atmosphere. I have been asked to teach and lecture at several prominent universities.

On the people-side, I have been able to help thousands of engineers with their careers. I have given career guidance seminars in nearly every state and at major engineering universities. I have published over 60 papers in trade journals, newspapers, and conference proceedings. I have been a director of engineering with the responsibility of campus recruiting, interviewing and hiring, career development programs for employees, approving raises and promotions, layoffs, and conducting hundreds of employee evaluations. This experience has given me wonderful insight into both the business and human aspects of career advancement.

This book allows the reader to draw upon the career advice and tips that I have assembled over my lifetime with input from universities, company executives, successful engineers, and human resource departments. The book is designed to help the engineer deal with career issues throughout their engineering career lifetime. I hope my advice assists you in achieving the career you dream of and I welcome feedback from my readers; I may be reached at j.hoschette@ieee.org.

JOHN A. HOSCHETTE

Minneapolis, MN

ACKNOWLEDGMENTS

I would first like to thank my wife Linda for her unwavering encouragement and support in writing this book. She is the most loving, caring, and wonderful partner. She is a fantastic mother who has always made our family the top priority. She is amazingly talented and has the gift to inspire beyond belief, which I have drawn upon throughout our life together. It is because of her support, sacrifices, and inspiration that I once again have been able to write another book. Linda-Lue, thank you for supporting my career and my dream to help engineers with their careers.

I would like to thank my children, Tina and John, for their love and support as well as the many ideas and words of encouragement. My parents, Veronica and Vernon, who gave me the opportunity to attend college and told me I could be whatever I wanted to be. And thank you to Linda's parents, Florence and Roy, whom were always of great support.

I would like to thank all the senior engineers, managers, executives, and mentors who have provided me with career guidance throughout the years and in doing so, helped me develop some of the material for this book. Specifically, John Miller, Frank Ferrin, and George Hedges, who provided guidance in the early years of my career. Dr. Edwin Thiede and Dr. John Glish for the experience of lecturing with you both, throughout the United States and Europe. Professor Eziekel of MIT, for the valuable opportunity to work with you. Tom McGrath, Vice President of the THAAD program, and Dr. Vance Coffman, former CEO of Lockheed, for the opportunity to work on one of the most exciting missile programs of the century and for your personal guidance. Paul Kostek, former President of IEEE USA, for his support by sponsoring my classes over the years. The IEEE Society for allowing me to be an officer of the Career Maintenance, and Development Committee as well as sponsoring my workshops. The IEEE USA Executive Engineering Council for honoring me with one of their highest achievement awards, the National Citation of Honor Award. Professor Arthur Winston, former IEEE President and Director of the Gordon Institute at Tufts University, for allowing me to teach for ten years as part of this career development program and for his improvements to my material. For Santa

Clara University, Stanford, University of Arizona, Brown University, Georgia Tech, and the University of Minnesota for allowing me to teach as an adjunct professor in their programs and believing in the necessity for engineers to receive career training. I would like to thank Honeywell, Loral, Alliant Tech Systems, Lockheed Martin, Raytheon, Draper Labs, US Army, and Northrop, for the support of my career development workshops. I would like to also thank the engineering societies that have also sponsored my workshops: the Institute of Electrical and Electronics Engineers (IEEE), Society of Women Engineers (SWE), National Society of Professional Engineers (NSPE), American Society of Mechanical Engineers (ASME), Society of Manufacturing Engineers (SME), and the International Council on Systems Engineering (INCOSE).

I would also like to thank the Evans Scholars Foundation which allowed me the opportunity to become an engineer and attend graduate school.

J. A. H.

ABOUT THE AUTHOR

John Hoschette is a Technical Director with Lockheed Martin in Eagan, Minnesota working in the Tactical Avionics group. His work encompasses developing the next generation of super mission computers for the F-35, F-36, and F-22 jet fighter aircraft. His area of technical expertise is optical data networking. Over the past 34 years John's career has covered such areas as developing infrared sensors for night vision, laser sensors for weapons, helmet-mounted displays, and optical fiber channel networks.

John has experience in all aspects of an engineer's career. This experience covers designing, testing, and fielding of advanced electronics, as well as twenty years of managerial experience. In his management role he has been responsible for the career development of employees, employee appraisals, recruiting, hiring, campus recruiting, and downsizing. John has held an executive staff position at Lockheed Martin, Sunnyvale, California assisting in the career development and management of approximately 5,000 engineers at this campus. John is a certified Master Black Belt in Six Sigma. The previous Fortune 100 companies John was employed with include Honeywell, Northrop, Alliant Tech Systems, and Loral spanning from the East to West coast.

John holds BSEE and MSEE degrees from the University of Minnesota as well as a Business Administration Certificate. John has been an adjunct professor at Santa Clara University and the Gordon Institute of Tufts University for almost a decade. In addition he has been a guest speaker for many universities including University of Arizona, Brown University, Georgia Tech, University of Massachusetts—Amherst, University of Minnesota, Stanford University, and University of California—Berkeley. As a member of the National Speakers Association, John has presented the keynote address for engineering conventions such as Sensor's and societies such as IEEE, SWE, NSPE, SME.

John is a senior member of IEEE and has served the IEEE organization in various leadership positions. He has held such positions as the Vice Chair of the Career Maintenance and Development Committee under IEEE-USA. He is a strong supporter of the IEEE PACE organization and has conducted many workshops for them throughout the United States. He has notably received a Citation of Honor Award by IEEE-USA.

John has published over 60 articles dealing with engineering career development. Special feature article have appeared in *Today's Engineer* and *Spectrum Magazine*. John continues to consult for many companies, universities, and engineering organizations.

PART 1

SUCCESSFUL CAREER PLANNING

CHAPTER 1

SUCCESSFUL CAREERS START WITH YOU

Successful careers start with you. It's about your abilities to plan and control your career advancement. Do you have a career plan? Are you in control of your career? If you ask most people these questions, they usually respond with "yes," when, in fact, they have no plan and no control. They proceed along, day after day, assuming their next promotion is just around the corner but when nothing happens, they don't understand why. "I deserved it," they say, "Why does everyone else get promoted and not me?" "Why did they get the promotion and I'm still waiting?" The answer is simple; they do not have a good plan and they were not in control of their career.

Controlling your career takes a great deal of planning, hard work, and the ability to manage circumstances [1–3]. Knowing the criteria it takes to get promoted in your company and constantly working toward meeting and exceeding the criteria is essential to achieve career advancement.

Often, there are things going on behind the scenes of which you must be aware. As shown in Figure 1-1, there are multiple factors that could be limiting your career growth. Key among these are obsolescence, your personality, education level, your supervisor, and maybe even the company or department structure (just to name a few).

Always knowing what is happening and managing circumstances to the best of your ability is essential to reduce the time frame between raises and promotions. Are you in control as you should be? To illustrate the point, answer the questions below.

1. Was your last pay increase smaller than expected or your performance rating lower than expected?
2. Do you feel your hard work often goes unrewarded? Does it seem like other people are always getting awards or recognition?

The Engineer's Career Guide. By John A. Hoschette
Copyright © 2010 John Wiley & Sons, Inc.

FIGURE 1-1   What is stopping your career growth?

3. Have you ever been passed over for a promotion and don't understand why?

4. Do you feel like you are stuck in a dead-end, thankless job with little hope for advancement?

If you answered yes to any of these questions, it reveals you are not in as much control as you think you are. If you wait until your job review with your supervisor to discuss your career plans, it is too late.

▶ ***Career Tip.*** You must accept responsibility for your career and take control; no one is going to do it for you. You must make things happen for yourself!

Do you need to be in control at all times? Yes! Yes! Yes! If you are not in control and planning your next career advancement, it means you need to start planning and stop letting others control your advancement.

> Failing to Plan
> is Simply Planning to Fail.

Not planning your career will most likely result in no advancement and possibly opening the door to failure. If you have no career plans and have not identified specific goals, where are you going to be a year from now? The simple answer is, doing exactly what you are doing now. Most promotions take one year once you start planning. So if you want a promotion next year, you better start planning now.

> Who has the majority of responsibility for making
> your career growth happen? You do!

The perceived share of responsibility for the engineer's growth is shown in Figure 1-2. As shown in the top portion of Figure 1-2, most people mistakenly

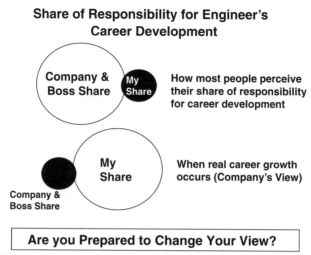

FIGURE 1-2 Share of responsibility for career development.

believe the company has the major responsibility for career development of the engineer. They sit back and expect the "company" to have a career plan for them and offer advances when the appropriate time comes. They think they only need to show up at work and complete the assignments given to them.

Real career growth generally occurs when engineers accept the responsibility for their own career development. This is shown in the lower portion of Figure 1-2, where the large circle surrounds the engineer and the smaller circle represents the company's share. In many situations, the company only provides the opportunity for advancements but it is the engineer's responsibility to make a promotion happen. Real career growth starts when engineers realize the responsibility for career advancement ultimately rests with themselves and not the company.

As Caela Farren points out in the book *Who's Running Your Career*, many factors may be trying to control it but ultimately the individual must develop their own career plan [3]. My experience and training confirms this to ultimately be true also. Take positive career advancing actions and be proactive in demonstrating to the company that you are ready for advancement.

Getting control of your career starts with you. There is no one going to make it happen for you. It starts with having a positive attitude. With a positive attitude you can make changes and better your career. Experience has shown that people with a positive attitude advance and those with a poor attitude stand still or digress.

Success in your career comes with the thoughts and actions of "I can." Failure usually results from the attitude of "I can't." Few people are promoted or advance when they are constantly responding to problems with "I can't" or "It will never work." Companies and CEOs promote people with positive and proactive attitudes who think, "I can" or "We can find a way to make it work."

Success comes in   "I Can!"
Failure comes in   "I Can't!"

It is going to take hard work and effort on your part and you are going to need to be determined and strive for the best all the time. Can you do it? Of course you can. This reminds me of an inspirational quote from Robert Schuller that has inspired me and I know it has inspired others as well.

Great People
Are
Ordinary People
Who Simply Do Not Quit!

This saying summarizes the basic component of successful career advancement. Engineers who ultimately reach goals do so simply because they never gave up. They may have changed their approach but they never quit. I have seen engineers bounce back from failure and achieve even greater accomplishments.

In addition to taking control, you must have a strategy for career advancement. Before you develop a strategy, you must understand how to play the game. You must know what, when, and how to do it.

THE BEST STRATEGY FOR CAREER ADVANCEMENT

Your fellow employees will tell you simple things you can do to get promoted: flatter the boss; it's who you know; work overtime and be a hero. The list goes on and on. In reality, most people have not done their homework and can only guess at what actually works. You must develop a strategic career plan based on how the game is played. In order to be promoted you must know how the promotion game is played and how to score points. The following example illustrates this point (Figure 1-3).

FIGURE 1-3   Learn the rules of the game.

FIGURE 1-4 Learn how your company keeps score.

Let's assume you are the key player of the team and everyone is counting on you. The situation is this, it is the second period of the third quarter, the teams have lined up, and the goalie is calling the play. On the previous play the putt was good for three points, but the right wing was penalized when the fifth base was stolen. They have the option to bowl for a strike or fast pitch for a slam dunk. What do you recommend?

Everything sounds somewhat familiar, right? But you really don't know the rules, the players, or how the points are scored. Therefore, your chances of making the right decisions are very slim.

If you don't know the rules of the game, how to keep score, or who the players are, how can you expect to play the game?

Obtaining your next career advancement is similar; you are the key player for this game. The game sounds very familiar; do a good job, flatter your supervisor, and work hard. These all sound like great things to do, but will they lead to career advancement?

If you don't know how your company promotes people or how they keep score, how can you expect to get ahead?

The bottom line is this: if you don't understand the game, you can't expect to score points or do well. Similarly, if you do not know how your company plays the career advancement game, you cannot expect to get ahead. Learn the game, learn how your company keeps score, and start calling the plays that will score your career advancement (Figure 1-4).

This book will provide you with career tools and guidelines on how to play the game and score points. It will identify how to determine the obstacles limiting your career advancement and what you need to do to overcome them.

DEVELOPING A STRATEGIC CAREER PLAN

Developing a great strategic career plan takes time and is an iterative process [4]. A simplified process for implementing a strategic career plan is shown in Figure 1-5. The first step is for the engineer to develop a strategic career plan. Next, the situations at work need to be analyzed to determine the best career actions. To analyze the work situations, career tools shared in this

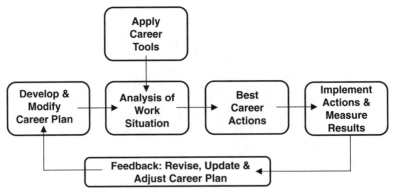

FIGURE 1-5   Process for implementing a career plan.

book need to be applied. These tools allow the engineer to identify what is happening behind the scenes and determine the best potential actions. After careful research and analysis, the best career actions should be selected and implemented. Next, the engineer needs to measure the success of the actions. If the actions are successful, keep going with the plan. If, however, the actions are not successful, the plan needs to be modified. The engineer needs to identify changes and make revisions to the career plan as necessary or adjust the career plan.

The process in Figure 1-5 is only a process; it does not identify the specific situations that need to be analyzed. The situations that need to be analyzed are shown in Figure 1-6 in pictorial form. These situations are shown as switches in a simple electrical circuit. The engineer needs to turn on the light of career advancement. They need to close ALL the switches and get the circuit operating to obtain career advancement. In the analogy, closing the switch

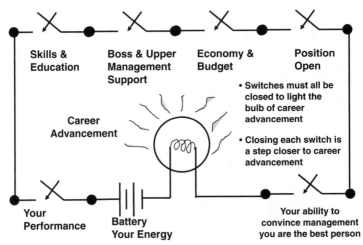

FIGURE 1-6   You must close all the switches to achieve career advancement.

is like controlling the work situation and getting the outcome desired per the strategic career plan.

The battery in the circuit represents the engineer's energy and drive. If the engineer has insufficient drive and energy to make things happen, it is like a dead battery.

However, simply having energy or one or two switches closed is not enough. The engineer must have energy and control ALL situations, or close all switches to turn on the light or get career advancement.

The first switch is the engineers' performance. You must have top performance ratings. Second, you must have the skills and education required for career advancement. Even having a PhD or advanced degree is no guarantee of success as Peter Feibelman points out in his book, *A PhD Is Not Enough: A Guide To Survival In Science* [5]. The third switch is the support of upper management, their backing, and belief you are ready for advancement. The following switches are concerned with the economy, profitability of your company, and a position available for you to move into. Finally, the last switch you need to control is the ability to demonstrate to management you are the best choice.

The chapters in this book have been arranged to show you actions you need to take for real career advancement. They will identify career tools to apply to a multitude of work situations and help you analyze the various situations in order to identify the optimum actions to take.

I leave the definition of career advancement up to you to define in your personal strategic career plan. Career advancement means different things to different people; to some people it means a promotion, to others it means a pay raise. And still to others, it means more challenging assignments or more responsibility. Defining what "Career Advancement" means to you and taking the actions necessary to make it happen is the essential ingredient for success.

Career Advancement = Promotions, Pay Raises, Better Assignments

Following the techniques presented in this book and applying the tools will shorten the time to your next career advancement. Remember, your next promotion is at least a year away from the time you start planning.

Plans Lead to Actions
Actions Lead to Advancement

As you read through the book and study the examples, I hope you will begin to accumulate actions you need to take for your particular career situation. To help you collect and organize these ideas and actions, I have created a notes section at the end of the book to record your thoughts. Simply record the actions you wish to take as you read the book. Later on you can use these ideas to generate your career plans.

So for now, let's move on to getting control of your career and hopefully realizing your career aspirations.

REFERENCES

1. Barkley, Nella, *Taking Charge of Your Career*, Workman Publishing Co., 1995.
2. Ficco, Mike, and Laplante, Philip A., *What Every Engineer Should Know About Career Management*, CRC Press, 2008.
3. Farren, Caela, *Who's Running Your Career*, MasteryWorks, 1997.
4. Mattiuzzi, Cici, *The Ultimate Career Planning Manual for Engineers and Computer Scientists*, Kendall/Hunt Publishing, 2006.
5. Feibelman, Peter J., *A PhD Is Not Enough: A Guide To Survival In Science*, Basic Books, 1993.

CHAPTER 2

DEVELOPING A SUCCESSFUL CAREER PLAN

The first step to getting control of your career is developing a strategic career plan. A good strategic career plan should identify the path to your ideal job and identify the immediate and long-term actions that will lead you to your definition of career success.

A strategic career plan is a map guiding you along in your career so you know the path to your ultimate goal. It provides direction and guidance at each bend in the road so you always stay headed in the right direction.

Engineers who fail to develop a strategic career plan are like a person driving to a place they have never been before without directions or map. The analogy is this, a person who believes that as long as they drive along the freeway at 60 miles per hour, they are making progress. They may not know which exit to take, but as long as they are traveling at 60 miles per hour they are making great progress. Sometimes, going fast isn't as important as knowing where you are going to get the results you need. Spend time developing a strategic career plan and don't waste a lot of time and energy on activities that may hinder your career advancement progress.

PURPOSE OF CAREER PLANNING

There are several important reasons you need to develop a career plan. These reasons include:

- Means to reach your goals
- Sense of purpose for daily activities
- Sense of fulfillment and accomplishments
- Shock absorber during setbacks
- Develop backup or recovery plans

The Engineer's Career Guide. By John A. Hoschette
Copyright © 2010 John Wiley & Sons, Inc.

Developing a plan provides the map to guide you. Career planning also provides one with a sense of purpose for all those mundane daily activities. It is quite easy to lose focus on your short-term goals and long-range career objectives when you have long daily commutes and an endless stream of, what appear to be, meaningless tasks to accomplish.

While working in Los Angeles, Boston, and San Jose, I would sit in traffic for hours and often find myself asking, "Why am I doing this again?"

One of the most satisfying reasons for career planning is that it provides the individual with a sense of fulfillment and accomplishment as the goals and career advancement opportunities are realized. This provides a renewed sense of energy and pride in your work and keeps you working toward the next goal. "Success Breeds Success."

Finally, career planning is absolutely essential in the event of setbacks. Career setbacks include such things as being laid off, career plateaus, demotions, lower than average raises, and job obsolescence due to technological advancements. One recent career threat has been job outsourcing to foreign countries. With the rapidly changing technology environment and global economies, lifelong employment at a single company is becoming a very rare event.

The average engineer can expect to change jobs multiple times in their lifetime. Therefore, a good career plan can serve as a shock absorber and provide backup or recovery plans in the event you need it.

A graphical representation of an engineer's career is shown in Figure 2-1. The x-axis is time, and the y-axis is salary/responsibility. In an ideal situation, the engineer's career grows exponentially with few setbacks until, at the end of the career, the engineer retires. However, the reality of life is shown in the lower curve.

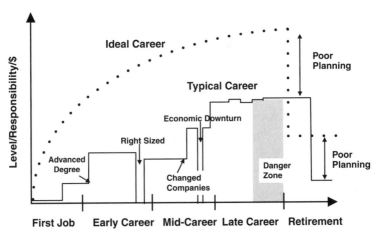

FIGURE 2-1　　Ideal career versus typical career.

In the typical career, the engineer's growth is characterized by a series of ups and downs. Careers are aided by returning to school and receiving an advanced degree, job changes, excellent performance reviews, and growth through seniority. Most often it is slower than expected. There are circumstances when the engineer may be laid off due to economic downturns in the industry. A good career plan will aid the engineers through the good times as well as the bad. However, as shown in the lower curve, most often the typical engineering career is far less successful than originally hoped for, and a leading factor contributing to this is poor planning.

In recent years, a new danger zone has risen for engineers between the ages of 55 and 65; they are the leading candidates for layoff during company downturns. The reasons being they are usually the most expensive salary-wise, cost the companies the most in medical benefits, and are often technically obsolete if they have not returned for training in the latest technology.

What is constant during an engineer's entire career? Two things: first is the need for lifelong career planning to deal with all the changes. Career plans, like technology, quickly become obsolete due to all the changing circumstances. Products come and go, technology changes, supervisors change, and even companies come and go. Career plans need to be updated yearly and changes made when required.

The second constant during the career of an engineer is the need for technology updating. This is necessary now more than ever since the half-life of technical knowledge is now down to roughly five years. That is to say, half of the technology you learned only five years ago is now obsolete or overcome by new technologies. People ask, how can this be true? These questions may illustrate the point, what do you call a 5-year-old PC? Most answer, obsolete, outdated, no longer used while some even claim they qualify as antiques! With technology changing so rapidly, the only hope an engineer has to remain successful is to make lifelong learning part of their career plan. Do not become obsolete like your PC, make lifelong learning part of your career.

DANGER SIGNS INDICATING CAREER PLANNING IS NECESSARY

There are definitely danger signs where career planning is necessary. These danger signs are

- Career plateau, your career growth has stopped or you are receiving poor ratings
- Excessive hours at work, 50–60 hours per week is the norm
- Self-worth measured in terms of "Company"
- Work is no longer fun, conversely staying at work to avoid going home
- Stress—quick to anger—sarcasm—absenteeism—inflexibility
- Economic downturn in your industry, downsizing, and layoffs

▶ *Career Tip.* If you find yourself experiencing any of these warning signs, then it is time to do some serious career planning. Recognize the signs and take the appropriate actions now.

CAREER ASSESSMENT AND PLANNING DECISION TREE

A simplified career assessment and planning decision tree is shown in Figure 2-2. The decision tree starts with asking the basic question, "Are you happy with your present career?" Answering this question leads to three different choices. The first is, Yes I am happy with the way my career is progressing and I like my job. Then keep doing what you are doing and congratulations. Make sure you have backup plans just in case. The second choice is, I am not happy with my present job, but I do not want to leave my employer. In this case, you need to develop career plans to explore your options.

I have listed three options for this leg of the decision tree: look for new assignments within the department, leave the department, or get new training to qualify you for better assignments.

If you are unhappy with your present employer and want to leave, you need to assess what it would take to make you happy and what you want to change. From here, the next step is to generate a career plan, update your resume, and start interviewing.

As you look at the options on the decision tree and decide which actions you should take, you might want to consider actions in several of the branches at once. If you look closely, all branches involve generating or updating your career plan. Therefore, this is a good career action regardless of the decisions you make. Also, you may start down one branch of the decision tree only to discover this is not what you really want and switch to another branch. And as always, things will change, so you will need to observe circumstances over a period of time to get a true reading of where things are headed. It is most important that you spend the time thinking about and planning your career options. You can always make adjustments and change your approach. As a matter of fact, career engineers with 30 or more years of experience have found that constantly assessing the situation, defining adjustments, and changing plans are the norm.

THE PROCESS OF GENERATING A CAREER PLAN

Stephen Covey, in his book *The 7 Habits of Highly Effective People*, defines a simple process for an individual to follow to generate a career plan [1]. This process is shown in Figure 2-3 and contains three simple steps. These steps are self-evaluation, generating a career plan, and taking action.

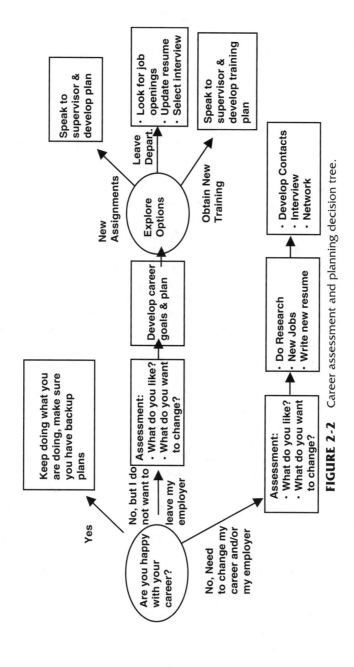

FIGURE 2-2 Career assessment and planning decision tree.

Process of Developing Career Plans

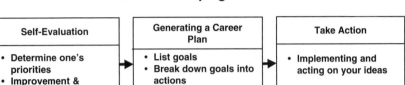

Self-Evaluation	Generating a Career Plan	Take Action
• Determine one's priorities • Improvement & change areas	• List goals • Break down goals into actions • Develop a schedule for actions	• Implementing and acting on your ideas

Going Through the Process Is as Important as the Product

FIGURE 2-3   Process of developing career plans.

The first step in the process is to determine one's priorities in life. Simply put, we do not have time to do it all, so we must choose the most important things and prioritize the list. In addition, one must identify improvement areas. Another way to look at this is to determine our weaknesses and come up with an improvement plan. It is not usually our strong areas that hold our career back but our weak areas, so the best return on effort is to improve areas where we are weak and turn them into strengths.

Once we have performed a self-evaluation, the next step is to generate our goals list. Clearly identify your career goals and then break down those goals into actions. As we identify career actions, we need to determine a schedule for the actions. Set a deadline for each action and make it happen. Think of it as an assignment from your supervisor, one that is important and needs to be accomplished on time.

For the final step in the process, we need to take actions, great plans are useless without action. It is action that translates plans into accomplishments. In this step of the process, we realize the benefit of our planning.

Some people I have coached feel that generating a perfect plan is the key. And they get all hung up in the way the plan looks and how it is formatted. A neatly formatted plan is not a guarantee to successful career planning. The most significant benefit behind planning is simply going through the process and forcing yourself to think about your career. Don't worry if your career plan is not perfect and there are holes or undecided items. Just trying to generate a plan and recording your thoughts puts you ahead of the majority of engineers.

CAREER WORKBOOK—HELPING YOU PLAN YOUR CAREER

To help you organize your thoughts and guide you in the assembly of your career plan, the next sections of this chapter will identify typical engineering career goals and the corresponding actions to accomplish these goals. As you read through the remaining sections in this chapter and start to identify your goals and actions, you can record them in the career workbook provided at the end of this book, in Appendix A.

It is very important you take the time to write out your goals and actions in order to accomplish them. Experts all agree, writing them down, discussing, and formulating action plans will lead to success.

The workbook in Appendix A, will help you formulate your own personalized career plan. Please take a moment and review the workbook before starting the next sections of this chapter. Briefly study the workbook and identify the questions and data it will be asking of you. Then return here and we will start the career planning process.

Quickly review the career notebook in Appendix A and then return here!

Career planning involves each individual doing assessments of what is important to them and where they ultimately want to go with their career. To help you identify and select your personal career goals, I have identified some typical career choices you need to make. These career choices can be categorized into the following:

- Technical versus management career plan
- Typical goals during an engineering career
- Skills assessments (strengths and weaknesses)
- Examples of great career goals
- Balance in your career and great personal goals

To develop a strategic career plan, you need to look down the road of a typical engineer's career. Where do assignments lead? What do you need to be doing to accomplish your present short-term career goals and subsequent long-term goals. Let's look at the typical career of an engineer as they progress from leaving school to senior levels.

TECHNICAL VERSUS MANAGEMENT CAREER PLAN

A fairly typical career path for most engineers is shown in Figure 2-4. Engineers leave school and join a company in an entry level position. These positions are identified in Figure 2-4 as levels E1, E2, and E3. Generally, a Bachelor's degree with no experience will join the company as a level E1. If the engineer has some experience or a Master' degree, they enter at level E2. If the engineer has a PhD, they may enter at the E2 or even E3 level.

At the lower levels (E1–E2), the engineer is expected to be a good team player and take directions from lead engineers and follow orders. In addition, the engineer should be further developing their technical skills and knowledge base. They should be learning to work on teams, interface with other engineers of different disciplines, and learning the company structures and processes. The quickest way to advance is by getting their assignment correct and done ahead of schedule. This clearly shows they are ready for more challenging assignments and ready to move up.

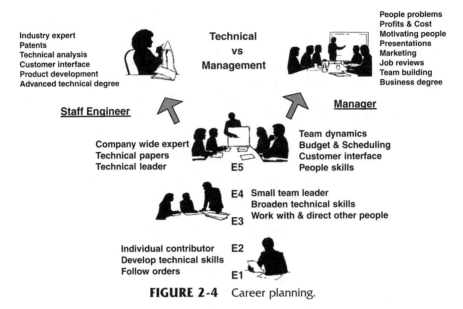

FIGURE 2-4   Career planning.

As the engineer moves into the middle ranges (E3–E4), they become a small team leader. They must continue to broaden and expand their technical knowledge base and also lead the efforts of a small team of engineers. Their assignments continue to include doing technical work in their area and expand to learning how to get other people to support team efforts accomplishing the goals. They work with other people and get assignments done by others.

In the top levels (E4–E5), the engineer usually becomes the leader of large teams and often is responsible for the team's technical success as well as the financial success. As a project lead engineer, they may be directing the activities of several large teams at once. On the technical side, the lead engineer is responsible for the team meeting the technical objectives of the projects. They are often considered industry experts, may hold several patents, and may be the technical director for the project. They lead and are often responsible for the team's engineering design, build, test, and product production. They direct the technical activities of the team, including making technical decisions and approaches.

On the management side, these lead engineers (E4–E5) are responsible for the financial success of the project. They work to ensure the project is completed on time and within budget. They are required to have a variety of management and people skills. They must be able to organize teams, plan projects and resources, set objectives, interface with customers, and resolve conflicts in the best interests of the team and company. They often become involved in personnel issues such as hiring, layoffs, raises, and promotions. The lead E5 engineers need to develop great business skills in addition to keeping their technical skills updated. Generally, it takes about 10–15 years for an engineer to progress from an E1 to the E5 level.

At the E5 level, the engineer has to make a major career choice. The engineer has reached a fork in the career roadmap. This career choice is shown in Figure 2-4 as becoming a "Technical Staff Engineer" or "Manager." If the engineer decides to focus on becoming a Technical Staff Engineer, their assignments usually become strictly technical in nature. They focus on the technical aspects of the project and leave all the business decisions to the managers.

The technical staff engineers are responsible for the technical success of the products, projects, and company. They usually hold several patents, publish technical papers, and often return for advanced technical training and degrees. They interact with the customer and the customer's technical experts and are considered industry experts in technological development.

If the engineer decides to become a manager, then they need to develop an entirely new skill set. The engineer takes a major turn in their career to focus their energy on the business success of the company. They handle company finances, banking, people problems, profit and loss, marketing, and customers. They are responsible for the financial success of the company. The shift to a management career requires the engineer to return for business schooling and develop an entirely new skill set.

Deciding on whether to remain technical or change and become a manager is entirely dependent on your preferences. If you find yourself having no interest in dealing with people, organizing teams, and dealing with the business side, your career choice best lies along the path of becoming a technical staff person. If, on the other hand, you like to lead people or wish to develop good people skills and enjoy organizing teams, then the management path is your best choice.

If you are unsure, the best thing to do is talk over your choices with your mentors, supervisors, and others in the company who are presently in those positions you hope to reach some day. Ask them what they like and dislike about their jobs, what they find challenging, and how to best prepare for a position similar to theirs.

As you move up the engineering levels, there is a constant trade-off between acquiring more technical skills or management skills and you will need both to advance. However, the timing of acquiring these skills is key to successful and rapid growth. The recommended guidelines for allotment of time and effort you spend on technical versus managerial planning and training for the various engineering levels are shown in Figure 2-5.

I have witnessed junior engineers (E1–E2) announce they are returning for an MBA because they want to become a manager. They falsely expect by the time they receive a degree in three years the company will simply make them a manager. This is good planning but unrealistic goals. They must also prove they have the technical skills and demonstrate a good technical base.

It is not likely that junior engineers will ever move up two to three levels in a single promotion. Therefore, when others around the engineer hear their expectations, it may cause negative reactions. You must develop a sound technical base as well as developing your leadership skills.

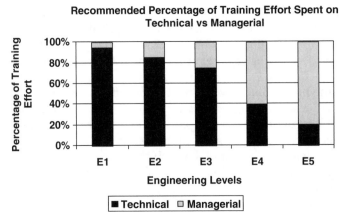

FIGURE 2-5 Time and effort spent on technical versus managerial training and planning.

On the opposite side of the issue, I have observed junior engineers return as advanced technical masters and expect to skip levels up the corporate ladder simply because they know more than anyone else. They develop their technical skills and knowledge base but neglect their leadership and people skills. After they graduate with a advanced degree and are put in charge of a team, the team quickly disintegrates because they lack the skills to resolve problems and manage teams.

Over the years, I believe, or have been convinced, the best approach for rapid career growth is to concentrate more time and effort on the technical skills in the lower levels (E1–E3) and more time and effort on the leadership skills in the upper levels (E4–E5).

Now that you have looked at the typical paths available to engineers, what are your career plans? If you're in the lower levels of engineering, your major career planning should include broadening your technical skills and developing your team leadership skills. If you're in the more senior levels, you have a major decision ahead to either remain technical or go into the management side.

▶ *Career Tip.* Regardless of your level, you should be looking ahead to levels you one day aspire to reach and develop career actions to ensure you will reach these levels.

Go to the career workbook and fill in the data requested for developing your long-range career plans. You should identify where your ultimate career path will lead, technical or management. Once you have made this choice, it is time to identify the actions needed to accomplish these long-range career objectives.

An example of a multiyear career plan is shown in Figure 2-6. The goals are identified for the present year, three years, five years, and 10 years.

This Year	3 Years	5 Years	10 Years
Goal: Outstanding rating E3, Senior Engineer	Goal: E4 level, Lead Engineer	Goal: E5 level, Project Leader	Goal: Manager
Actions required to meet goal:	Actions required to meet goal:	Actions required to meet goal:	Actions required to meet goal:
1. Discuss performance and future plans with supervisor	1. Secure training in making presentations	1. Return for MBA	1. Complete Master's in Business Admistration
2. Identify weak areas	2. Further technical training	2. Lead major project	2. Return for technical updating
3. Training in technical area	3. Get leadership assignment	3. Publish papers	
4. Team training	4. Talk to leaders about career paths	4. Complete leadership assignments successfully	
5. Excellent performance		5. Solve major technical issues	
	Actions for 3-year goal:	Actions for 5-year goal:	Actions for 10-year goal:
	1. Training in meetings	1. Identify project lead opportunities	1. Gather data for MBA program
	2. Training in budgeting	2. Get assignments in leading small efforts	2. Enroll in MBA program

FIGURE 2-6 Example of a multiyear plan.

FIGURE 2-7   Discussion with supervisor to identify criteria for "Excellent Rating".

For each goal, career actions have been identified. For this example, the engineer has identified the career path to becoming a manager within the next 10 years.

The engineer's goal for the present year is to obtain an "Outstanding" or "Excellent" rating. To accomplish this goal, the engineer identified the actions of career advancement in a discussion with the supervisor to determine the criteria for an "Excellent" rating (Figure 2-7). Other actions include identifying weak areas and developing plans to strengthen these areas. In addition, the engineer has identified both technical and team training to be completed.

Also identified are the career actions to be accomplished within three years. The engineer has identified the 3-year goal of becoming a lead engineer. The engineer must work on the actions for this year as well as actions for development for the following years. The engineer has identified the career actions of learning how to run meetings and how to do department budgeting as needed for the 3-year goal.

For each of the engineer's major career objectives for the next 10 years, actions have been identified to accomplish these goals. The engineer has set up a career plan to make sure their performance at the present level is "Excellent" while at the same time learning new job skills that allow them to be ready to handle job assignments at the next level. The simple rule for my engineers is: "Top Performance Today is Training for Tomorrow."

▶ *Career Tip.* Top performance today is training for tomorrow.

If you have selected your long-term career path, you must start now to break your goals into a series of actions you need to accomplish this year, 3 years, 5 years, and even 10 years from now. Turn to the workbook in

Appendix A and fill in, as best you can, the multiyear career goals and actions you will need to take.

If you do not have enough data or a clear understanding of the steps leading to your goal, it is time to talk with your supervisor and mentors about what you need to do. When you have completed this worksheet, return here and we will continue the career planning process.

TYPICAL GOALS DURING AN ENGINEERING CAREER

A matrix of typical career goals as a function of the engineer's age is shown in Figure 2-8. The goals have been separated into various age groups as well as personal and company goals.

This example matrix was developed from materials received from the IEEE, AIAA, and SME engineering societies. These societies have published various lists of career goals for their memberships based on age.

The matrix identifies typical career goals throughout an engineer's life. As the engineer matures and advances, the goals change. Therefore, career planning is a lifetime activity. In the first five years after leaving the university, the immediate goals are for the engineer to learn the company engineering/business methods. In parallel, the engineer should be developing a long-range 5–10-year career plan and expanding their technical skill base.

Years from Graduation	0–5 Years First Job	5–10 Years Early Career	10–20 Years Mid Career	20–30 Years Late Career	30–40 Years Retirement
Age	22–30	30–35	35–45	45–55	55+
Company/ Technical Career Goals	Adjust to work environment Learn company ropes Enhance technical training Career planning	Focus on technical specialty Develop team skills Higher levels of responsibility Publish papers Lead product development	Tech vs Business Develop leadership skills Update training Supervisor Return for MBA Publish papers Patents	Continue leadership development Technical update Upper management Mentoring junior people Senior role in company/staff	Leveling of career and responsibility Consulting role Teach classes
Personal and Family Goals	Payoff college debt Have fun Financial plan for retirement Elder caregiving New car	Marriage Purchase home Start family Elder caregiving Plan for retirement	Family vacations Child development Child school activities Elder caregiving Plan for retirement	Family vacations Children college Elder caregiving Plan for retirement	Very late for planning of retirement Wedding of children Grandkids Less pressure

FIGURE 2-8   Typical lifelong career goals.

In the next stage of early career, 5–10 years after graduation, the engineer should focus on a technical specialty and start to develop team skills. The engineer will be accepting assignments with more responsibility and could even lead development teams. In this stage, the successful engineer is enhancing their technical area of expertise, broadening their technical background into new areas, and working on team skills. At this point, engineers return for some type of advanced training, either technical or business in nature.

At the mid-career stage, 10–20 years after graduation, the engineer is usually in a leadership position directing either the technical or business aspects of a product or team. The engineer has risen to the point where they need to decide if they are going to remain technical or become more of a business leader. Once again the engineer needs to acquire new skills and talents.

Engineering careers generally peak in the next stage from 20 to 30 years after graduation. If the engineer has been successful and accomplished their career goals, they are usually in a senior level technical or business leadership position, making significant contributions to the company's success. They are the upper level leaders determining the technical approaches and directing the engineering workforce. They are doing many career actions simultaneously, running departments, returning for training, mentoring junior people, and continuing their leadership development.

In the last stage, the engineer has generally reached a plateau and can start to give back by teaching classes and mentoring their juniors. Oftentimes, senior engineers start a consulting career based on their acquired knowledge. And finally, if the engineer has done the proper planning, they may retire in comfort to enjoy life.

There is an equally important set of goals to your professional goals and these are your personal goals. Do you know what your personal goals are? For you? For your spouse? Children? Personal goals in the first years after college deal with establishing a solid career plan, paying off college debt, doing some financial planning, and possibly giving elder care to parents. As the engineer ages, these goals may change to marriage, starting a family, and buying a home. Then as the children grow, the goals might include becoming involved in the children's school activities and planning family vacation. Finally, in the later years, a typical engineer may be supporting children in college or planning weddings. A constant planning activity throughout an engineer's lifetime should be financial and retirement planning.

After you have reviewed Figure 2-8, turn to the career workbook and fill in the section on career goals based on your age. Remember to identify your personal as well as work goals.

Engineers have asked, "What are some great career goals and actions?" Brainstorming with other engineers, we have tentatively identified the following list of great career enhancing actions you can take.

EXAMPLES OF GREAT CAREER ACTIONS

- Join an engineering society
- Host a society meeting at your company
- Visit the university bookstore
- Have a feedback session with your supervisor
- Write a technical paper
- Attend a conference
- Attend a seminar
- Apply for a patent
- Write "lessons learned" memo
- Do a hardware demo
- Write an article for company paper
- Do a new product demo
- Submit a team or coworker for an award
- Volunteer to improve something in your company
- Return for more schooling and/or advanced degree
- Learn a software program
- Write a report
- Read new technical journals
- Learn about the job level above yours
- Help organize engineers week
- Update your resume
- Take a time management course
- Get your work done early and ahead of schedule
- Develop a new simulation/model

Are these the only great ones? Absolutely not, they are provided to stimulate your thinking and get you thinking outside of the box. Go to the career workbook and select some career goals for your plan.

EXAMPLES OF GREAT PERSONAL/SOCIAL ACTIONS

Similarly, we generated a list of great family/social actions you can take.

- Plan family vacations
- Sunday night family meetings
- Date night with spouse
- Plan birthdays
- Plan anniversaries

- Plan volunteer work
- Attend your kids' school functions
- Coach a sports team
- Exercise (stress relief)
- Make it home for dinner more often
- Spend "special time" with your kids
- Financial planning

Have you ever considered these types of goals in your career planning before? Why not? I am sure your family members will have their own ideas about great activities. Have you ever stopped to ask them? Maybe you could run a brainstorming session at home just like you do at work and generate new family activities that everyone enjoys. The important considerations are getting their involvement and making good plans even better. People tend to support and help implement actions they helped develop and enjoy.

Go to the career workbook and select some personal/social goals for your plan.

KEEPING BALANCE BETWEEN WORK AND PERSONAL LIFE

One of the most important things I have found in my 30 years as an engineer is to have a balance between work life and your home life. Every time I let things get out of balance, it has resulted in trouble for me either at work or at home.

An oversimplified example of what I am discussing is shown in Figure 2-9. The teeter-totter is like your life. On one side are personal goals and demands and on the other side are work demands. It requires a constant effort to balance both and there will be times when you get out of balance and restoring balance is no easy task.

Try to monitor your daily activities at work and at home. How many days do you make it home on time? How many days do you work overtime? If you are out of balance, how do you plan on getting back in balance? Restoring balance will not occur naturally, you must work hard at it.

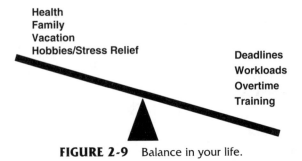

FIGURE 2-9 Balance in your life.

I have known many engineering managers who traded their family for a promotion, only to regret it. To have balance in your life, you must have a family/social plan as well as a career plan and work constantly at both.

How do you get things back in balance once they are out of balance? The best answer I can give is, time management. You simply need to take control and limit your time at work. One recommendation I can make here is to take time management classes to help you organize your time and be more productive in a shorter time frame. Another recommendation is to schedule going home just like you do regular work meetings. Block out your calendar and schedule a meeting called "Time To Go Home" at the end of each day, the meeting notice will pop up and remind you. No excuses for showing up late for dinner again!

I have completed research and attended many seminars on getting and keeping balance in your life. The experts in this area all indicate you need to set priorities in your life and constantly keep these priorities in mind as you grind through the daily hassles of life. These experts recommend you set your priorities in the following order:

1. Religion (moral beliefs)
2. Family
3. Self
4. Work

From my 34 years of experience, I found I felt the best and was the happiest when I kept my priorities in this order. When I felt the worst, I had set priorities based on the reverse order with work holding the number one position. I have also observed many other engineers and managers who told me they felt the worst about their lives when they set their priorities in the reverse order. How do you have your priorities set?

CAPITALIZE ON YOUR STRENGTHS AND MINIMIZE WEAKNESSES

Each of us has a set of strengths and weaknesses. The key for a successful career is to capitalize on your strengths and eliminate your weaknesses by turning them into strengths. A partial list of skills that engineers will use during their careers is given in Table 2-1. Please read through this list and do a self-assessment on whether these skills are a strength or weakness. By definition, if it is not one of your strong points, then it is a weakness.

As you go through the list, think of some type of accomplishment or action that indicates this is a strength or weakness for you. If this list does not meet your needs, then I suggest you design your own list.

After you have reviewed the list, go to the career workbook and fill in your personal career assessment. I have given more blank lines in the workbook for you to fill in skills you know of and are not on the list.

At this point, I am asked, "What are the most important skills and what are other skills needed to advance?" To this I reply, see your supervisor and company job descriptions. These are an excellent place to start. Another excellent person to check with is your mentor. A good mentor will provide you with an objective opinion on the importance of these skills and help you rank them. Mentors can also identify skills missing from the list. Then prioritize your list in the workbook. Make sure you focus your energies on the highest priority skills (Table 2-1).

Go to the career workbook and select some career goals for your plan. Return here after you are finished selecting goals.

TABLE 2-1   List of Engineering Career Skills

Type of Skill	Strength or Weakness	Accomplishment Demonstrating Strength or Weakness
Technical		
Product design		
Technical knowledge in your field		
Technical knowledge in other fields		
Product building		
Product integration		
Laboratory test		
Laboratory research		
Technical publications		
Computer modelling		
CAD design and modeling		
Analysis and modelling		
Experimental research		
Patents		
Technical awards		
Programming		
Manufacturing		
Project Management		
Planning		
Budgeting		
Organizing		
Developing policies		
Developing procedures		
Cost tracking		
Schedule planning		
Customer interface		
Team formation		
Salary administration		
Department budgeting		
Presentation skills		

TABLE 2-1   *(Continued)*

Type of Skill	Strength or Weakness	Accomplishment Demonstrating Strength or Weakness
Interpersonal Skills		
Motivating		
Team leadership		
Conflict resolution		
Work relationships		
Meeting skills		
Versatility		
Team dynamics		
Communication style		
Customer relationships		
Social abilities		
Mentoring		

TRANSLATING GOALS INTO ACTIONS

Translating goals into action is the heart of career planning. For each goal you establish in the workbook, you should identify the actions and dates you need to accomplish them. An example of the career goal "Getting a Master's Degree" and the corresponding actions to accomplish this goal are shown in Table 2-2.

The first example is for returning to pursue an advanced degree. Here, the goal is broken down into simple tasks that are easily accomplished with deadlines set for each task. The first step to realizing your career goal of obtaining an advanced degree is doing some research on the types of degrees and the programs offered at various universities. The next step is talking with your supervisor and Human Resources department about your plans and getting their support and approval. Next, you should generate a degree plan and complete the admission forms and finally start. Please note the example time line shown. From start to actually taking your first class could be six months. So if this is your career goal, get going now.

I have dedicated an entire chapter of this book to going back for your advanced degree. I identify the options available, what they mean to your career, and advise on how to overcome roadblocks during this process. This material is provided in Chapter 45, "Getting the Most From Your Company's Education System." The steps shown in Table 2-2 are presented for guidance in formulating an action plan to achieve this goal. When you actually start down the road to this degree, you will quickly realize that there are a significant number of other tasks and subtasks not listed. In fact, if you come up with a detailed plan of all the actions, you will quickly realize you will be busy doing something every week to accomplish this goal.

TABLE 2-2   Career Goal and Actions for Getting A Master's Degree

Career Goal	Actions	Date to Complete Action
Get Master's degree (MBA)	• Contact university on programs available and get information	• January
	• Talk with supervisor	• January
	• Talk to HR about tuition refund	• February
	• Generate plan for degree	• March
	• Fill out application form	• April deadline for starting in fall semester
	• Apply	• April 25th
	• Start	• September 15th, first class

Another example of a career goal might be getting your next promotion. This is no simple task, and there are no guaranteed ways to accomplish this career goal. However, there are steps you can take to significantly reduce the time to your next promotion. The career goal to obtain a promotion and a simplified example list of the corresponding recommended actions for accomplishing this goal are shown in Table 2-3.

The remainder of the book is dedicated to the actions you can take and career tools that are available for your use. The first steps are to determine how you are perceived by your supervisor and how far away you actually are from getting promoted. Therefore, the first step recommends reviewing your last

TABLE 2-3   Career Goals and Actions for Getting a Promotion

Career Goal	Actions	Date to Complete Action
Promotion	• Review your last job appraisal and performance rating	Start
	• Determine your weaknesses and strengths, action plans	1 month
	• Research the criteria for your present position	2 months
	• Research the criteria for the next level up	3 months
	• Talk with supervisor about your desires and potential for promotion	1 month
	• Develop an action plan leading to promotion	2 months
	• Demonstrate you are ready!	6 months to 3 years
	• Make sure there is an opening	Constantly
	• More career discussions and feedback with supervisor	Every 3 months
	• More demonstration of performing at the next level up	Constantly
	• Doing a performance review	Yearly

performance appraisal and determining your improvement areas. Next, you must come up with a plan that demonstrates, when executed, you have mastered the requirements/criteria for your present level.

The following action is to determine the criteria for the level of promotion you wish to receive. This will require doing a significant amount of research. This research will include finding out the formal and informal criteria for promotion and determining your supervisor's criteria, what Human Resources has set as standards for the next level, and work performance you must demonstrate. This is followed by doing months and even possibly years of consistently outstanding performance at your present job. Along the way, you must be in sync with your supervisor and obtain feedback on your performance improvement.

The most significant point here is that your promotion is a minimum of a year away from the time you decide to accomplish this goal. So if you do nothing other than your normal job, where can you expect you will be a year from now? Doing the same job you are doing now.

▶ *Career Tip.* Your promotion is a minimum of a year away from the time you decide to accomplish your goal.

The typical time to move from each level of engineering in most larger companies is 3 years if you are a star performer and 4–6 years if you are an average performer. If you count the number of levels in your company, and multiply by the years per level for average performers, you will quickly realize that, unless you become a star performer, it is difficult to reach the upper levels of most major corporations. To be a star performer, you must have a clearly defined career plan and consistently execute critical career actions. The career plan must be a well-formulated plan, capable of handling roadblocks and setbacks. Therefore, having a career plan is absolutely necessary if you wish to move up in the company.

Please turn to the career workbook again and look at the career goals you wish to select for you. Then start to fill in the actions you feel are necessary to accomplish these goals. Make sure you identify a date for the action and the priority level.

If you feel you do not know enough to generate actions for each of the goals, don't panic. You can leave them blank for now. As you read through the remainder of the book, I will be providing you with career tips and actions for these various goals. So for now, fill out what you can, and as you read through the remainder of the book and find something you would consider as a great career action, quickly turn to the back and record it.

The next step to formulating a career plan is to fill in the career goal worksheets provided in the career workbook. The worksheets are formatted like Tables 2-2 and 2-3. Identify your goal, then determine and record the actions necessary to complete this career goal. This is a very iterative process and will require an extended period of time to complete these worksheets. You will identify goals and actions and as you think through the process, you will more than likely change or modify your approach. Take your time and complete the worksheets over several weeks.

The next step is to get feedback on your plan. Go over the plan with your mentor, your spouse, or anyone else with whom you feel comfortable. Get feedback on your thoughts and ask for ideas on how to improve it. Your mentors should have some great ideas and experience they can share with you that should help your planning. Your mentors may also have good feedback on the timetable you have estimated. You may be overly optimistic or too pessimistic for your estimates of the time it will take to accomplish these career actions and goals. One good person to get feedback on your plan from is your mentor; please refer to Chapter 7 on this.

Make sure you prioritize your career goals and personal goals from highest to the lowest. You are not going to have time to do everything, so making sure you have a prioritized list is essential. When you have prioritized career goals and action lists, you are ready to move on.

Before we continue, do you have balance? Have you considered your personal/family social goals? Make sure you identify these goals as well.

You are now close to completing a career plan but there is one final, very important step, that is critical to getting started. Your task is to generate a timetable or calendar for career actions.

ORGANIZING CAREER ACTIONS INTO A CALENDAR OF EVENTS

The last page of the workbook is a simple chronological listing of your career and personal goals/actions by month for the year. This should be very easy to fill out from your completed workbook sheets. Make sure your highest priority actions are listed. When the calendar is completed, you will have a prioritized list of career actions you need to take each month. The company has goals for the year and now you have your own goals. You may even consider combining both. A sample calendar is shown in Figure 2–10.

Once you have completed your personalized career calendar, the next step is to post it where you can see it daily. I have seen a format in Power Point and Excel spreadsheet and both work nicely. Posting your completed calendars behind one's computer is another excellent career management technique. This way, when you are working at your computer, you are reminded of the actions you planned to take. You might even add notices to your electronic calendars that pop up and remind you. Another good technique is to use it as a screen saver. I add check marks as each task gets accomplished.

January
- Update plan
- Plan year
- Wife's birthday

February
- Kick-off new project
- Take time management class
- Kid's birthday

March
- Plan family vacation
- Attend financial planning workshop
- Spring break holiday

April
- Career discussions with supervisor
- Major program review
- Taxes

May
- Launch new product

June
- Anniversary
- End of second quarter financial report

July
- Family vacation
- Complete major test
- Parent's birthday

August
- Update plan to year-end
- Major design review with customer
- Kid's soccer

September
- Kids start school
- Run special testing
- Job review

October
- Third quarter technology planning
- Customer meetings in New York
- Raises and promotions

November
- Trip to parent's home (out of state)

December
- New Year department budget and planning
- Celebrate holidays

FIGURE 2-10   Sample personal and career goals calendar.

HOW TECHNOLOGIES, PRODUCTS, AND INDUSTRIES IMPACT LONG-RANGE CAREER PLANNING

Excellent career plans are not a guarantee for advancement and may even fail miserably. The reasons the career plans can fail are generally due to engineers neglecting to monitor and adjust their career plans for changes in the job market, industry ups and downs, and rapidly changing technology. Sometimes engineers end up leaving their engineering field or worse yet, leaving engineering completely at a significant cut to their salary and not to mention a large time investment.

As shown in Figure 2-11, the optimum career success occurs when you are at the intersection point of all three factors (technologies, products, and industries), and all are thriving. If you find yourself in thriving industries,

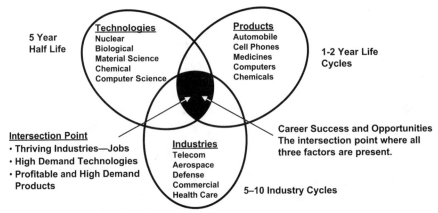

FIGURE 2-11   Career intersection point of technologies, products, and industries.

and your technical background is in demand and you are working on profitable high-demand products you are in a very good career situation.

▶ *Career Tip.* The purpose of long-range strategic career planning is to keep you at the intersection point of technologies, products, and industries.

As you do your career planning, have you considered the effects that technology obsolescence, product life cycles, and industry ups and downs will have on your career and ability to stay employed?

Plots showing the relative number of jobs in various industries and technology fields over the past 40 years are shown in Figure 2-12. These graphs show the cyclic nature. In the upper graph of Figure 2-12 is the number of jobs that cycle up and down for various industries.

In the early 1960s, NASA and space were the hot industries that all the engineers wanted to get into. The space industry was where the new engineering jobs were in the 1960s. However, with the Vietnam War reaching its peak in the late 1960s and with the successful landing on the moon, jobs turned from space to the defense industry. Then, with the end of the war in Vietnam, defense spending fell and jobs moved into the private sector. In the 1980s and Reagan years, the Cold War was at its peak and defense was once again a hot industry. Defense has a characteristic 10–15-year cycle of ups and downs. In the early 1990s, the Berlin Wall fell and with the end of the Cold War, the defense industry did another downturn and lost a significant amount of jobs.

In the late 1980s, the telecom industry was one of the hottest industries but this quickly died out by the mid-1990s. A very quick burn industry in the 1990s was the DOT.COMs. Companies and millionaires were literally made overnight, and just as quickly as they came, they disappeared.

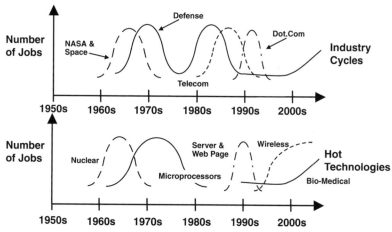

FIGURE 2-12   Industry and technology effects on number of jobs.

The important point to consider in your career planning is what stage your industry is in today. Is it growing, peaking, or decaying? Your career plans should include provisions and adjustments for these swings in industry. Your objective is to make your career resilient to downturns.

Another factor to consider is the technology with which you are working. Technologies and your technical training go through periods of high demand and then obsolescence. As shown in Figure 2-12, in the 1960s the hot technology was nuclear. The nuclear engineers were in high demand and this demand quickly died off when elimination of nuclear power plants and atomic weapons became the concern of many nations and political groups.

Then with the push to get to the moon, the semiconductor industry and microprocessor technology became the high demand. Employment forecasts indicated that there would not be enough engineers to ever supply this demand. However, as the technology matured and the number of transistors per chip quickly increased, the demand for these digital designers fell off. Then in the 1980s, and 1990s, the server technology and web page technology became hot. This quickly fell off leaving many web page designers looking for new employment. Another hot technology, and still growing, is the wireless technology. Cell phones, remote controls, and computers are all going wireless. This has recently created a high demand for wireless engineers. A new and emerging technology has been the biomedical field of the healthcare industry.

One unforeseen event has been the 2009 economic downturn and its impact on engineers' careers.

Does your technical training support the development of a growing technology field or are you working in a decaying technology field with obsolete technical skills? You need to assess your conditions. If your technical training and skills are becoming obsolete and the technology field you are

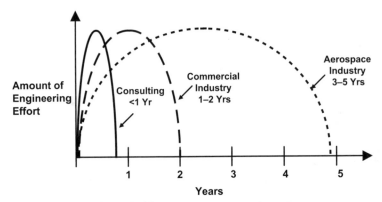

FIGURE 2-13   Typical product life cycles.

working in is getting replaced by new and better technology, it is time to do some serious career replanning and retooling.

Another equally important factor to consider in you career planning is the product life cycle. Many engineers plan their career solely around a single product and act surprised when the product is phased out after a year or two and so are their jobs. A typical product life cycle for various industries is shown in Figure 2-13.

The shortest is the consulting product life cycle; many engineers try to become a consultant hoping to provide their services to many companies. However, the typical consulting contract (in my experience) is usually less than 1 year. This means the engineer must be looking for a new job on a yearly basis. In the commercial industry like automobiles, TVs, computer, and personal electronics, the typical product life cycle is 1–2 years. This means the engineer must be willing to change products and incorporate new improvements and technology every 1–2 years. The aerospace and defense industry typically have a longer product life cycle ranging from 3 to 5 years. However, the products in this industry are tied to government spending and funding changes yearly. So, the engineer must be attuned to government budgeting and spending on a yearly basis.

In all cases, the product life cycle should be considered in your career planning. What part of the product life cycle is your present job supporting? What are the future products of your company? Does your product have funding and plans for updating? Is your product just one in a family of products and can you work on others if the product goes away? These are all questions you should be exploring during your career planning.

STRESS RELIEF PLANNING

One of the major factors in successful careers is being able to handle all the stress that comes along with the job and home life. Do you have a stress relief plan? To be successful and handle all the stress that is going to be thrown at you, a stress relief plan is essential. A graphic depicting the combinations of stress you are going to face during your career and family life is shown in Figure 2-13.

Along the bottom are shown your age and various stressful conditions that normally occur. Along the left-hand side are shown the job levels and on the right-hand side the age of your children.

The key in studying this chart is to determine when you may be a great candidate for overstressing and making sure you have a plan in place to deal with this stress in a positive manner.

In studying the diagram, you can see that at various times in your life you are going to be subjected to an unusually large amount of high stress. The high stress job levels are related to first and second line supervisors. During your lifetime (age axis), stress is high when you first leave school, start a new job, and possibly get married. Later on in life, another high stress condition is in your 30s and 40s when you may return to school. The final stress generator is your family or children. A high stress time is when your children are first born and when they reach the teenage years. You can see from the graph that at certain times you will have three to four high stress conditions all occurring at once.

The most stressful conditions seem to occur all simultaneously. The worst case is when you have recently been promoted to supervisor, starting a new

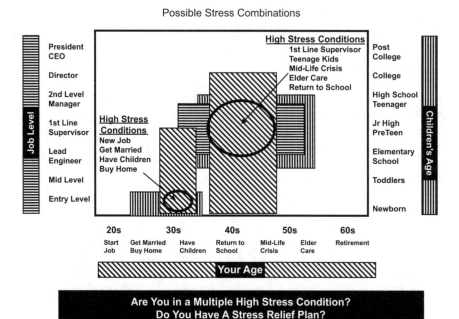

FIGURE 2-14   Possible stress combinations.

project, returning to school for an advanced degree, and your children have reached their teenage years and are still living at home.

Are you prepared for these conditions? Are you presently in a high stress situation and don't realize it. Do you have a stress release plan? Check in your local bookstore and purchase a book on how to relieve and handle stress. Do you exercise regularly? Do you take time out for yourself? Do you reward yourself after each significant milestone? Having a great stress relief plan will allow you to perform better, not take problems out on people, and keep you calm and in control when all others are not.

▶ *Career Tip.* Having a great stress relief plan will help you perform better and accomplish more under high-pressure situations.

Leave your personal problems at the door when you go to work in the morning and when you leave at the end of the day try your best to leave work problems at the office. If you do not have a stress relief plan, start working on one. How about taking your significant other out for dinner and putting a stress relief plan together.

DEALING WITH CHANGING SUPERVISORS—REBOOT TIME

One final aspect (I recommend) you should take into consideration while you are executing your career plans is to make sure your career plans are not solely dependent upon your supervisor. Many engineers will develop career plans entirely dependent on their supervisor promoting them into the next level. Then halfway through plan and with the promotion nearly in sight, the supervisor leaves and a new one is assigned to lead the department. When this happens, the engineer must start all over again from the beginning to prove to the new supervisor they are ready for the promotion. It is like entering all the data into your PC and before you complete the analysis the system crashes and you lose all the data. You must reboot and start over. This is exactly what happens when your supervisor changes. You start all over trying to prove you are ready for the next promotion.

In my 34 years in industry, I have had over 65 different supervisors. Typically, over a 30-year period it is not unusual to have 50 or more different supervisors. You must be prepared for this when you develop career plans. Sometimes supervisors can change as often as every 3 months.

Your supervisor will play a significant role in your development and promotion, but you must be prepared in case your supervisor suddenly leaves. It is called contingency career planning and is absolutely necessary if you are going to survive. Your career plans should include your supervisor, but not solely dependent upon him or her.

If your supervisor does have to leave due to an assignment change, make sure they acknowledge that you are on track for a promotion before leaving. Ask if he or she will discuss with the new supervisor your expectations on promotion. The supervisor who is leaving can brief the new supervisor on your progress and promotion plans. This will bridge the transition from the old supervisor to the new supervisor and hopefully minimize the impact on your promotion plans.

Another career move is to discuss the changing of your supervisor with the manager who is the next level up from your supervisor. If the manager agrees you are on track for a promotion, then you have protected yourself against the supervisor leaving and having to start all over with the new supervisor.

▶ *Career Tip.* Your promotion is a year away—get started now!

Get started on your career plans now; most significant career accomplishments or promotions take years to accomplish. You can significantly reduce the time to your next promotion with a great career plan.

Experience has shown that when engineers finally start to work for a promotion in earnest, it is usually a year or more away before it actually occurs. So if you want to be promoted a year from now, get started today.

Who is going to do your career planning? Only you can. Career planning is not hard and the more you do it, the easier it becomes.

SUMMARY

Career planning is a never-ending task. It is personal to everyone and only you can do it for yourself. Not having a career plan is like driving to a place you have never been without directions or a map. A good career plan guides you along, provides you with a sense of purpose, and even acts as a shock absorber during bad times. Career planning takes time and it is an iterative process. Your plans will need to be updated yearly and even sooner if you are in a highly dynamic environment.

You should have short and long-term career goals. Your career plan should include work as well as personal goals. The best plans have balance in them. When you are out of balance it will take effort to get things back in balance, they will not go back naturally.

Get started on your career plans now; most significant career accomplishments or promotions take years to accomplish. You can significantly reduce the time to your next promotion with a great career plan.

Have you identified any career actions you want to take as a result of reading this chapter? If so, please make sure to capture these ideas before you forget by recording them in the notes section at the back of the book.

ASSIGNMENTS AND DISCUSSION TOPICS

1 How long should it take to develop a career plan and how often should you update it?
2 What are the benefits of generating a career plan?
3 Name some danger signs that indicate you should be doing some career planning.
4 How do technologies, product life cycles, and industries affect your career planning?
5 Is stress relief important?

REFERENCE

1. Covey, Stephen, *The 7 Habits of Highly Effective People*, First Fireside, 1990.

CHAPTER 3

CAREER STRATEGIES
WHAT WORKS AND WHAT DOESN'T?

In this chapter, you will begin to map out a career strategy to obtain the goals identified in the previous chapter. There are many career strategies discussed in this chapter. Your challenge is to identify which career strategy you are going to use for achieving your career advancement. All of these strategies work and have been utilized by engineers to achieve their career goals. These strategies only work when combined with hard work and excellent performance. The strategies fall apart when the engineer does not fill the basic requirement of excellent performance on the job. Even though you may have an excellent strategic career plan, it is no guarantee for advancement. Good career plans can sometimes lead to failure; in this chapter we also discuss what doesn't work, so you can avoid these costly mistakes.

One of the first things to consider in developing your career strategy is your overall philosophy on your employment. Your options are to develop a career strategy based on the philosophy to work for the same company for your entire career versus a philosophy of moving from company to company to attain career advancement. The career option of remaining with the same company, in the same department for your entire career is what I refer to as the "One Department for Your Career Strategy."

DEVOTED TO ONE DEPARTMENT FOR YOUR CAREER STRATEGY

This career strategy is shown in Figure 3-1. As you grow and mature and are promoted, you always remain in the same group or department of the same company. This is becoming a very rare event lately due to all the company buyouts, workforce reductions, and short product life cycles. However, I have also witnessed this career strategy work successfully. I have

The Engineer's Career Guide. By John A. Hoschette
Copyright © 2010 John Wiley & Sons, Inc.

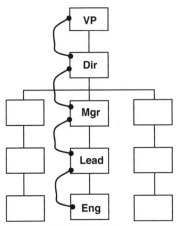

FIGURE 3-1   Remain in one department for career strategy.

- Remain in the same company and group for your entire career

- Becoming a very rare event

- Best retirement benefits

- Requires patience and ability to handle ups and downs of product and technology cycles

- Depends upon group attrition for advancement

- Requires one to constantly outperform outsiders

met people who have been with the same group and product area for over 30 years; they are forever loyal to their group and usually enjoy their work greatly.

From a retirement benefit point of view and vacation accrual, this is the best strategy. Your retirement benefits are best since most companies offer higher retirement benefits in exchange for longer years of service. In addition, many companies in the defense industry offer the 85 points plan that allows early retirement. This benefit is simply that you can retire early when you have reached 85 points. You receive one point for each year of service and one point for each year of age.

This means that if you started with a company when you graduated from college at age 22 and worked for the same company your whole career, you will have 85 points when you become 54 years old in this example. In this example, engineers who have stayed with the same defense company their entire career often elect to retire early at 55 when they have their 85 points. However, this benefit has become pretty much obsolete by many companies due to financial reasons with the elimination of pensions.

Another great benefit of the forever loyal strategy is that you accrue the maximum allowed vacation which, in most companies, can be as many as 6–8 weeks per year. This means you work only 10.5 months out of the year and even less when you consider all the paid holidays. You often reach this level after 20+ years of service. Thus for a 35-year career, you only work 10–11 months of the year for the last 10 years of your career.

Many engineers capitalize on this large amount of vacation when it comes to retirement. In the last few years before retirement, engineers only use a portion of their vacation each year and bank their unused vacation. They do this since most companies allow you to carry over any unused vacation from year to year. After several years of banking unused vacation, the engineer can retire with 8–10 months of vacation in reserve. The company pays the engineer for all the unused vacation at retirement, which is like getting a bonus of 8–10 months salary on the day of retirement.

There are special challenges to being successful with the forever loyal strategy. One of the special challenges is being able to handle the up and down cycles of the group. The group will go through periods of growth and decay as well as products will come and go. The engineer must adjust and be open to these changes to remain in a viable and productive career.

As the old products phase out, new ones will be introduced and the engineer must have the skills and training for the new products. Contracts will also come and go. There will be periods when virtually no work exists and the company must carry the salary of the engineer. The decision to continue to pay the engineers even though work is scarce requires the backing of upper management. Few companies do this since the emphasis is so great on the bottom line. So if you plan to stay with the same group, you should explore what will happen during downturns and loss of contracts. Make sure your management is committed to paying your salary and will support you during these times.

The next challenge for the engineer in the "Forever Loyal to One Group Career Strategy" is that their career advancement be tied to attrition in the group. The engineer's only opportunity for advancement comes when someone above them leaves the group and an opening becomes available.

The engineer should monitor how often or how long before the people in the group above them leave? If the average turnover rate for people in the group is 30% in 3 years, then according to the mathematics, the engineer should be the senior person in the group within 9–12 years, based on the assumption that they move up every time someone leaves. This is a great career growth rate.

If, however, the employee turnover rate is very low for your group then chances of advancement seriously diminish. How does one find out the turnover rate? Simply ask around how long each person in the group has been in the group and what happened to the people who left and the reasons for leaving. If most of the staff have been in the group 5 years or less, you are in a very high turnover rate. If most have been there 10–15 years, you are in a very low turnover rate group.

There may be other career barriers within a group. The following chapters will discuss typical career barriers within a group and methods to get over them.

What if the group is new and there is no history? You need to observe the group's operation over time. Relax and monitor the products, technology, and turnover rate of staff. If senior workers start bailing out, this could be good or bad. If the products and technology are solid, then remaining in the group is a good career move. If the products are becoming obsolete and customer demand for the products is diminishing, then maybe you should consider moving to other groups. However, people who are loyal will stay until the last opportunity is gone or help turn the situation around. For their loyalty, and choosing to stay, they are often rewarded with senior level positions.

However, as I have found, simply staying is not a guarantee that things will automatically turn around. I tried staying to the end believing the market and economy would turn around, but it did not and found myself laid off. It was the early 1990s and my division had over 700 employees laid off. The decision to stay or bail is strictly an individual one based on specific circumstances. There is no guaranteed right or wrong decision, it is your call on whether to stay or move. I learned the hard way. Be extremely aware of your company's profitability and future business plans.

▶ **Career Tip.** The best career move is to continuously gather all the data you can about the company, the group, its people, the customers, and future business outlook.

Another special challenge to consider, if you choose the forever loyal career strategy, is how are you going to keep up with product and technology changes? You will need to do technology updating and have the ability to introduce new, successful products to replace the older obsolete ones. In fact, being the engineer assigned to introduce a new technology to the group is a great career move. Once you have obtained the necessary training for the new technology, you become the group's senior person, recognized expert, in this area. These engineers are the ones that often get promoted to lead positions. Just make sure the new technology is going to significantly enhance the product performance. If the new technology fails, then you are in a poor career advancement position.

To remain on top of the same group over an extended number of years, you will need to constantly prove your worth and outperform others in the group. Remaining loyal does not mean you get to sit back and rest. You must continue to improve and expand your horizons. You must be the one introducing new methods and technologies into your group.

MOVE DEPARTMENTS, BUT STAY IN THE SAME COMPANY

The next strategy and more common, but still rare these days, is the "Move Departments, but Stay in the Same Company" career strategy. This strategy is shown in Figure 3-2.

For this strategy, the engineer remains in the same company, but moves from department to department following the best contracts and jobs to obtain promotions. This type of career path is very typical for most people staying in the same company their entire career.

The engineer may start in one department, get a promotion then move to another department and another job to get the next promotion. For this strategy, the engineer must constantly be monitoring all the other groups in the company and determining the next move. This is a great career strategy

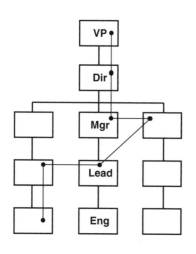

- Remain in the same company and follow the best program, contracts, and jobs

- Typical career of people remaining in the same company

- Working for the same company entire career is becoming rare event

- Best retirement and vacation benefits

- Requires ability to move within the company (special skill)

- Isolates your career from dependence on one product

- Looked upon favorably by upper management since engineer has experience across many product lines

- May require moving to other divisions in other states or countries

FIGURE 3-2 Move between departments, but stay in the same company.

when you are up against department barriers and getting a promotion in your present department is highly unlikely. It also makes you a more valuable employee since you can work in multiple groups.

This career strategy has excellent vacation and retirement benefits since the engineer stays with the same company and accrues the maximum benefits. Another great aspect to this career strategy is that it isolates your career from dependence on a single department or product to be successful. When one product or department has a downturn, you simply shift over to another department or product line hopefully without losing seniority or pay. Most often if you coordinate well enough in advance, the new group may offer you a promotion to make the transition.

This career strategy may or may not have some negatives depending on if you want the opportunity to live in and see different parts of the world. With some companies, moving departments may require a move to another state or even another country. Many large corporations have operations in foreign countries and require their upper level managers to work in these foreign countries to gain the experience necessary to direct the division one day. Accepting a position in another division may mean moving away from family and friends. Are you prepared to do this?

PICKING MAINSTREAM JOBS AND DEPARTMENTS

If you select to stay with the same company, your next strategic career planning exercise is to map out the career path flow within the company. A typical large company pyramid structure is shown in Figure 3-3. At the bottom of the pyramid are the engineers. Moving up the chain are the lead

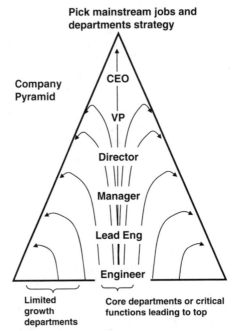

FIGURE 3-3 Core department career strategy.

engineers, managers, directors, VPs, and CEO at the top. Career paths are represented by the arrows.

In all companies, there are mainstream career paths leading to the upper levels. These mainstream or core departments or critical functions become the supply chain for the upper levels in the company. As shown in Figure 3-3, up the center of the pyramid are the core departments that feed the leadership of the company. The centerline departments perform the critical functions that become the backbone of the company. Without their successful operation, the company would not survive.

The career strategy here is to always work for a core or critical department. This career path leads to highest levels in the company. A good example of this might be in a chemical company, where you are a chemical engineer working on the most successful product and you control the key processes to make the product.

To either side are the noncore departments that support the organization but their career paths often stop short of the top. The engineers work in departments that help support the company and are critical to its success. A good example of this might be a software network engineer working in the information services (IS) department of a chemical company. The engineer supports the network but it is not a critical department that generates revenues for the company. In fact, this department is just the opposite and considered an expense that most upper level executives want to minimize when expenditures are high. It is highly unlikely that the manager of the IS department is going to be promoted to VP or CEO of the chemical company.

How do you map out the career path flow in your company? Easy, just ask around the senior level people which departments they have worked in during their rise up the corporate ladder. If you can, find out what departments the previous four or five CEOs worked in prior to becoming the CEO. Once you do this, you will quickly learn what the feeder departments are for the upper levels.

▶ *Career Tip.* Before moving into new departments check to see if they are a core business department.

The condition of career peaking is represented by the arrows moving outward and eventually peaking and turning downward. I have shown the arrows peaking and turning downward since this is what research shows happens to most engineers' careers. The engineer's career reaches a peak. If they are capable, they will remain at that level until the end of their career. However, if the engineer is not successful at the peak, they are often replaced and must accept a lower level position at which they are capable of being successful.

Another career phenomenon is that people are promoted on the basis of their skills and at some point they reach a level where the next level up requires more skills than they possess. However, this is not found out until they are promoted to a level they are not capable of performing. This is often called the "Peters Principle" of management after it was recognized by Tom Peters in his book *Thriving on Chaos*. Sometimes people refer to a manager and ask, how did he or she get that job, they are totally incapable of performing at that level? And the answer usually comes back, "You have heard of the Peters Principle: haven't you?"

As the engineer approaches retirement, one career move is to accept a lower level position to reduce the responsibility and stress in their life. It is also part of the replacement plan in which you step aside and train your successor. Your career strategy should include a growth and development strategy as well as what is referred to as an exit or transition to retirement strategy.

LEVERAGING THE RESEARCH GROUP

Another career strategy when working for the same company includes leveraging the research group. This strategy is shown in Figure 3-4. The leveraging research group strategy is basically one of working in the company research group or advance development group, then transitioning to an upper level position in the division following the natural product development flow into the division. This career path is shown by the dotted lines.

**Work your way up the chain or
leverage research groups**

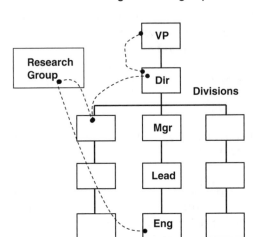

FIGURE 3-4   Leverage the research group career strategy.

Most corporations have a research or advance development group that is responsible for developing new products and getting them into production. Engineers who work in these groups are naturally the most knowledgeable about the product and would make the best manager when the product is transitioned to production. Therefore, a good career strategy is to leverage an assignment into the research group that later leads to a leadership position in the division. Your career path lies along getting an assignment in the research group, developing a new successful product, and then following it into production as the leader.

Having a successful career and remaining with the same company your entire career requires you to understand the career paths within the company. Learning about these career paths, will require you to do research about the company's product development flow and engineering structures your company utilizes to create and sell products.

FOLLOWING HOT COMPANIES AND PRODUCTS

The next and most common among engineers' career strategy is what is called "Following Hot Companies and Products." This career strategy is shown in Figure 3-5. For this career strategy, the engineer moves between several companies and departments to gain promotions. For example, as shown in Figure 3-5, the engineer realizes that in his present company the opportunities for advancement do not exist, so he or she makes a lateral career change from company A to company B that has more opportunities. Once in company B he switches departments and gains a promotion to lead engineer. The next move again involves a lateral move to a new company C. However, once in

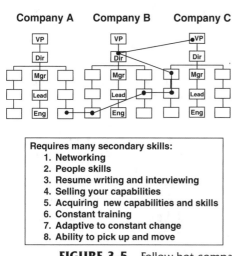

• Change companies to follow the best
program, contracts, and jobs

• Becoming most common (two to three
company changes is very common)

• Reduced retirement and vacation
benefits

• Isolates your career from dependence
on one product and one company

• Looked upon favorably by upper
management since engineer has
experience across many product lines
and companies

• Constantly learning new products,
technologies, and industries

• May require moving to other companies
in other states or countries

• Too many changes and becomes liability

Requires many secondary skills:
1. Networking
2. People skills
3. Resume writing and interviewing
4. Selling your capabilities
5. Acquiring new capabilities and skills
6. Constant training
7. Adaptive to constant change
8. Ability to pick up and move

FIGURE 3-5   Follow hot companies and products strategy.

company C he gets promoted to the manager level. Then the engineer returns to company B to receive a promotion to the director level. Finally, the engineer returns back to company C for a vice president level position. In this example, the engineer is changing departments and companies to follow the best programs, contracts, and jobs.

This type of career is becoming the most common due to the rapidly changing technologies, products, and companies. It is becoming the norm for engineers to have worked at several different companies in their lifetime. This career strategy isolates your career advancement from being dependent on one product or company.

I have personally changed companies twice without changing location, job, or even desk. The old company was acquired by a new company and everything remained the same except for the name of the company.

Changing companies is often looked upon favorably by upper level managers since the engineer has been exposed to the best practices of several companies. Hiring managers look favorably on the company changes since the engineer brings along all this experience to the job when they are hired. To survive and advance using this career strategy, the engineer must know how to handle and adapt to change.

The engineers who have followed this career path have said they enjoy the change since they are constantly learning new techniques, new technologies, and developing new products. They say they would stagnate if they had to work on the same product their entire career.

How do you find the hot companies and products? Simple. You network with people at conferences, conventions, or engineering society meetings. Ask people if their company is hiring and hand out your business cards.

Another way is to search the web for companies hiring. Have you attended a job fair recently? These are usually the hottest companies since they are often willing to extend second interviews or even make job offers on the spot.

▶ *Career Tip.* Networking with people at conferences, conventions, and engineering society meetings is great for your career!

There are also some disadvantages to this career strategy and the most obvious one is having to move your family to new cities, states, and even countries. If they are unwilling, this strategy has the pitfall of losing your family for your job. Also, if the engineer changes jobs too often, employers might interpret the large number of moves as an indication that the engineer is a problem employee and is the reason behind the engineer's changing jobs so often. To get around this image make sure your resume clearly indicates that you moved because of a better opportunity.

Additionally, regions in the United States have totally different attitudes regarding whether or not staying with the company or moving is good or bad. For example, many employers in the Midwest portion of the United States view changing your job every 5–7 years as an indication that you are a problem employee. These Midwest companies also look at employees who have stayed 10–15 years at a company as very desirable employees since they are not going to jump ship any time soon.

However, for companies in the large cities on either coast, moving every 5–7 years makes you a more valuable employee since you are probably working on the hot contracts and exposed to the latest methods of multiple companies. These companies view the engineer who has remained at the same job for long periods of time as stuck in their old ways and stagnant. They view the engineer who has only worked at one company as undesirable rather than a valuable asset. You must be able to explain your decision to remain at one company and the benefits you have for doing this.

Another negative aspect of this career strategy is the lack of, or reduced, retirement benefits, and vacation benefits. Changing your employer every few years results in rebooting or starting over on your retirement benefits and vacation accrued unless you negotiate this to your advantage when moving. Nonetheless, the engineer should make sure they get the maximum possible increase in salary each time they move.

I have done this analysis several times in my career at which time I have evaluated the career options on whether I should move or not. And I keep coming up with the same answer. For a move to another company to be profitable and a real improvement, the company must be offering 10% or more than your present employer is offering. That is a 3% raise to stay at your present job and employer is equivalent to a 10% raise from a potentially new company where you have to move your family. The reason for this is all the hidden costs with making a career change involving changing your residence.

When you start to add up hidden costs you incur when moving, real estate fees, moving costs, buying a new home, obtaining new licenses, permits, and so on, you quickly find moving to a new company may not be in your best interest. Perhaps you may even have to consider the spouses' loss of employment and income. When you consider all these hidden factors, staying with your present employer and accepting the 3% raise is not so bad after all. If you decide to make the move, try to negotiate that all of these hidden costs be covered and get help in obtaining employment for your spouse. But you must also be prepared for the event where the new company does not support reimbursement of all your hidden costs.

"Following the Hot Products and Contracts" career strategy requires the engineer to acquire some new career skills as listed in Figure 3-5. The engineer must have excellent networking and people skills to learn about these new opportunities. In addition, the engineer must be able to obtain interview offers as well as have excellent resume writing and interviewing skills. Constant change and moving becomes a way of life and the key to survival is being adaptive to change. Finally, the engineer must consider the impact on the family and obtain support from the family members for each move.

Our family has found that with each move it took about one year before things returned to what might be considered "normal" again. My wife and I made new friends, the kids accepted their new schools.

When I lived in Los Angles, Boston, and San Jose, fellow engineers would tell me to take a map of the city and mark the locations of all the potential companies that I might work for. Then draw a line from one to another and where all the lines intersected would be the best place to purchase a home. Purchasing a home close to the intersection point would minimize the chances of having to change your residence if you should happen to change your employer.

UP AND DOWN THE SUPPLY CHAIN CAREER MOVES

This career strategy is very common and utilized by most engineers at some point in their career if they move from company to company. The basic strategy is shown in Figure 3-6. Movement up and down the supply chain makes good career sense when considering a company change for your career move. As shown in Figure 3-6, the top of the product supply chain is what is referred to as the prime contractor. The prime contractor usually supplies a product to the customer made up of lower level products integrated together to form the final product sold to the customer. Good examples of a prime contractor might be a car, aircraft, or PC manufacturer.

The first level below the prime contractor is what is referred to as first level tier suppliers who provide major assemblies to the prime contract. Good examples of these types of products might be engines, radios, or monitors. The next level down is referred to as the second level tier suppliers and these

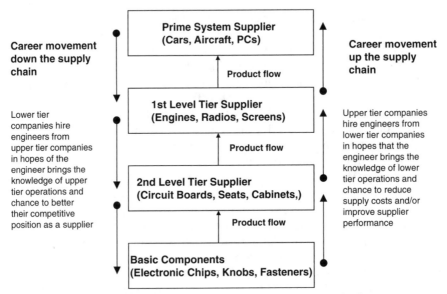

FIGURE 3-6   Career moves up and down the supply chain.

suppliers provide basic subsystems to the above level. Examples of these products might be circuit boards, seats, and cabinets. Finally, at the lowest level is the component supplier who may supply such things as electronic chips, knobs, and fasteners.

Career movement up and down the supply chain is a good career option and used quite often in the industry. Movement either way, up or down the supply chain, can be beneficial to your career.

Career movement down the supply chain can be beneficial for several reasons. Companies down chain like to hire engineers from up the chain in hopes that the engineer will bring valuable insight into their customer operations. Hopefully, the engineer brings valuable insight that can be translated into better performing products that the customer will purchase more of. The engineer coming from up chain has the opportunity to see the products in use as well as their defects. He or she will bring a wealth of technical knowledge about the use of the product down chain. The engineer is in a good technical leadership position. The supplier also hopes that by hiring engineers from up chain, hopefully they will have a better competitive advantage with the customer. The engineer is hired in hopes that they had a great personal relationship with the customer and these relationships can be capitalized upon for the benefit of the company.

Career movement down the chain is not a good career move when you leave the upper level on a bad note and go to one of the suppliers. The upper level company may in fact look at the move as bad and it may harm the lower level suppliers' relationship. It could ultimately lead to loss of business and ruining your career. Correspondingly, moving up the supply chain and leaving the lower level supplier under bad circumstances may also turn out to be a bad career move. If for some reason the supplier loses business with the upper level customer, it could be blamed on you and potentially lead to lawsuits.

Career movement up the supply chain can also be a very beneficial career strategy. Customers often hire engineers from the suppliers down chain to help with technical issues and problems encountered in the products they buy from suppliers. These engineers are considered a valuable asset in dealing with suppliers. They understand the technical issues of the product best since they came from the supplier.

The down side of moving up the supply chain is the engineer no longer does the same type of product development. He is responsible for procuring the product and not developing. The engineer is typically put in a broader assignment involving integration of multiple products.

The key for good career movement up and down the supply chain is to leave the company under good conditions. Go through the effort to make sure you leave the company on a good note; make sure they understand this is a good opportunity for your career.

SELECTING THE BEST CAREER MOVE FOR YOU

Selecting the best career move for you is not a simple task. It will take considerable thought and analysis every time you decide to make a career change or move. To aid you in the process and help you determine what the best career move is for you, the simple principles of systems engineering can be applied.

Shown in Table 3-1 is what is referred to as trade study matrix. Down the left column of the matrix is a list of the key factors you want to compare for the various career moves you are contemplating. Across the top row of the matrix are the options you have identified.

This is an example to help you construct your career options matrix. For the example shown, some key factors are identified when considering a career move. These include chances for promotion, better job assignments, office and lab facilities, impact to family, supervisor, interesting work, expanding and growing workforce, and long-range impacts. The second column lists the weighting of factor in scoring the final decision. Each factor is allocated a percentage of importance in the total. All the factors weight add up to 100%.

For this example the highest weight factor is chance for promotion weighted at 25. This is the most important factor for the engineer and carries the heaviest weight. When constructing your matrix assign the highest weights to the factors you consider the most important and lower weights to the ones you consider the least important. For this example, the lab facilities and long-range impacts were rated the lowest.

TABLE 3-1 Career Options Rating Matrix

Factors in Consideration	% of Total Importance	Option 1, Stay in Job	Option 2, Change Departments	Option 3, Leave Company
Chances for promotion	25	5	20, job postings in HR 1 grade level up	10
Better job assignments	15	2, same old assignment	15, new and more challenging	10
Office and facilities	5	5	4	3
Impact on family	15	15	15	10, move to new town
Supervisor	10	2, disliked supervisor	10	6
Interesting work	10	2	8	4
Expanding and growing workforce	15	5	12, new contract in other departments	6
Long-range impacts	5	5	5	1 (unknown)
Total score	100	41	89	50

The next three columns to the right are the options available to the engineer. They have identified three options, stay in the job, change departments, or leave the company. The next step for the engineer is to assign a score to each of the factors for the various options. We can see from this example, the engineer has scored all three options. Staying in the present position (option 1) had a total score of 41 which was lower than changing departments (option 2) score of 89, and leaving the company (option 3) score of 50. The engineer ranked his best career move would be to change the departments. Note the reasons for the various rankings entered into the cells.

Using an options matrix with weighted factors is a very powerful technique that allows the engineer to see the sum of all the factors being considered at once. It is highly recommended that you use this technique when considering career moves. It has helped me make some very difficult career decisions during my career.

My family was faced with a very difficult decision when I lost my job. We had to decide whether we should move out of state and follow the jobs or should we stay in state and I change careers to something other than engineering. To help the family make this decision I utilized the option-rating matrix. I called the family together and we discussed and listed key factors. After the key factors were identified and agreed upon we all put in weighting factors. Then each of us ranked the factors for each option. All things considered, our best option was to move out of the state and the choice was more readily accepted by the family.

COMPANY HIGH TALENT AND LEADERSHIP DEVELOPMENT PROGRAMS

Most major corporations have realized that they need to be constantly training their future leaders. To this end, they develop with the help of their managers and Human Resources department what is referred to as the "High Talent List." Are you aware of this list at your company? If not, ask your supervisor if a list like this exists. If it does, ask if you are on it. If the answer comes back yes, you are on the list, then you are in great career shape. Why? Because the people on this list are usually the star performers and recognized future leaders of the company.

▶ *Career Tip.* Investigate to see if your company has a leadership development program for engineers and if so join!

Another great career move is going to your Human Resources department and asking if the company has any leadership development programs. These are specially designed programs to provide training to high talent junior engineers who show a potential for becoming future leader of the company.

The programs are often structured to provide career guidance and opportunities to engineers. The programs may call for the engineer to return to school to obtain an advanced degree. Other programs have the engineers in a job rotational assignment. The engineer rotates jobs throughout the company every 6–12 months. Now this is the way for career advancement, having the Human Resources department sponsor you and get you highly desirable job assignments every few years.

It is recommended that you look into your company's high talent or leadership development programs. If you are not in a leadership development program, then find out the criteria to become part of the program and work as hard as you can to meet these criteria and join. This is what is referred to as a career accelerator! Or for you Star Trek followers "Moving at Warp Speed!"

SUCCESSION PLANNING—YOUR KEY TO MOVING UP

One of the key activities every manager has to perform is succession planning. This is simply identifying a potential replacement for key individuals in the group including themselves. If you are a junior engineer, are you on the succession list for the lead engineer spot? If you are a lead engineer, are you on the succession list for the manager spot? One good career move is to explore with your supervisor if you are a succession candidate. If you are considered a succession candidate, keep performing and be patient for when the next opportunity arises you may be the leading candidate.

SUMMARY

All successful career strategies are based on hard work and excellent performance on the job. There are many career strategies you can follow that span the spectrum from staying in a single company for your entire career to continually changing companies to advance. All these strategies work and your task is to select the career strategy you are most comfortable with. Most engineers employ a combination of these strategies to advance and you may find that you need to use different strategies throughout your career in order to advance.

Changing companies and jobs is not a simple decision and involves many factors like future promotions, salary, job assignments, impacts on the family, and impacts on your long-term benefits. When you are considering a job change, give consideration to more than just the salary increase. Other companies may offer you a larger salary but it may not cover all the expenses you are going to incur to make the move. Keeping your present job with a smaller raise may be the best when all the other factors are considered. Finally, take advantage of programs your company have which offer special advanced training or leadership training; taking advantage of these types of programs are always excellent career strategies.

Have you identified any career actions you want to take as a result of reading this chapter? If so, please make sure to capture these ideas before you forget by recording them in the notes section at the back of the book.

ASSIGNMENTS AND DISCUSSION TOPICS

1 Discussion topic, "Is it better to stay with one company or move from company to company to obtain career advancement?"

2 How can you find out about the career paths in another company before you make the decision to leave your present company and go to a new one?

3 What are the core departments in your company which lead to executive management?

4 How do you leave a company on a good note?

5 The final homework assignment is to determine what your career strategy is? Is this good for all time? How often should you review your decision?

CHAPTER 4

SUCCESSFULLY MAKING A JOB OR CAREER CHANGE

KEY FACTORS IN MAKING A JOB OR CAREER CHANGE (COMPANY, TECHNOLOGY, INDUSTRY)

Tightly coupled and often interwoven with the previous career strategies is the type of career change you are making when changing jobs, departments, or companies. There are three major factors that the engineer must consider when making a job or career change. These factors are the company, the technology, and the industry. The first factor when contemplating a career change is the company. The obvious consideration for this is whether or not to stay with the same company or move to a new one.

The next factor is the technology or degree you must have for your new career choice. Are the technology and the corresponding degree the same or you must return for further training and obtain a new degree. An example is making a career move that requires you to be a software engineer when you are presently a chemical engineer. To be successful in your transition to a new career you will have to return for further training and may even have to obtain a new degree.

The final factor to consider is the industry in which you will be employed. Is the career choice in the same industry or will your career change move you to a new industry? An example might be moving from the automobile industry to the food industry.

A model to help you think through your career change choices is shown in Figure 4-1. The career choices have been mapped into a cube. Along one axis are the company choices, same or new. Along the next axis are the technology choices, same or new. And the final axis shown is the industry choices, same or new.

The Engineer's Career Guide. By John A. Hoschette
Copyright © 2010 John Wiley & Sons, Inc.

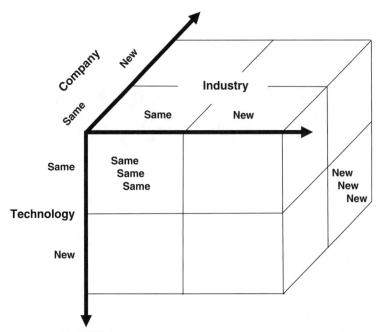

FIGURE 4-1   Career change choice cube model.

The upper left cube at the origin is where you are presently located. Making career changes and moving to different locations in the cube will involve varying degrees of risk and different types of career actions to successfully support the change.

The safest move and lowest career risk is normally changing your job where you remain with the same company, the same industry, and the same technology. This career move is shown in Figure 4-2, where a career change of this type keeps you all within the (same, same, same) cube.

This is considered the safest career move since you minimize the impact and amount of change to your career. You stay with the same company, in the same industry, and are working in the same technology field. It is not risk-free but risk has been greatly minimized. If you want to minimize risk and keep the amount of change in your life as small as possible, making a job change of this type is what you should seek.

The next type of change involves moving along a single axis where you only change one factor and keep the other two factors the same. I refer to this as a one-dimensional change. You move in the model in one axis only, so you have two factors the same and only one factor changing (same, same, new). An example of this is shown in Figure 4-3 where you move to a new company and stay with the same industry and same technology.

The next level of risk-taking when changing your career is doing a two-dimensional move or diagonal move in one axis. This is shown in Figure 4-4. This type of career move involves changing two factors at once and represents

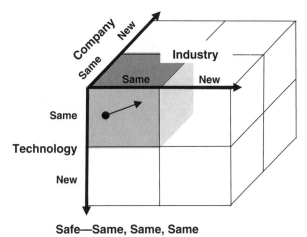

Safe—Same, Same, Same

FIGURE 4-2 Safest career change.

a high-risk situation (i.e., new, new, same). These types of career moves have to be well thought out with encouragement and help from your mentors and career counselors.

The highest risk career change you can make is when you change all three factors at once and move in three-dimensional diagonal (i.e., new, new, new). This career move is shown in Figure 4-5.

For us, probability engineers, the total number of possible combinations of career changes is equal to 2^3 or 8 possible combinations. These combinations have been ranked in Table 4-1 from the lowest to highest in risk.

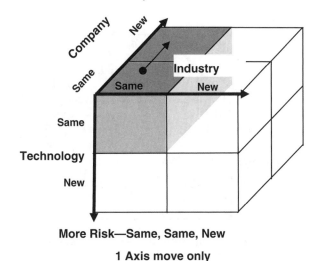

More Risk—Same, Same, New

1 Axis move only

FIGURE 4-3 Career change involving one factor.

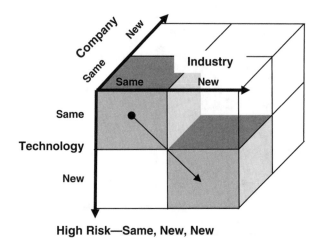

High Risk—Same, New, New
Diagonal move in 2D

FIGURE 4-4   Career change involving two factors.

In Table 4-1, the left-most three columns identify the factor changing and the fourth column identifies an associated career risk level for the type of combinational change. Rated the lowest risk and the safest career change is when you make a career change where you stay in the same company, with the same technology, and the same industry. Although totally not risk-free, this type of career change usually has minimum impact on your work, residence, and your family.

The next level up of increased risk is changing just one factor (company or industry). This level is rated more risk. If you are contemplating a career change of this type, please be aware that significant planning should go into making this type of change.

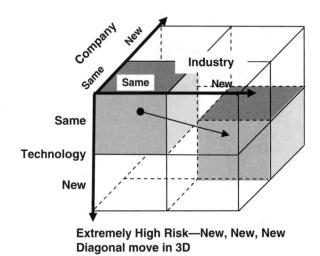

Extremely High Risk—New, New, New
Diagonal move in 3D

FIGURE 4-5   Career change involving three factors.

TABLE 4-1   Career Change Options

Company	Industry	Technology	Risk Consideration	
Same	Same	Same	Safest - Lowest Risk	Safe, but do planning
Same	New	Same	More Risk	Beware, significant planning required
New	Same	Same	More Risk	
Same	Same	New	High Risk	
Same	New	New	High Risk	Caution! Intensive planning required
New	Same	New	High Risk	
New	New	Same	High Risk	
New	New	New	Extremely High Risk	Dangerous! Get help

Are you prepared to do risk management?

Changing two factors or the special case where you change the technology or engineering degree you must have is rated high risk. Changing two factors at once or the single factor of your technology base represents a significant change in your career direction. It is not something to be taken lightly and extensive planning should be involved. Again, work with your mentors and career counselors if you are considering this type of high-risk change.

Finally, the highest risk or "total makeover" approach is to change all three factors at once. This is an extremely high-risk career move and the engineer should have professional help when making this type of change. A good example of this might be a nuclear engineer deciding to become a software engineer and leaving his company to go back to school to get a degree in computer science.

PLANNING FOR SUCCESS WHEN CHANGING YOUR CAREER

Key to making it through any change is identifying the risks upfront and have a risk mitigation plan for each one. The positives, negatives, and potential actions to reduce risks with making a career change for each of the factors are shown in Tables 4-2 through 4-4. The positive reasons for making a career change are generally the same for all three factors. People make career changes generally for better opportunities, better pay and benefits, and more interesting or challenging work. These are all very good reasons to consider changing your career.

Changing your company is the first factor we consider. The negatives for changing your company include potential home relocation. Others are loss of vacation benefits and retirement benefits. Also included in the negatives is loss of your people network. You are leaving all the people at your old company behind and the support they provided. Through years of experience you developed friends and technical points of contact that you could go to for

TABLE 4-2 Positives, Negatives, and Actions when Changing Your Company

Positive	Negative	Actions to Reduce the Impact or Risk
+ Better opportunity for advancement	− Potential home relocation	• Assuming you are still in the same industry utilizing the same technology:
+ Better pay and/or benefits	− Loss of vacation benefits	− Select a company close to home so you do not have to move
+ More interesting or challenging work	− Loss of retirement benefits	− Select a company you have dealt with previously and know about
	− Loss of people network	− Network and meet as many people as you can once you arrive at the new company
	− Must start over proving yourself; new person in the company	− Bring along examples of your work portfolio to share with people
		− Spend extra time learning the new policies and procedures of the new company
		− Learn about the products and customers

help. In the new company, you will have to start over developing your support network.

There are several actions you can take when changing your company to minimize the risk to your career. These risk reduction actions assume that you are only changing your company and not changing your industry or technology at the same time. Top of the list is to select a company close to your home so that you do not have to move or change your residence. You may also select a company you have dealt with in the past, maybe a supplier or customer and know about their products. Network and meet as many people as you can when you first arrive at the new company. Bring along examples of your work and take the time to show key people in the group you are joining some of your past accomplishments. Share your portfolio if you have one.

**TABLE 4-3 Positives, Negatives, and Actions when
Changing Your Industry**

Positive	Negative	Actions to Reduce the Impact
+ Better opportunity for advancement	− Loss of experience base (people, customers, and methods of doing things)	• Assuming you are still working for the same company and in the same technology
+ Better pay and/or benefits	− New industry standards	− Network and meet as many people as you can
+ More challenging work	− Loss of people network	− Learn about the products and customers
	− Must start over proving yourself	− Obtain training on the new industry standards
		− Look for visible ways to stand out
		− Highlight things you bring from your old industry that may improve things
		− Attend conferences and engineering society meeting in the new industry

One of the key actions is spending extra time at work to learn the new company policies and procedures. This may include signing up for special training and attending more than your fair share of training courses in the first year.

Finally, but certainly not least, find out about the company products and customers. Get yourself invitations to customer meetings; learn about the customer hot buttons. Attend program reviews and company quarterly status meetings whenever you can.

The next career change factor to consider is changing your industry. This discussion assumes you are still with your same company and you will be required to use the same technology background. A good example of this might be changing from the car industry where you are a mechanical designer to the food industry where you handle mechanical processing and packaging of food. The negatives, positives, and recommended actions when changing industries are shown in Table 4-3.

The positives for changing your industry remain the same: better opportunity, better pay, and more interesting work. The negatives change,

TABLE 4-4   Positives, Negatives, and Actions when Changing Your Technology

Positive	Negative	Actions to Reduce the Impact
+ Better opportunity for advancement	• Return for new education, costly tuition	• Assuming you are still working for the same company and in the same industry
+ Better pay and/or benefits	• Loss of experience base; re-entering workforce like a new graduate	− Obtain training or degree before you make change
+ More challenging work	• Loss of people network	− Learn about the products and customers
	• Must start over proving yourself	− Obtain training on the new technology standards
	• Potential pay reductions	− Highlight things you bring from your old industry or company that may improve things
		− Attend conferences and engineering society meetings in the new technology
		− Network and meet as many people as you can

however, from the previous example. Changing industry causes you to lose your experience base that includes people, customers, and methods of doing things. To counter this and minimize the impact on your career you remain working for the same company and can easily go back to your former group should you need them and ask for help.

In the new industry, you are going to be subjected to all new standards and methods for doing things. Hopefully they will not be that different so you can easily adapt. An example of this might be changing from knowing all the crash standards in the automobile industry to health regulations for packaging in the food industry.

Again, you will lose your people network and spend time re-building a new network of coworkers, customers, and suppliers. In addition, you will be the new kid on the block so you will have to prove your technical skills all over again to your team.

Actions you can take to minimize the career impact for changing industries include networking and meeting as many people as you can. Putting in the extra effort and spending extra time learning about the products and customers in the new industry is a must in order to survive in your new position. You should read old test reports and get training in the new industry standards as soon as possible. Look for visible ways to stand out and highlight your accomplishments. You could also highlight or share with coworkers the technology and methods of doing things that you bring from your past employer. Finally, a good way to get up to speed quickly is by attending industry conferences where you meet all the vendors and can obtain training by attending seminars.

The last career change factor to consider is changing your technology or degree. This discussion assumes you are still with your same company and you will be in the same industry. A good example of this might be a mechanical engineer who is considering returning to school to become a software engineer. Another example is an electrical engineer returning to school for a biomedical engineering degree. The negatives, positives, and recommended actions when changing technologies are shown in Table 4-4.

The positives for changing your technology remain the same: better opportunity, better pay, and more interesting work. Changing your technology is probably the hardest of the three factors to change and requires the most effort to accomplish. Returning for a new degree is not easy, to say the least. It could be very costly and require a significant multiyear investment of your time.

When you change your technology you lose part of your experience base. It is like re-entering the workforce at the college graduate level. Another negative is that you lose your people network and must start all over proving yourself. It may even include a salary reduction since you are starting over at the most junior level.

Even though there are very compelling reasons not to do this, it can be a very good move for your career if your technology base has become obsolete. Frequently, engineers who return for new degrees and return to work have very successful careers. Part of the reason is that they found new work where their old degree and new degree were both utilized, making them a highly educated and highly desirable employee.

To minimize the risk while making this change, one can obtain the training or degree before leaving your present company. Have your present company pay the tuition and hopefully help you find a job when you complete the degree. Another good move is to learn about the products and technologies that will utilize your new training. You can also minimize risk by learning the new industry standards. Once you start your new job make sure you highlight all the experience you bring from your previous job.

Network and meet as many people as you can in the new technology area. You have the option to network with the professors teaching the class, other students, and even customers and suppliers. Utilize your HR department and have them do a search for people in your company who already have the degree you are trying to acquire.

Making a technology or degree change is going to be very difficult and you need to call upon all the resources you have available: Human Resources department, fellow employees, professors, mentors, and career counselors.

SUMMARY

There are three major factors that the engineer must consider when making a job or career change. These factors are the company, the technology, and the industry. When making a job change, the safest move is staying in the same company working on the same technology for the same industry. Changing one of the three factors is higher risk and changing two or all three factors is considered the highest risk moves for your career. There are positives and negatives associated with any career change and knowing what to do to minimize the negatives is key to making a successful change. A highly risky career change can be successfully made when the proper actions are taken. These actions include networking with people who already have made significant career changes, educating yourself about the new area that you wish to change into, and getting professional help through career counselors.

For further information about making a career change, I recommend you search on the Internet using "Career Changes" or "Engineering Career Changes."

Have you identified any career actions you want to take as a result of reading this chapter? If so, please make sure to capture these ideas before you forget by recording them in the notes section at the back of the book.

ASSIGNMENTS AND DISCUSSION TOPICS

1   Which of the three factors (company, technology, industry) is the most difficult to make? Which is the easiest?
2   Do companies help you make a technology career change?
3   How long does it take to make a significant career change?
4   Is returning for a Master of Business a one-factor change only?

CHAPTER *5*

STRATEGIES FOR GETTING YOUR NEXT PROMOTION

There are multiple viable strategies employed by engineers to obtain a promotion and in this chapter we discuss the most commonly used ones. The first and most obvious one is getting a promotion in your present department.

GETTING A PROMOTION IN YOUR PRESENT DEPARTMENT

The first place to look for a promotion is in your present department and with your present supervisor. The best way to do this is to have an open and direct communication with your supervisor about the subject. In other words, have a career discussion. Pick a time and place that is comfortable for both of you. Start out by establishing that you want to discuss your career options within the department. Ask open-ended questions and listen to your supervisor's answers. Here are some good questions to ask:

- How do you see my career coming along?
- What would you recommend I do to get a promotion in the future?
- How close am I to fulfilling all the criteria for promotion?
- Am I demonstrating that I have the capabilities to be promoted to the next level?
- Do you have any specific criteria or capabilities you are looking for in order to be considered for a promotion?
- Are there any barriers or personal improvement areas that I should be aware of?
- What can I improve upon to demonstrate that I am ready for a promotion?
- How long do you think before someone of capabilities could be promoted?

The Engineer's Career Guide. By John A. Hoschette
Copyright © 2010 John Wiley & Sons, Inc.

These questions are structured, so they are not threatening in hopes that your supervisor will open up and have some direct communication with you on your chances of being promoted. If the answers come back that you are very near to getting promoted and they are seriously considering it, you are in good shape. You need to keep up the excellent performance and make sure your work continues to support the justification for your promotion.

If the answers come back that you are a long way from being considered for promotion then you need to identify what you are lacking and what you need to improve upon to be considered. The best thing to do is identify some specific area you can improve upon and discuss with your supervisor an action plan leading to your next promotion. If you feel that getting a promotion in your present department is not likely, then you come to the next strategy.

CHANGING DEPARTMENTS FOR A PROMOTION STRATEGY

Going to another department for a promotion is a viable immediate near-term career move you can make and probably the most common one used. New openings in other departments are constantly occurring due to new work coming in, rotational assignments where employees leave the department and even retire. If you use your networking skills, you can find out when these openings are going to occur in advance of them being announced. Fellow employees will tell you so and so just quit, or someone moved, or share their department plan to hire several new people. Knowing in advance when these openings will occur gives you a better opportunity to be ready and sell yourself when the official interviewing starts. Another good indication that openings are becoming available is the retirement party notices and "leaving the company luncheon" signs that are posted in the workplace.

Changing your department may be a good career move even when your present supervisor is happy with your work *and* you are growing and performing but they have given no indication of promoting you. Then what? Your career advancement has come to a grinding halt in the department. In Chapters 8 and 14, I discuss career tools you can utilize against this situation and how to find out the reasons for your lack of advancement.

Another good time to consider this option is when you are not selected for the promotion that exists within your department and someone else in your group is selected. This is a clear indication that your career advancement in this group may be limited and it may be the time to move on or you need to continue to improve your performance before you will be considered.

If you are performing on the job and getting good reviews, and you have not been promoted in the past 5 years, then it is also another good time to consider moving to another department.

Changing departments is not a good career strategy to follow if you make it the only way you get promoted and use it every couple of years to get the next promotion. Engineering managers look at how long you have been in a department when it comes to the interview for a new position in their department. If you are jumping to a new department too often (multiple department changes in less than two years), you might get labeled as a problem employee who is just getting shuffled around and your chances of being selected are significantly diminished. So this career strategy may help you, but also hurt you if you use it incorrectly.

LEAVING THE COMPANY FOR A PROMOTION STRATEGY

Going to another company to get the promotion is often referred to as jumping ship. This is a good near-term career move if the opportunity at the new company is a significant improvement in your career, or offers a greater long-term advancement potential than simply staying at your company.

Another time to utilize this technique is if, for some reason, your reputation becomes tainted at your present company and it is obvious you are not advancing because of it. Another reason may be that you made a big mistake and because of it, you are going to be passed over for near-term advancements. There is a saying in the industry I have heard many times over the years and at different companies by a multitude of engineers. The saying goes something like this: your great successes are only remembered until the next panic, which is usually only several weeks, but your mistakes are never forgotten and usually take 4–5 years for your career to overcome.

Engineers can use going to another company very effectively to get a promotion. To successfully execute this strategy, the engineer must leave their company on a good note and take a position that is a promotion in another noncompetitive company. I have also witnessed engineers returning to the original company after a couple of years with another promotion. The engineer picks up the two promotions within a short period of time. What was really astounding was that, in several cases, the original supervisor hired back the engineer. It turned out that the original supervisor did not realize what a valuable employee they had until the employee left. Once the employee left, the manager realized the mistake and in one case actually pursued the engineer to return. If you decide to engage in this strategy, make sure you check the company policies about leaving a company and then returning. Some companies allow you to come back within 5 years and retain your seniority and benefits!

▶ *Career Tip.* Be knowledgeable about your employer before job switching!

This strategy sometimes fails miserably when the engineer leaves the company for a promotion, only to have the new company go out of business shortly thereafter, leaving the engineer unemployed or having to take a significant pay cut just to have a job. Be knowledgeable about your employer before job switching.

GOING TO THE COMPETITOR FOR A PROMOTION

Going to the competitor is another career strategy that is commonly used and may work successfully and also fail miserably. The idea here is the competitor is willing to hire the engineer since they know all inside information on the company business, who the customers are, and may even bring customers and contacts along. Making this career move is very stressful. Usually within an hour of the employee announcing they are going to a competitor, they along with their belongings are marched out the door by security, never to return. Some engineers believe since they did all the work, they can bring whatever knowledge and technology they wish to the competitor. This violates nearly all company rights. If the engineer brings unauthorized information to the new company and they benefit, there is usually an investigation by the legal department and if anything shows up, lawsuits are filed. The situation can turn ugly rather quickly.

This strategy might also be short-lived as the new employer may just be using the engineer to gain whatever knowledge they can. After the engineer has shared all the information from the old company with their new company and has nothing left to offer, they are suddenly let go for some nebulous reason. The engineer was simply terminated because they did not perform up to expectations during the probation trial period. The engineer finds themself unemployed and searching for a new job.

On the positive side, competitors may quickly realize the value of the new engineer and move them into higher levels of responsibility. The engineer's career advances and they become a driving force in other companies.

One of my most interesting observations is that this strategy, although highly risky, is utilized by engineers at all levels of the company. I know of engineers from junior levels to the most senior level of vice presidents who have utilized this strategy of going to a competitor to get the next career advancement.

SUMMARY

There are multiple viable strategies employed by engineers to obtain a promotion. The first place to start looking is in your present department. Your supervisor is key to this and having a frank and open discussion on the subject of your promotion is highly recommended. If you determine this is not an option you can look at other departments within your company or consider moving to other companies. Going to a competitor is included in looking outside the company. Whatever your strategy is for getting the next promotion it is going to require a high-energy state on your part and tact in dealing with people and situations. Promotions do not happen just because you think you are ready; they happen because others think you are ready also. Others will only consider you are ready for a promotion after you clearly demonstrate your ability to handle more and perform at a level above your present one in all aspects of your job.

Have you identified any career actions you want to take as a result of reading this chapter? If so, please make sure to capture these ideas before you forget by recording them in the notes section at the back of the book.

ASSIGNMENTS AND DISCUSSION TOPICS

1　Why should you start with your supervisor when looking for your next promotion?
2　Should you share what you are doing with others in your department? Or others in your company?
3　Can your mentor help you?
4　What is a reasonable time to wait for a promotion?
5　Do you know the criteria for the level above yours?

CHAPTER 6

CAREER DISCUSSION GUIDE FOR EMPLOYEES, MANAGERS, AND MENTORS

Hopefully by this point in the book you have a tentative career plan and mapped out a strategy to accomplish your goals. The next recommended step is to get feedback on your plan from your mentor and manager through engaging in career discussions.

What are career discussions? Career discussions are simply a series of conversations between the engineer and their manager, leader, career advisor, or mentor, about your career goals. There are multiple reasons for making it a series of conversations. First, it is a conversation to explore and discuss in an open and non-threatening fashion the engineer's career plan and goals. Second, it is an opportunity for the engineer and manager to review mutual expectations and set goals together. And third, it provides a forum for the engineer to check and validate the career goals they identified to be realistic, obtainable, and lead to future advancement. It is really the start of a journey between the manager and the engineer leading to benefits for both (Figure 6-1).

In this chapter, we first provide tips and guidance for the engineer to maximize the benefits from career discussions. This is followed by tips and guidance for the manager. The career discussion guidelines and recommended actions are provided to help both the engineer and manager be productive during the exchange. Questions are proposed that each should answer in order to stimulate forward thinking and goal setting. Preparation and homework are recommended to facilitate better career discussions.

The career discussion process is beneficial to both the engineer and the manager. For the engineer it provides meaning behind day-to-day activities and a long-range vision. It gives the engineer something to look forward to

The Engineer's Career Guide. By John A. Hoschette
Copyright © 2010 John Wiley & Sons, Inc.

FIGURE 6-1   Meet with career advisors about your goals.

and aspire for. Career discussions often motivate and stimulate both the engineer and the manager to strive for better performance.

For the manager, career discussions are means of communicating the company standards, expectations, and organizational goals. It is an opportunity for the manager, in a non-threatening way, to provide input to the engineer career plan and make adjustments when needed. Engaging in career discussions is beneficial to the organization because it provides a valuable opportunity to maximize the engineer's skills and competencies while enhancing career satisfaction. Career satisfaction is a critical factor in employee retention. In addition, career discussions followed up with coaching may help attract, develop, and retain a highly skilled workforce. Career discussions have been shown to positively impact employee retention and are a "win-win" action for the organization.

I have found that as a manager, when I take the time to have career discussions with employees that it motivates them to perform at a higher level and reduces the tensions while helping eliminate misunderstandings.

OBJECTIVES OF CAREER DISCUSSIONS

There are multiple key objectives of a career discussion. These include:

- Development and review of an effective career plan
- Defining short- and long-term goals
- Identifying and analyzing career choices by clarifying ideas, goals, and expectations
- Evaluating career alternatives and choices

- Review the engineer's skills to determine strong and weak areas
- Sharing resources to aid in career pursuits
- Jointly develop a realistic and meaningful learning plan
- Setting organization expectations and challenging the engineer

It is important that career discussions are open, two-way conversations with both the employee and the manager sharing and exchanging ideas. The engineer or leader can request the career discussion, which should be scheduled at a mutually convenient time and held privately without inter-ruptions. It is equally important that both parties are committed to periodic follow-up after the initial discussions and not let the effort die out after one meeting.

CAREER DISCUSSION GUIDE FOR ENGINEERS

The majority of the responsibility for making career discussions happen belongs to the engineer. Here are several items and recommended preparation actions for the engineer to consider when preparing for a career discussion:

- Have you completed a career plan?
- Have you taken advantage of the career development resources avail-able in the company and Human Resources department? Does the company have an online career planning site or career planning tools?
- Do your career goals align with the overall company goals?
- Are your goals realistic?
- Have you shared your career goals with another business leader or mentor prior to sharing with your supervisor?
- Have you contacted and discussed career plans with someone in the company who holds a position you are interested in aspiring to reach some day?

The first step to preparing for the career discussion is to complete a career plan. If you have completed the career workbook you are in good shape. If not, this will be your first task to complete in preparation for the career discussions.

Next, stop by your supervisor's office and ask if they are open to participating in career discussion with you. If the answer is yes, then ask for a good time. I have found the best time for career discussions is later in the day and not early in the morning. Generally, most supervisors are too busy early in the day handling critical issues to suddenly stop and shift gears to do a career discussion. Their body may be in the room, but their mind is usually off solving that latest panic. The best times I have found are during lunch or later in the day when things have settled down.

▶ *Career Tip.* Don't wait for your supervisor to initiate a career discussion; take it upon yourself to make it happen.

If the supervisor refuses the request to participate in a career discussion, you might ask why? Maybe they have never done it before, or simply do not feel they have the time. If they have never conducted a career discussion before, you may have to engage them by sharing this section of the book with him or her. Share with the supervisor the recommended actions and ask them to reconsider.

If they say they are too busy and it's not a good time, push it out. However, do not let it drop. Ask your supervisor when in the future would be a good time and get a meeting on their calendar. Let them know you are open to pushing it out to a better time and would like to reserve that time. Don't give up until you get a date.

The day of the career discussion you may be very nervous. Make sure you have a plan on how the discussion should go. Make an agenda of items to discuss and stick to it. Don't get side tracked. Start the meeting out with small talk and some socializing. These are called ice breakers for meetings and good topics might be the weather, latest sporting events, or similar other light topics. After a few minutes of socializing, it is time to get to work. Here is a recommended list of items to aid your discussions.

1. Review career plan; short-range and long-range goals, actions to reach goals
2. Your strengths
3. Your weaknesses or improvement area
4. Other people you should contact for career discussions
5. Summarize agreed upon actions

Remember this career discussion is a meeting where you are responsible for conducting. Like any meeting there is a set of actions you can take to ensure it goes smoothly. Here are some additional suggestions for conducting successful career discussions:

- Plan for your meeting a week or two ahead of time to give you time to prepare
- Keep the meeting time short, 30–60 minutes
- Know your career discussion objectives and be ready
- Bring copies of documents to share and make notes on
- Speak openly
- Schedule a follow-up time, remember this is a journey with several meetings planned along the way
- Relax and start out the meeting by making small talk

- Follow up with a "Thank You" email
- Write up meeting notes and identify agreed upon actions with a closure date. Start out the next meeting by reviewing the notes from the previous meetings; this will stop you from re-hashing previously discussed items.

▶ *Career Tip.* Career planning and discussions are an ongoing process that takes commitment and effort. Your career goals will change with time since career opportunities available to you are constantly changing. It is up to you to take control and determine the path of your career.

CAREER DISCUSSION GUIDE FOR MANAGERS AND MENTORS

As an engineering manager it is one of your job functions to stimulate and conduct career discussions with your employees. Are your employees asking for career discussions? If so, are you making time for them? I hope you are! Making time for employee career discussions can really benefit your organization. If your employees are not asking for career discussion, it is in your best interest to stimulate the process.

Now before you shout at me the famous "I have no time!" Think for a minute, we are talking about only about 45 minutes on a periodic basis and the benefits are substantial. In addition, the employee does the majority of the work; you provide your honest opinions and helpful guidance. This is definitely something worth doing.

Career discussions follow along much like other work directions that you give. Let's review some basic interaction guidelines to ensure a meaningful discussion and exchange between you and the employee.

- Open the discussion with small talk and make sure the employee is at ease prior to starting heavy career discussion. Remember the employee is feeling they are bearing their soul about what they consider are their very private goals.
- Open the meeting with a small discussion of the objective and what you both expect to accomplish.
- Make sure you ask to see their career plan. Do they have something written down and does it make sense? Are their long- and short-term goals identified?
- If they do not have a career plan then educate them on what a career plan is and help them by starting a plan. Refer to Chapter 2 for help with this.
- Get agreement on actions, who will do it, and by when.
- Close by expressing your support and encouragement for the employee's development.

- Be honest and open, point out good as well as improvement areas.
- Identify specific actions, including resources available to the engineer.
- Are company guidelines available that identify skills and criteria for various engineering positions in the company?
- Any company web sites you can direct the employee to? If not how about an engineering society web site? IEEE has an excellent one. Go to www.IEEE.org and search on "career" for further assistance.

To facilitate the discussion, you as a supervisor should use your communication skills and keep in mind that your task is to ensure you are meeting the expectations and personal needs of the employee. Your employees want to feel they are valued, listened to, understood, and involved. As a result of conducting career discussion and meeting the personal needs of employees you build trust and a relationship that makes employees more dedicated to you and your organization. Here are several key meeting skills you should be utilizing during the meeting and potentially some good responses you can ask to show you are engaged in the process.

1. Listen and respond with empathy
   - Career planning and discussions are not easy.
   - This is going to take work, but you can do it.
   - It will be well worth the effort!
2. Complement and enhance self-esteem
   - Your plan has some very good points.
   - This is a "good" or "excellent" effort.
3. Promote and encourage their involvement
   - I can see you succeeding with this plan.
   - How can we make this plan even better?
   - What other things might we try?
   - What can we both do to improve this meeting?

Active listening is a great skill to utilize during career discussions. Active listening shows you are hearing what the employee is communicating and improves mutual understanding. By using active listening techniques, you clarify your understanding of employee expectations and at the same time can create a results-oriented and realistic career development plan. You might consider the following tips for active listening:

- Avoid interruption and don't jump to conclusions.
- Don't tell them every little thing to do. Discuss options.
- Avoid judgment responses like: "That's a bad approach," "Will not work," "Unrealistic ideas and goals."

- Restate or paraphrase what the engineer just said to check for understanding.
- Use probing open-ended questions to explore interests and desires: "Any other career actions you could possibly take?," "Do you see any issues or roadblocks to your approach?"

Holding meaningful and regular career discussions are key ingredients to a successful career development process. As the engineers' manager you will have a better understanding of where to focus developmental employee activities by conducting career discussions. You will also have insight into the future activities and opportunities that your engineers are looking for.

One of the most satisfying accomplishments in my career was to hold career discussions with engineers, identify goals, and then see engineers accomplish these goals. I truly felt that I had made a significant difference in furthering their career.

SIMPLE STEP-BY-STEP PLAN FOR CONDUCTING CAREER DISCUSSIONS

In the previous sections of this chapter, we discussed guidelines for the engineer and manager to follow in preparing for the career discussions but not an actual step-by-step plan. In this section, we identify a simple step-by-step plan to guide the engineer and manager through the career discussion process.

The career discussion process is simply a series of conversations between the engineer and manager to help guide the engineer in a meaningful way to set career goals and establish a career development plan. This series of conversations can follow a simple process. This process is shown in Figure 6-2. A recommended list of actions for the manager and engineer is shown for each meeting.

In the first meeting the primary objective is simply to get to know one another. Oftentimes, a manager may work daily with an engineer and assign work but never really have the time to get to know the engineer's hopes, desires, and career plans. The first meeting is planned to be an ice breaker where both the engineer and manager share their backgrounds, work experience, and career aspirations. In preparation for this first meeting the manager may notify engineers they are open to, and available for career discussions. A good tool to facilitate the start of the process is to hand out or email career planning forms and solicit the engineers to request a career discussion meeting. Naturally, the manager should accept meeting invitation.

As part of the first meeting, the manager is encouraged to discuss their background as well as the company goals and objectives. This gives the engineer a better understanding of the manager's perspective.

1st Meeting Getting to Know Each Other	2nd Meeting Reviewing Career Plan	3rd Meeting Setting Actions	Follow-up Meetings
Objective: First priority is getting to know each other. Sharing of backgrounds and career desires.	**Objective:** Identify a career plan with specific goals	**Objective:** Determine the actions and options to obtain goals. Document actions with completion dates.	**Objective:** Monitor progress toward goals and adjust plan.
Manager Actions: – Notify engineers you are open to and available for career discussions – Hand out/email career planning form – Accept meeting invitation – Discuss your background – Discuss company goals and objectives	**Manager Actions:** – Request career plan prior to meeting – Discuss career paths available to engineer in company (Tech vs Mgmt) – Identify potential goals – Feedback to engineer on strengths and weaknesses	**Manager Actions:** – Request meeting if engineer does not set up meeting – Request updated career plan prior to meeting – Get agreement on career actions going forward – Challenge engineer to set higher career goals	**Manager Actions:** – Request meeting if engineer does not set up meeting – Request updated career plan prior to meeting – Review progress on actions – Identify new actions as old ones are completed
Employee Actions: – Schedule meeting with manager – Fill out draft career plan – Discuss your background and career goals – Discuss actions and document	**Employee Actions:** – Schedule meeting with manager – Complete career plan – Identify goals—near term and long term – Discuss actions and document	**Employee Actions:** – Schedule meeting with manager – Update career plan – Report progress on actions – Discuss actions and document	**Employee Actions:** – Schedule meeting with manager – Update career plan – Report progress on actions – Discuss actions and document

FIGURE 6-2 Recommended career discussion process.

As part of the first meeting, the engineer has the responsibility of doing most of the effort. This includes scheduling a time with the manager, filling out a draft career plan, and coming prepared to discuss their background and career goals. The engineer has the responsibility to document the agreed upon actions and set up the second meeting.

The objectives for the second meeting are to further refine the engineer's career plan by coming up with specific actions and completion dates for each action. The employee can simply declare "I'm returning to the university to get my Master's degree." Declaring a career goal in this manner is too nebulous. The manger should help the employee break down the goal into obtainable steps with a completion date for each. For instance, the steps leading to a Master's degree would be selection of a university program, application, acceptance, course planning, tuition reimbursement planning, and finally registering for the first class. The manager's responsibility in employee discussions is to help the employee discover the steps leading to the goal, guiding the employee through the steps, and helping them set realistic dates to achieve these tasks. If the manager does this, they are truly having productive career discussions.

Also at the second meeting, the manager should identify the career paths within the company available to the engineer and especially the choices of staying technical or moving into management. The manager might want to identify some specific goals the engineer needs to set and accomplish. This second meeting is also an excellent time for the manager to discuss the strengths and weaknesses of the engineer. The manager should be recommending other people in the organization that the engineer should also have discussions with; preferably people in the organization who presently hold the position the engineer hopes to aspire to obtain one day.

The engineer should come prepared for the second meeting with a fairly complete draft of a career plan with goals identified and potential completion dates. These goals should address immediate and long-term desires. The engineer has the responsibility to document the agreed upon actions and set up the third meeting.

The objective for the third meeting is to break down the career goals into specific smaller actions and options leading to the ultimate goal. Then document these actions with completion dates. The manager should encourage the engineer to set goals that will test/challenge the engineer. Some goals may be easily obtainable while others may significantly challenge the engineer. The engineer should come to the third meeting with an updated career plan from the second meeting and be prepared to discuss progress on any actions. Naturally, all agreed upon actions at the third meeting should be recorded for future meeting.

By the fourth meeting the engineer and manager will probably have a well-established routine if they follow this process. The purpose of further meetings is to track progress on the goals identified in the career plan and create new goals as the older ones are accomplished.

Do career discussions go as smoothly as the process seems to indicate? They may not. Many things may interfere with this process, namely, work deadlines, unplanned setbacks, and simply daily work assignments. However, if the engineer and manager can stick to some type of routine and continue to hold meetings on a regular basis, both are going to benefit. It may take several meetings just to establish a plan versus the two meetings suggested in the process. It may take more meetings, but the important part is getting the engineer to develop their career plan, not the number of meetings it takes to do it.

Once the engineers have established some type of plan, they are going to be asking, "Is this a good career plan? Will it lead me to be successful? Simply put, is this a guaranteed formula for success?" The manager is very prudent if they stress during the career discussions that just having a plan does not guarantee you will be successful. However, with a career plan, the outcome is more likely to be successful.

Next, the manager might want to discuss options for going around roadblocks to career goals and help the engineer think about alternate methods for reaching goals. Exploring other options opens the engineer to the realization that there may be several means of obtaining the engineer's goals. The manager can emphasize to the engineer not to stop at the first failure or setback; there is always another means to accomplish your goals.

▶ *Career Tip.* There are multiple ways around every career roadblock; you have to be open to alternative methods.

The final challenge faced by the manager and engineer is what happens when career plans do not result in advancement. Do they change plans completely, abandon the process, or even go so far as to throw everything out and start over? To these questions I respond with: use your engineering analytical skills and apply them to your career planning. Should the plan or process not work out as planned, identify disconnects, analyze the actions planned, results obtained, and finally determine why the desired objectives were not accomplished. The next section is dedicated to helping the engineer identify and overcome career disconnects and barriers.

A final note to the managers remember when you developed your career plans and how much they changed? Make sure you share this with your engineers so they do not become discouraged with the constant changing and updating they will be doing. Here is a saying I would like to share with managers and hopefully they will share it their engineers.

First you dreamed it
Then you believed it
Next you worked at it
Lastly you achieved it

SUMMARY

Career discussions are simply a series of conversations between the engineer and their manager, leader, career advisor, or mentor about your career goals. It is a conversation to explore and discuss in an open and non-threatening fashion the engineer's career plan and goals. Career discussions provide a forum for the engineer to check and validate career goals. It is really the start of a journey between the manager and engineer leading to benefits for both.

The key objectives of a career discussion include the development of a career plan, defining short- and long-term goals, identifying and analyzing career choices, reviewing the engineer's skills to determine strengths, weaknesses and set expectations on job performance. The majority of the responsibility for making career discussions happen belongs to the engineer. One of the engineering supervisor's job functions is to stimulate and conduct career discussions with employees. Career discussions do not happen in one meeting but with a series of three to four meetings and good career plans evolve and change over time.

Have you identified any career actions you want to take as a result of reading this chapter? If so, please make sure to capture these ideas before you forget by recording them in the notes section at the back of the book.

ASSIGNMENTS AND DISCUSSION TOPICS

1   Name three objectives of career discussions.
2   When was the last time you had a career discussion with your supervisor?
3   What resources are available to you from the Human Resources department?
4   Are career discussions valuable to the manager? The company?
5   Describe the steps in the process of making career discussions.

CHAPTER 7

THE MENTORING PROCESS AND VALUE TO YOUR CAREER

The dictionary defines the word mentor "as a person who is a wise and trusted counselor or teacher." At all stages in our lives, each of us depends upon and uses mentors. These mentors are usually referred to in different terms. When we are young our mentors are most often our mothers and fathers. As we mature, other mentors come into our lives. These mentors may be teachers, coaches, or guidance counselors, or may even be older brothers or sisters, and aunts or uncles [1]. As we continue to mature, our mentors might be called monsignor, rabbi, or professor. The point here is that each one of us calls upon and asks for the advice of mentors in our personal lives. Are engineering mentors necessary for successful career development? Yes. You need mentors at work just as you do in other areas of your life.

UNDERSTANDING THE BENEFITS OF MENTORS

My experience and that of others has shown that finding and developing the right mentor can significantly increase your chances for career success. The Institute of Engineering and Technology has a website dedicated to helping its members understand the value of having a mentor [2,3]. If your mentors are well placed and well thought of in the organization, they can be your guide to the top, provided of course, you can meet their expectations and perform on the job. Mentors can help you develop the necessary technical and business skills you need to progressively rise to the top. Kevin Hoag in his book *Skills Development for Engineers* published by the Institute of Electrical

The Engineer's Career Guide. By John A. Hoschette
Copyright © 2010 John Wiley & Sons, Inc.

FIGURE 7-1   Having a mentor is very valuable for your career.

Engineers identifies mentoring as a key component in skills development [4] (Figure 7-1).

A good mentor who is strategically placed in the organization can introduce you to the inner circle of executives who make the decisions as to who will be moving up the corporate ladder.

Correspondingly, a mentor who is not strategically placed or not very well thought of in the organization, or offers poor guidance, can severely hurt your career. Your challenge is to find the right mentor and develop an amiable relationship. This relationship will significantly enhance your career development. In this chapter, we will identify all the benefits a good mentor can provide you, and then we will suggest ways you can find and develop good mentors.

A good mentor will help sponsor you, provide coaching, protect you, recommend you, and even help you get the more challenging and consequently more rewarding job assignments. Through these helpful mentoring steps, you will be exposed to the shortcuts and hopefully avoid the pitfalls and setbacks.

▶ *Career Tip.* Having a great mentor can make your job significantly easier.

Sponsorship from a mentor is essential to getting the challenging assignments that provide you with the opportunity to perform and clearly demonstrate you are ready to move up. A good mentor is usually in a position to defend your abilities should they come into question by management. He or she is the person who interacts with management to vouch for your suitability to handle difficult assignments. Your mentor is there recommending you for the assignment, clearly identifying you as the best candidate to successfully complete it. It is this type of sponsorship that will make you stand out from the crowd.

▶ *Career Tip.* A mentor can help sponsor your next promotion.

Coaching is an equally important function a mentor will perform. A good mentor can provide the coaching or polishing necessary for advancement. Good coaching can help overcome a multitude of problems that you might otherwise have been unable to handle. Good coaching may come in the form of learning how to deal with an overly domineering supervisor or coworker. It may also come in the form of technical advice as to how to solve problems and how to avoid technical failures.

Good coaching may come in the form of challenging assignments causing you to expand your expertise, or it may also come in the form of recovery from failure or getting advice on what to do if you fail. A good mentor is a good coach, always there challenging you, inspiring you, and demanding the best. Protecting is another function that a good mentor performs for you.

Often people will try to find a cause or a person to blame when the engineering project fails. A good mentor will stand up for you in times of need and defend your actions, provided of course they agreed with them in the first place. Mentors can easily deflect arrows or blame away from you and can be a shield in troubled times.

▶ *Career Tip.* A mentor can protect you in troubled times.

By developing a good relationship with a powerful mentor you assimilate power just by being associated with them, you send signals to other people in the organization that you are a member of a powerful team. You acquire some of your mentor's influence and will have resources behind you, which may make it possible for you to obtain inside information or new organizational power to cut through the bureaucracy and red tape.

FINDING GOOD MENTORS YOU CAN RELY UPON

The best place to begin looking for a mentor is outside your department; any senior supervisor in a position to influence your career positively is a potential candidate. Choosing someone inside your department may be quick and easy but will cause too many problems. Having a mentor inside your group may cause jealousy from other group members, innuendoes of favoritism, and even the alienation of other group members.

Finding a good mentor is not an easy task and may take quite a long time to cultivate [5,6]. When looking for a mentor you should be looking for someone with whom you are comfortable and compatible in ideology. The mentor should share similar views to yours about company strategies and success in the corporate world. The mentor should appear to you as a successful role model whom you would like to emulate. You should find the mentor stimulating, challenging, and an inspiration to you to perform to the top of your ability. This type of relationship cannot be planned but must be cultivated over a period of time.

The mentor should be a person with whom you can share your triumphs as well as your defeats. You will need him or her for guidance, nonjudgmental listening, and constructive criticism. They should be a person with whom you feel comfortable trying out your new ideas and you should respect and value your mentor's honest opinion. They should also respect your intelligence and capabilities.

▶ *Career Tip.* You should be comfortable to discuss problems with your mentor.

Unfortunately, mentors do not walk around the company with badges identifying themselves. So you must be aware of the subtle hints potential mentors give. When you are working with senior people, do they take the time to explain to you everything you should know? Do they spend extra time making sure you get the assignment right? Do you find yourself sharing similar strategies on the best approach to the problem? Is there an unexplainable chemistry between you and your mentor? Do you enjoy discussing difficult problems and tasks with him or her? Does your mentor like your style and compliment you on your work? Similar outside interests are other things you might have in common with mentors. People who like your work and spread the word are good candidates. These are all good signs the person you are working with would make a good mentor for you.

▶ *Career Tip.* You have to actively seek out mentors. Join a mentoring group if possible.

Once you have found a good candidate to be your mentor, the challenge becomes one of developing your relationship. Just as in any other relationship, you must invest your time. A good way to do this is by sharing lunch hours. Stop by your mentor's office on a periodic basis just to get an opinion about the project you are working on. Pass by your mentor's office on the way home at night. Volunteer your help on a project they are working on. If your mentor is like most people, they could always use a little extra help. Look for special projects around the company that are pet projects of your mentor and get involved in them.

As you start to develop the relationship you can cultivate it by letting your mentor know how much you value their opinion. Everyone likes to hear how valuable their opinion is. You might even ask if they would mind if you considered them as a mentor. The point here is that you must be aggressive about finding a mentor and maintaining a relationship. However, you need to be cautious when doing so.

First, being overly aggressive can be misinterpreted as being too pushy or give the impression your foremost interest is empire building. This can trigger resentment from your coworkers and earn you the reputation of being a "brown nose" or "apple polisher." Second, the relationship with a mentor must be beneficial to both you and them because you do not want to become just a "gofer" or "yes man." If your mentor is only using you for their own benefit, it's time to move on and look for another one.

▶ *Career Tip.* Pick mentors who are a minimum of two levels above you and have 8–15 years more experience than you.

Age difference is an important factor in picking a mentor. If your mentor is within 4–8 years of your age, they are more of a close friend than a mentor. The best age difference is between 8 and 15 years of age. More than 15 years of age, the relationship may turn into a parent–child relationship.

Don't put all your eggs in one basket. In other words, don't rely on only one mentor. If you spend all your time developing a relationship with one mentor and he or she leaves the company, you are suddenly left stranded. This is where having more than one mentor can be beneficial. Even though mentors may appear to know it all, one mentor simply cannot provide all the guidance you need. For engineers, it is good to have at least one technical mentor, one political mentor, and a mentor with great business acumen. This way you gain experience in dealing with technical, political, and business moves within the company.

▶ *Career Tip.* Find several different mentors, each with different backgrounds and experiences.

For example, you may be working on a tough technical problem and require assistance from a senior technical person, in this case your technical mentor may be able to provide the best guidance. On the other hand, a project you are working on may be a dead-end project as far as company management is concerned. A smart business mentor can forewarn you of the futility of working on the project and even assist you in moving on to more rewarding projects.

HOW TO UTILIZE A MENTOR

You must be aware that utilizing mentors can be both over and underdone. You have to strike a balance, which will come with time and experience. If you are constantly seeking out the advice of your mentor for every conceivable problem, then you are overusing them. Your mentor will soon come to realize you are incapable of making decisions and view you only as coming to them so they will do your work for you. On the other hand, if you only go to your mentor after you have solved everything and there is no real need to take advantage of their wisdom and advice, you lose the benefit of coaching. In either case your relationship with the mentor will not work. Here are some guidelines for when to utilize your mentor.

Times of trouble are probably the most obvious. When you are having difficulty and do not know where to turn, see your mentor. However, you should not "dump" your problems on your mentor. Coming to the discussion with some possible solutions indicates that you are thinking of potential solutions and really trying to work the issues. Be ready to discuss your options and highlight the pros and cons.

Ask your mentor for advice and find out what they think your options are. If you have missed anything helpful, your mentor will point this out for you. Ask for their guidance and be ready to act on it afterward. Nothing is more discouraging for a mentor than to recommend a course of action and have the employee fail to try it out.

Reporting back to your mentor on your progress is also important. Provide feedback on your activities; this shows you are really taking and utilizing their advice. Utilize your mentor for trying out new ideas and approaches to problems. A good mentor will have years of experience and should be able to assess your chances for success. Nothing helps sell a new idea faster than when your mentor is pushing for it along with you. In expressing new ideas, let them make constructive criticism, then implement any suggested changes. You may not see the need for the changes, but he or she may know of hidden barriers in the organization of which you are unaware. Your mentor's suggestions should help you to overcome these barriers.

▶ *Career Tip.* It is always a good time to utilize mentors.

A good time to utilize your mentor is at the beginning of the project. Meet with them and discuss your plans for the project, for example, how you have set things up, your planned actions, and any problems you anticipate. Your mentor has years of experience and should be able to identify, in advance, your problem areas.

Approximately halfway through the project, discuss the problems you are encountering and the steps you are taking to solve them. Try to present problems with potential solutions, ask for an opinion on how the project is coming along, and ask them to do research on how other superiors perceive the progress of the project. It's always good to know whether management thinks you are doing a good job or not. Remember, mentors are great resources for finding out how to overcome barriers in the organization.

Another good time to seek out the guidance of your mentor is when the project is coming to a conclusion. What is the best way to end the project and present the results? How can you make management aware of the fine job you did on the project? What steps can you take to determine the next project you will be working on? Does your mentor have any recommendations? Can he or she sponsor you on another project? Who are the people you must contact? These are all important questions a good mentor should be able to help you with at the end of the project.

Another good time to utilize your mentor is when you feel that your position seems to be stagnating or when you feel you're in a dead-end job.

Good mentors will be able to tell you about things you may not be aware that are going on behind the scenes. It is quite easy for them to sit down with your supervisor or other people in the organization and get any information to which you may not have access. A good mentor will show you how to get out of the situation or just weather the storm.

When you have failed on a project and need to recover, never be afraid to ask for help. Some people react to failure by trying to hide it, failure does not have to be a career limitation. In fact, a good mentor can show you how to overcome failure and actually make it an opportunity for advancement. Nothing is more impressive to management than when you can show them how you sought out the help of others, identified a solution, and made corrections after a failure. Doing this is not career-limiting but part of career advancement. Shawn McCarthy in his book, *Engineer Your Way to Success* published by the National Society of Professional Engineers, points out overcoming difficulties or failure is all part of being successful [7].

Developing and utilizing mentors is a skill you need to develop for career advancement. Knowing the best time to go to your mentor will depend upon your relationship with them. The above-mentioned times are suggested as guidelines for helping you. Gordon Shea in his book, *Making the Most of Being Mentored*, points out many excellent times and actions you can take to get the maximum benefits from your mentoring experience [8].

HOW TO BE A GREAT MENTOR

Being the mentor means you are the leader and will guide the protégé or mentee through this process. The key words are "lead" and "guide." Lead the mentee through the process. They will be looking to you as having a clear vision of where they need to go and know how to best get this accomplished. You are not to drive or tell the protégé what to do, but let the protégé discover and learn solutions to problems. You are a guide along their trail to career success.

The first step for the mentor is to ensure the relationship gets off to a good start by setting a tone of enthusiasm, excitement, and genuine interest in their career. Mike Pegg in his book, *The Art of Mentoring*, suggests some great ways to start out your program [9]. Make sure you allow enough time socializing and exchanging personal information about each other's careers. Try meeting in more relaxed areas like the cafeteria or company lounge.

Regular meetings will establish a rhythm and foundation to construct successful career plans. Plan to meet for an hour, preferably two, at least once a month. Have a plan for each meeting and ensure your time together is fruitful and each of you walk away with identified actions.

GREAT QUESTIONS FOR MENTORS TO ASK

Initially, as a mentor, you should assess the condition of the protégé's career plan. Does he or she have a career plan with goals identified? Refer to Chapter 2, Developing Career Plans for Guidance. Have short-term and long-term goals been identified? What is most important to the protégé and what gets their excitement up? Ask open-ended and probing questions.

Where do you want to be in 5–10 years?
What do you like to do the most?
How would you describe your ideal job?
What are you really good at?
What areas do you want to improve?
How can I help you?
What organization are you in and what would change?

MENTOR ROLES AND RESPONSIBILITIES

To fully engage your protégé, take them to meetings you think might benefit them to attend. Copy them on leadership articles.

This is one of the challenges for you as a mentor—describing how you learned what you know so your protégé can learn from your experience. One way to accomplish this is through examples. By thinking through specific experiences, you can gather the gems that will be the most beneficial to your protégé.

The following are some suggestions for activities that may help you in this process:

1. Share your weekly activity report. Explain what the "hot" issues are and why your group has been concentrating on them.

2. Discuss your resume. Share your career decisions, and review the process you went through to make them.

3. Copy articles you find in business or industry publications that may be particularly helpful to your protégé.

4. If possible, ask your protégé to attend one of your staff meetings. Discuss the impact of some of the decisions made, the group dynamics and any follow-up activity that will be necessary.

5. Review a list of volunteer organizations you are participating in, and the role you play. Discuss how involvement in those groups has helped your career.

▶ *Career Tip.* Mentors open doors and windows to new insights, by sharing their experiences and knowledge.

By simply exposing your protégé to alternative ways to accomplish goals, you are opening doors and windows they might not have considered before. Great door-opening activities are lab tours or production area tours and attending meetings with you.

Connecting your protégé with colleagues that can become informal mentors is another door to open. As you get to know your protégé, you will think of others who may be able to provide them extra guidance. Establish meetings with these individuals and help your protégé think of questions they might ask.

Learning from ones' mistakes is often the most difficult part of being a mentor. Allow your protégés to make their own mistakes, and help them learn from them. Also, help your protégés learn from your mistakes, so they do not have to experience the same pain.

- Think through a time when things did not go so well. What happened? Why did it happen? What could have prevented it from occurring?
- Have you had an opportunity to apply what you have learned from that experience? If so, what happened? What did you do differently from the first time?

Have them practice a presentation before giving it to a group. Ask the "tough" questions and make sure they are prepared. Ask them to profile the audience for you and see what questions are anticipated. Give them feedback on their presentation style and format.

Be a sounding board. Let your protégé talk through strategy with you so they can define it clearly.

▶ *Career Tip.* Sometimes the best thing a mentor can do is simply listen.

What we learn from our mistakes is more important than making the mistake in the first place. In fact, if we don't make some mistakes, we won't be able to learn as much, or as quickly. Help your protégé by sharing what you have learned, and by talking with them about what they have done or are about to do.

Mentors provide protégés with valuable knowledge and insight. They can also help connect protégés with others who are critical to their career advancement. This is particularly helpful for females and minorities trying to break through the glass ceiling.

The following questions are designed to facilitate a discussion about networks.

Who is in your communication network and how did they join it?
Who is in your expertise network and why are they in it?
Who is in your trusted network and what have they done to earn a spot there?

The following are some characteristics of successful mentors:

- Communicate politics within the organization
- Reveal unspoken corporate rules and values
- Point out areas necessary for advancement
- Show how to influence and persuade others in positions of authority
- Demonstrate how to earn the respect of peers and executives
- Share expertise
- Listen actively to protégé's goals
- Share vision and insight
- Provide support and encouragement
- Build protégé's self-worth and confidence
- Help protégé take safe risks
- Prepare protégé for dealing with setbacks and failure
- Provide honest and constructive feedback
- Make contact with protégé's manager—get them involved
- Establish a process for regular contact using phone, email, or other means
- Ask probing questions
- Require your protégé to document his or her lessons learned and accomplishments
- Provide visibility for protégé
- Set expectations of performance jointly
- Offer challenging ideas, raise the bar
- Provide direct feedback on negative behaviors and observations

Successful mentors do not:

- Agree to get a protégé a promotion or raise
- Take responsibility for protégé's actions
- Make promises on behalf of their management or the company
- Wait for the protégé to do everything
- Create situations that may cross the line into less than a professional relationship
- Forget that the perceptions of others are as important as the reality of the situation
- Use the protégé as means to get extra work done
- Forget to be sensitive to the differences between you and your protégé

PROTÉGÉ'S ROLES AND RESPONSIBILITIES

The protégé also has roles and responsibilities to ensure the mentorship yields benefits. The protégé should keep a journal of the meetings and document expectations and discussions with the mentor. It is the responsibility of the protégé to initiate and schedule routine interactions with their mentor.

The protégé should share job experiences to help the mentor gauge what would be the most helpful for them. The protégé should be open and sharing of personal performance on the job, and if possible, provide the last job review and rating as discussion aids. The protégé should be the one who creates a development plan and documents and monitors progress against that plan.

The protégé should not expect the mentor to solve problems for them but to help them work through the potential solutions and identify the best path forward. As a protégé, it is up to them to determine their game plan, focus, and priorities. The mentor is there to help as the protégé determines actions, but the ultimate decisions are up to the protégé.

THE PHASES OF THE MENTORING PROCESS

The actions and objectives of the mentor and protégé change as the relationship goes through each phase. The three major phases of the mentor/protégé relationship are shown in Figure 7-2.

The first phase is the "Getting to know and trust each other." Like any relationship, time should be spent at the beginning just getting to know each other. A good way to do this is to simply talk and share experiences. Some recommended things you might share with each other are your areas of technical focus, degrees, and where you went to school. Other good items to share are your likes and dislikes about the engineering fields. Looking at each

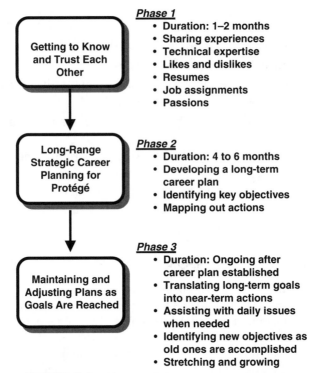

FIGURE 7-2   Phases of the mentoring relationship.

other's resume is an excellent way to spend some time getting to know each other. Asking the protégé about their job assignment will provide insight into their daily activities and the issues he or she is facing in the near term. Also consider sharing your passions and discussing the ideal job.

The first phase should last about 1 to 2 months during which you meet about 2 to 3 times. Take notes on your protégé answers, so you can refer back to them in future conversations. It also sends the protégé a very strong message that you consider what they are saying is important and you are taking notes.

An excellent action the mentor can do in the first phase is to contact the protégé's supervisor and spend time discussing protégé performance. This action will provide valuable insights for the mentor and should reveal the strengths and development area for the protégé.

▶ *Career Tip.* Meet with the protégé's supervisor to discover strengths and weaknesses.

The second phase involves generating a long-range strategic career plan for the protégé. This phase usually lasts 4–6 months. During this phase, the mentor guides the protégé in constructing a long-range career plan and determining the near and far-term actions. Have the protégé write up the plan and clearly identify the actions needed. Break down the goals into near and far-term actions. For example, simply establishing a goal of getting a Master' degree is too vague. Break this goal down into easy obtainable steps with dates. For example, this goal can be broken down into:

1. Identify major
2. Talk to people in company who have recently returned for a Master' degree and discuss actions they should be taking
3. Obtain the university criteria for entry into the Master's program
4. Identify the funding source and apply
5. Register and get accepted into Master's program
6. Develop degree plan
7. Start first class

Please refer to Chapter 2 "Developing a Successful Career Plan," for guidance with this.

The final phase is the "maintaining the plan" phase. In this phase, the mentor and protégé are meeting to discuss progress toward career objectives. Refining the plans as goals are accomplished and identifying new goals to replace them. In addition, the mentor may be helping with the protégé's job issues or more challenging tasks.

The mentor should be helping guide the protégé through difficult situations and the more challenging assignments. Not simply giving the answer to the problem but helping the protégé brainstorm solutions and evaluate the optimum solution to peruse. Sometimes the problems may seem trivial to the mentor and the best action the mentor takes is to just tell the protégé what to do to instantly fix it, but this is not mentoring. The mentor should only do this as a last resort and clearly identify that in the future this is not going to be a common practice.

YOU CAN MENTOR; IT'S EASY!

Setting up and running a mentoring program for your engineering society or company is easy and very rewarding. You can do this! It is very rewarding to see people develop and become leaders. Please remember that mentoring is not an exact science but a journey. You will travel on uncharted paths and discover new actions along the way and sometimes you will even go down dead-end paths.

All protégés are not alike and there is no one magic formula for all protégés. You, as a mentor, will have to experiment with various methods and techniques until you discover the best ones that a protégé responds to.

Your protégés will all have different career plans and objectives. Trying to mentor people into your exact likeness and desires will not work. David Clutter and Belle Ragins provide many excellent tips for mentoring people from diverse backgrounds in their book *Mentoring and Diversity* [10]. And finally, protégés are going to fail and make mistakes. You cannot take responsibility for their failures just as you cannot take credit for their success. You are like road signs along the journey of life guiding them as they go. However, it is the protégé who drives the car and determines which way to go and how fast. Good luck and please mentor someone!

A WORD OF CAUTION ABOUT FINDING MENTORS FOR WOMEN

The world of engineering appears to be dominated by men but every year more women are majoring in engineering. The chances still remain greater for most women to have men for mentors than other women. Unfortunately, good mentoring relationships between a male mentor and female are extremely hard to develop and maintain.

If you are a woman and develop a male mentor, be alert for the relationship shifting to something more than friendship. If this happens, then it is obviously time to change mentors. If you become involved it diverts your energies from your primary goal of career advancement. It can quickly ruin your mentor's career and yours. This type of involvement is to be avoided.

A good male mentor should challenge you and encourage you to make bold strides rather than timid little steps; his advice should be the same regardless of your gender.

Women looking for good mentors may want to take a different approach. Organizations such as Women in Engineering or Women's Engineering Societies are made up solely of women in engineering. Their goals and objectives are networking for women engineers, and therefore are excellent places to find mentors.

Women mentors can provide tips on what it is like to be a woman in the engineering field. They can provide the guidance and coaching necessary for women to overcome the "Good Old Boys" barriers, how to handle sexism, and deal with the "overly friendly" male supervisor. For a woman in a male-dominated field, having several female mentors is a necessity.

A FINAL NOTE ON MENTORING

Most people become squeamish when faced with the task of developing a mentor. To this I say what do you have to lose? The answer is absolutely nothing, with everything to gain. Remember, several good mentors are necessary for career development.

You will need to have many mentors throughout the course of your career; no one single person can provide all the guidance necessary. You may also have several different mentors at one time and as your career progresses mentors will come and go.

SUMMARY

At all stages in our careers, each of us should depend upon and utilize mentors. Finding and developing the right mentor can significantly increase your chances for career success. Mentors can help you develop the necessary technical and business skills you need to progressively rise to the top. The best place to begin looking for a mentor is outside your department; any senior manager in a position to influence your career positively is a potential candidate.

When looking for a mentor you should be looking for someone with whom you are comfortable and compatible in ideology. The mentor should share similar views to yours about company strategies and success in the corporate world. Finding a good mentor is not an easy task and cultivating a good relationship will take time.

Utilize your mentor during times of trouble, at the beginning of a project, at the end of a project, and nearly anytime you need advice or help on assignment. It is easy if you are a senior person to mentor people and the experience can be fun and rewarding. The mentor's roles are to fully engage your protégé, provide guidance but not tell them what to do, expand their thinking, and share your experiences. The protégé's role is to keep a journal of the meetings and document expectations, schedule routine interactions with their mentor, share job experiences, and development of a career plan. The protégé should not expect the mentor to solve problems but provide guidance. The mentoring process takes time and periodic meetings are recommended for effective mentoring.

Have you identified any career actions you want to take as a result of reading this chapter? If so, please make sure to capture these ideas before you forget by recording them in the notes section at the back of the book.

DISCUSSION TOPICS

1   Make a list of several people in your company who you think might make good mentors for you. Pick one and approach him or her about some problem you have and ask for guidance. Watch their reaction. Is it what you need in a mentor?

2   Can you identify a good technical mentor? How about a good business-oriented mentor?

3   What organizations outside of your company can you identify as good sources of mentors?

4 For women engineers, contact your local Society of Women Engineers. Can you think of any other organizations that might be a good source of mentors?

5 Topics of discussion:

Do mentors last forever?

How do mentors change as your career advances?

What are good qualities in a mentor?

REFERENCES

1. National Academy of Engineering, Institute of Medicine, *Advisor, Teacher, Role Model, Friend*, National Academy of Sciences, 1997.

2. The Institute of Engineering and Technology, Mentoring Books and Resources, 2008. http://www.theiet.org/careers/mentoring/resources/books.cfm.

3. Clutterback, David, *Everyone Needs a Mentor*, Chartered Institute of Personnel Development (CIPD), 1991.

4. Hoag, Kevin, *Skills Development for Engineers*, Institute of Electrical Engineers, 2001.

5. Johnson, Vern, "Career Mentoring 101," *IEEE Today's Engineer*, June 2006. http://www.todaysengineer.org/2006/Jun/mentoring101.asp.

6. IEEE, *The Balanced Engineer, Essential Ideas for Career Development*, CRC Press, 1998.

7. McCarthy, Shawn, *Engineer Your Way to Success*, National Society of Professional Engineers, 2007.

8. Shea, Gordon, *Making the Most of Being Mentored*, Crisp Publications, 1998.

9. Pegg, Mike, *The Art of Mentoring*, Management Books, 2000.

10. Clutter, David, and Ragins, Belle, *Mentoring and Diversity*, Butterworth-Heinemann, 2001.

CHAPTER 8

STUCK IN A DEAD-END JOB?
WHAT TO DO TO BREAK OUT

Not all jobs leads come with career advancing opportunities and it is highly likely at sometime during your career that you will find yourself in a dead-end. The job may lead to nowhere and if you stay in this job, your career will be on hold until you move. One good way to find out if you are in a dead-end job is to ask some simple questions about the position.

1. What happened to the person who had this position before me?
2. Where are they now?
3. What is the path to my next promotion if I stay in this job?
4. Is anything new planned for the job?
5. When does the present work end and what will I be doing next?

If you get the following answers to the questions, then it is a clear indication you are in a dead-end job and need to take some aggressive career actions.

1. The person who had your job left the department.
2. The person who had your job left and received a raise or promotion in doing so.
3. There is no path to a promotion or better job; there are no other job openings that you could be promoted into.
4. Nothing new planned for your job, the plan is to keep doing what you are doing now, only faster and better.
5. The job runs out when we stop making the product. The plan is to make the same product until the demand for it ends. Hopefully, this is for a couple of years.

If you received these types of answers or something similar to the previous questions about your job, then it is safe to say that you are in a dead-end job and it is time to make some career moves. Here is what you should be doing.

Have a Career Discussion with Your Boss. The first action is to have a career discussion with your boss. Find out what is planned for the department, the job, and your future in the department. Discuss how you want to advance and see if you both can identify an action plan to realize your goals. The boss may have other good things planned for you and simply has not had the time to discuss them with you. Discuss your performance and improvement areas. From these interactions you should be able to identify a path forward to new and better opportunities.

Investigate Obtaining New Training. New training to increase your skills can open up new opportunities for you. Maybe you are just one or two classes away from another career advancing job. Explore this option with your supervisor for moving out of the dead-end job. This may also include returning for further education in the form of another degree. Obtaining another degree is a sure method of opening up new opportunities to you.

Volunteer to Take on New Responsibilities and Tasks. Sometimes supervisors just assume you do not want to do any more work and never consider giving you more. If you volunteer to take on more work it shows you have more capabilities and are open to taking on new jobs. Volunteering to take on new responsibilities and work may open up new and better job opportunities.

Start Searching the Company Job Postings for Other Opportunities. On your own time, start searching the company job postings for new opportunities. Your department simply may not have any better opportunities for you and other departments just might. A simple transfer to another department in the company is an easy way to get out of a dead-end job.

Network with People in Other Departments. Another good career action to get you out of a dead-end job is to network with other people in the company and find out what departments have openings and possibly even better opportunities. Getting a coworker to recommend you for a better position in a new department is always a good career move out of a dead-end job.

Ask Your Mentors for Help. It is always good to discuss your job situation with your mentor and get advice on how to get out of a dead-end job. More than likely they have also faced and successfully overcame this situation in the past.

One note of caution: while you are looking around for a way out of the dead-end job, do not share what you are doing with your department coworkers. If you mention how you feel about your dead-end job and how you are looking, you are more than likely going to start the rumor mill running. People in your department will be talking about you and your actions. When this happens, it is only a short time before the boss finds out and then you have to do some real career damage control. Keep your feelings to yourself and be discreet about your actions and spend your time and energy on positive actions to help your career.

Have you identified any career actions you want to take as a result of reading this chapter? If so, please make sure to capture these ideas before you forget by recording them in the notes section at the back of the book.

CHAPTER 9

MY CAREER HAS PLATEAU, NOW WHAT?

A career plateau is reached when an engineer no longer continues to advance professionally or technically. Career plateau occurs when you are passed over for promotions, your job responsibilities do not grow, no salary increases, or your technical work is transferred to others deemed more capable. There are multiple causes of career plateau with some causes linked to the individual and others linked to the company or the economy [1,2]. The common causes of career plateau include lack of individual motivation, burnout, lack of update training, company business decline, poor economy, technical failures, and personal problems.

The impact of a career plateau can be devastating to the engineer as it can lead to feelings of being trapped, low value, low self-esteem, and no future to look forward to. If these feelings go unchecked they will certainly lead to poor performance and ultimately could result in job loss.

A simple example of career plateau is shown graphically in Figure 9-1 where an engineer's professional level is plotted versus age. For the example shown, the engineer has steadily moved up the engineering ranks and is now at level 5. The engineer moved a level every 3–4 years; however after achieving a level 5, the engineer has remained at this level without a promotion for 9 years. This data would indicate that the career has plateau. The normal reaction for anyone whose career has plateau is naturally to be concerned and wonder what to do next. The questions the engineer may be asking are:

Is my career over?
Is this as far as I am going?
What's next?

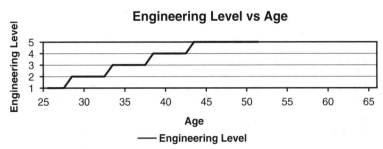

FIGURE 9-1   Plot of engineer's level versus age showing plateau.

These are all normal questions and if you find yourself in this position it is time to take some positive career actions. Here are some recommended career actions you can take if you wish to remain in your present job.

Look for Alternate Means of Job Satisfaction and Recognition. You can look for job satisfaction and recognition by volunteering for special assignments like training junior or new engineers, teaching a class at work, or leading a volunteer effort in the community on behalf of your company. You can mentor grade school and high school students in mathematics. Or you could organize and lead your local high school robotics team or Lego League team. Join an engineering society and volunteer to help others in their careers.

Take on the Challenge to Enhance Your Job. All jobs can be improved and taking on the challenge to improve things significantly is an excellent means to overcome career plateau. Can you identify cost savings or performance improvement projects you would be willing to participate? You can make a competition out of it and get others in the department or company involved.

Put Your Writing Skills to Work. Write a paper and submit it to a conference. Or if you have the energy and feel you have the knowledge, write a book on your experiences or technical knowledge. You could even file a patent on the idea you have.

Start a New Hobby. One way to get your mind off the problems at work and your career is to start a new hobby. A new hobby will give you something to look forward to and fill the void left by a career that has plateau.

Attend an Engineering Conference. Attending an engineering conference can be very uplifting with the exposure to the latest technologies and new products. Attend a training class at the conference in a new technical area.

Here are some actions you can take if you want to get your career growing again by leaving or changing from your present job.

Start Searching the Company Job Postings for Other Opportunities. Start searching the company job postings for new opportunities and network with others in the company to find out if there is anything new opening up. Even a lateral career move can provide new opportunities.

Have a Career Discussion with Your Supervisor. Find out what is planned for the department, the job, and your future in the department. Discuss how you want to advance and see if you both can identify an action plan to realize your goals.

Explore Returning to School for More Education. By supplementing your technical or leadership skills with more education you are increasing your net worth to the company. With more education you open new doors of opportunity.

Volunteer to Take on New Responsibilities and Tasks. Volunteering to take on new responsibilities and work can lead to career growth.

SUMMARY

A career plateau is reached when an engineer no longer advances professionally or technically. Career plateaus can lead to a feeling of being trapped, low value, low self-esteem, and no future to look forward to. If these feelings go unchecked and not turned around, they will certainly lead to poor performance and could ultimately result in job loss. You can overcome career plateau by taking positive career actions such as looking for alternate means of job satisfaction and recognition, taking on the challenge to enhance your job, putting your writing skills to work, starting a new hobby, joining an engineering society, or attending engineering conferences. The key to successfully deal with a career plateau is to take positive actions and implement changes in your job, surroundings, and lifestyle.

Have you identified any career actions you want to take as a result of reading this chapter? If so, please make sure to capture these ideas before you forget by recording them in the notes section at the back of the book.

ASSIGNMENTS AND DISCUSSION TOPICS

1   When is career plateau most likely to occur, at a young age or older age?
2   How do you determine if your career has plateau?
3   Name three actions you can take for career plateaus.

REFERENCES

1. Lee, Patrick C.B., *Career Plateau and Professional Plateau: Impact on Work Outcomes of Information Technology Professionals*, Nanyang Business School, Nanyang Technological University, Singapore, E-mail: acblee@ntu.edu.sg.
2. Sharma, Surajit Sen, "Stuck on a Career Plateau," http://www.hrcrossing.com/article/270059/Stuck-on-a-Career-Plateau/.

CHAPTER 10

HOW NEW GRADUATES CAN SUCCESSFULLY TRANSITION INTO INDUSTRY

If you recently graduated from college and joined a company, everything may seem quite different and not at all what you expected the industry job would be like. You may find it difficult to comprehend how your new company does business and what exactly your job is. If you are experiencing this, don't panic—you are not alone. Nearly every new graduate experiences this. The underlying reason for this is the fact that a university has a different set of operating rules than industry. And consequently the change in going from the university to industry is accompanied by learning a new set of rules, which often results in culture shock for a new graduate. The objective of this chapter is to share actions a new graduate can take to successfully leap into industry ahead of others.

HOW THE RULES OF THE GAME CHANGE IN INDUSTRY

The basic rule books for the university and industry are very different as Figure 10-1 suggests. To illustrate the point, let's look and compare some of the basic rules that govern university and industry and show the differences and how you need to adapt.

The first rule in industry is that your boss is not your advisor. Many students enter industry under the assumption that the boss is there for them just like their university advisor was. Don't assume the boss will be there looking out for your career and interests just like your university advisor did. This is simply not the case. The boss does not have time to guide and support your career like the university professors did. The boss is not there to make you successful, that responsibility is primarily yours.

The Engineer's Career Guide. By John A. Hoschette
Copyright © 2010 John Wiley & Sons, Inc.

FIGURE 10-1   University and industry rule books are different.

▶ *Career Tip.* Be prepared to learn a completely new operating system during your first years in industry.

The primary focus of a boss is to ensure products get developed and the company makes a profit doing it. If you are unable to perform, the boss should provide guidance and support to assist you. If you are still not successful, the boss is there to terminate your employment and get someone who can perform.

A critical change in going from the university to industry is how differently each operates. The analogy is similar to a computer operating system. The engineering knowledge you learned at the university is stored on your hard drive (mass memory) by the university operating system. University operates on a set of rules that dictate how you access the information and share it with others: how to register for classes, take tests, complete assignments, and which classes to take to obtain a degree.

The first thing industry does is replace your university operating system with a new industry operating system. Hopefully in doing so, the technical engineering information is not erased but only shows how you access this information differently. Industry has quite a different operating system with new commands and windows, most of which you have never seen before. Industry's goal is to get you quickly transitioned from the university operating system to the industry operating system. The faster you change over and become efficient with the industry operating system, the more you grow and your assignment becomes more challenging in nature.

Another very significant rule change between the university and industry is how work is accomplished. Most of the work in industry is done by teams of engineers and not individual efforts like at the university. This means in industry you rely on others to be successful, to perform tasks, so the whole team is successful. An analogy to illustrate the difference is to consider how an industry team would take a final exam. At the university, every student is required to complete a final exam solely on their own. In industry, the team would meet first and discuss the questions, then assign questions to members of the team to complete. If a person becomes stuck on a particular question, they would be expected to seek out graduate students to quickly help get the answer and report back. The team would complete the final exam together by sharing input from all members.

Another striking difference between how the university and how industry operates is how final grades are determined. In industry everyone on the team would receive a final grade equivalent to the lowest score of any individual of the team. In industry, if the team fails to deliver a good product and the company cannot sell it at a profit, the product and the entire team's effort will be considered a failure and everyone would receive a failing mark. This is equivalent to everyone in a university class getting the grade of the lowest test score in the class. One student may ace the final exam, however if other students failed, the student who aced the exam would still get a failing grade. In industry, the analogy would be one part of the product can be very successful, but if another part fails, then the whole product is considered a failure. For example, the electrical design of an IPOD may be outstanding but if the battery fails, the product is still considered a failure without a good battery. All team members must be successful in industry for the team and the product to be considered a success.

▶ *Career Tip.* Learn what it takes to be a good team player and how work gets done on teams.

In industry, sharing is rewarded and helping others is encouraged. To illustrate this point, let's compare what happens to a student who runs into a problem in a final exam at the university and a problem in industry. If the student is taking a final exam and does not know the answer to one of the problems, what would happen if during the exam, the student leaned over to a classmate in the middle of the exam and asked how they got the answer? Well, the university would probably expel the student from the class.

In industry, just the opposite is done. Industry rewards students who ask for help in order to get something done correctly and quickly. In industry, time is money. The sooner you get something done and correctly the first time, the less it costs. So getting help and correctly fixing problems is often rewarded in industry.

Another important difference to consider is that in industry, no one tells you the rules or exactly what should be done. At the university, the professor usually identifies for the students the number of quizzes, mid-term, final, and homework assignments required. A class syllabus is distributed showing the text and chapters to be covered. A class time and location is published. Nearly everything necessary to successfully complete the class is given in advance and the student needs only show up for class, study, complete the assignments, and pass the tests. In addition, student help is provided when problems arise.

▶ *Career Tip.* Learn the industry rules and what is expected of you.

Compare this to industry where the rules for completing an assignment are in constant flux and the engineer must determine what needs to be done to successfully complete assignments. Industry is a highly dynamic environment and the engineer must be able to successfully complete assignments while in a constant state of change.

As a boss once put it, "If I have to tell you what has to be done then I have the wrong person for the job. It is your job to figure what has to be done and get it completed successfully."

Other very significant differences between the university and industry, which new graduates often have a tough time handling, are the lack of feedback on performance and the long time for rewards or recognition to occur. In industry, employees receive very little feedback on their performance when compared to the feedback given by the university. Each week students receive feedback on their performance via quizzes and tests. Final examinations are often graded within a day, and several days later the final grades are posted. The reward or recognition for taking the course is given within days of completion and always before the next class is taken.

Compare this to industry where feedback on job performance is given usually once a year and sometimes even longer. The success of a project is not realized until the product has gone to market and is a great money maker for the company. This is often months after the product design was completed. Company rewards are usually given 6 to 8 months after completing the assignment. Can you imagine what students would do if they found out the grade for their class 8 months after completing the final? Or what would students do if they received feedback on how they were doing in classes only once a year? When a new graduate comes into industry, they usually expect weekly feedback and rewards immediately.

WHAT TO DO YOUR FIRST YEARS IN INDUSTRY TO GET AHEAD QUICKER

The first realization you must come to grips with is that you are no longer at the university and nearly all the rules have changed. You no longer have university professors and advisors helping you out and the burden for your success rests solely on your shoulders. You must take control of your career and develop a plan. Here are some survival tips for your first years in industry [1,2].

The very first day or week on the new job, ask your boss or lead engineer for a "Buddy." A "Buddy" is a coworker who you can go to when you need help, it is NOT your boss. Since everything is so new and different you will be asking questions most employees have figured out the answer to many years ago or seem obviously simple. You do not want to ask your boss these types of questions; it may not give the right impression.

Having a buddy to go to allows you to ask simple questions. The buddy should guide you through the rules and processes of the company similar to what the university advisors did.

▶ **Career Tip.** Ask for a "Buddy" or Human Resources department to help you learn the company rules.

The next major task is to meet all the people you can. You should meet at least all the people in your workgroup. These are people who you will be working with on a daily basis and who will be providing feedback to the boss on your performance. Find out what their job functions are and what tasks they are responsible to accomplish. Learn what is important to the department and how performance is measured.

The next step is to network at all the meetings you attend. Meet and greet everyone in the meeting. Make sure you introduce yourself and spend as much time as possible learning about what they do in the company. Let these people know you are new and appreciate any help or advice they might have. If necessary, follow up with key people after the meeting to learn more. Make sure you get a company business card and pass them out. Ask people for their business card or contact information, so you have a means of following up in case you have questions. Remember to make notes on the back of their card to remember them by.

▶ **Career Tip.** Meet and network with as many people as you can!

Make sure you visit your Human Resources department and complete all forms necessary for your benefits. If you don't understand how the benefit packages work, don't hesitate to ask. They're your benefits!

Go to lunch with your team members and others. Get to know the people you work with and what their interests are. Sit with different people at lunch so you meet more people every day. Two of the biggest career killers are going to lunch with the same people everyday or sitting at your desk during lunch and surfing the Internet.

▶ *Career Tip.* Eating lunch at your desk and surfing the Internet are two of the biggest career wasters.

One of the most important actions for one to take is finding out exactly what is required of you to successfully complete your job assignments. Your supervisor is a very busy person and will assume that if you have no questions you understand everything that is expected of you. This leads to misunderstandings and missed expectations about job performance. Spend a few minutes each week making sure your supervisor and you have a common understanding of your tasks for the week: what the order of tasks to be completed is, when the tasks are expected to be completed, and what is to be completed for a given task.

▶ *Career Tip.* Clarify and get agreement on your job tasks each week!

When clarifying your job assignments make sure to also clearly identify what expectations the boss has; this reduces misunderstandings and should help get your assignments done quicker than planned. This gets us to the next recommended action.

Get your assignments done sooner than expected. This is an indication to your boss that you are improving and ready to handle more complex assignments. One hard fact most new graduates fail to realize is that most assignments given to them could easily be accomplished by the boss or senior engineers in about one quarter the time. The assignments are designed to be easy and help the new engineer "learn the ropes" as they say. Therefore, when you complete tasks ahead of schedule it reveals your skill level is improving and you are becoming a more valuable asset to the organization. Correspondingly, if you are not able to complete the assignment in the time planned, it is usually an indication that you are not ready to take on more challenging assignments and may need special coaching. Something most supervisors do not have the time for nor want to do.

▶ *Career Tip.* Complete tasks ahead of schedule and ask for more work. Make sure your completed assignments are of high quality.

Another very good career action for a new graduate is to volunteer for assignments or committees. These may be technical in nature or more social like United Way or the company picnic. In either case, this provides you with an excellent opportunity to network and meet more people in the company. Working with people who are not in your department is beneficial since it can give you insight into other departments in the organization and possibly future promotion opportunities.

Learn the business your company is in. How does the company make money and what are the products? What are the key departments and what products do they develop? Visit the company websites and learn about the products they offer. Learn about their history, the good and bad times. Who were the past leaders and what their accomplishments were. How is the company stock doing? Review the annual report. Get a hold of an organization chart and identify the key departments in the company and their job functions. To put it simply, learn everything you can about the company.

SUMMARY

The rule book changes when you leave the university and leap into industry for your first job. Be aware of the changes and adapt to the new rules. Working in industry requires significantly more teamwork and the rules for rewards are quite different. Rewards are given to those who ask for help, share information, and help others become successful. Don't expect the immediate feedback and rewards you received while attending the university; industry moves much slower.

Your successful leap into industry requires you take control of your career and not wait for others to tell you what to do. Ask for a "Buddy" to help you learn the ropes and meet all the people you can. Get your assignments completed early and with high quality. Clarify your job assignments and make sure you know exactly what is expected of you and when. Learn everything you can about your company and its products. Determine how your department fits into the company organization. Finally, making the transition to industry is not an insurmountable task. Graduating from the university with a degree is much harder and you have already shown you are capable of that. Following the tips in this chapter should enable you to successfully leap into industry and hit the road running well ahead of others.

Have you identified any career actions you want to take as a result of reading this chapter? If so, please make sure to capture these ideas before you forget by recording them in the notes section at the back of the book.

ASSIGNMENTS AND DISCUSSION TOPICS

1 Why is it so important to get clarification on the expectations of your assignments?
2 What is the best thing to do when you run into problems on your assignment?

REFERENCES

1. Johnson, Vern, *Becoming a Technical Professional*, Casas Adobes Publishing, Tucson, AZ, 2000, http://www.dakotacom.net/~capublish.
2. Institute of Electrical and Electronics Engineers (Corporate Author), Gabelman, Irving J. (Editor), *The New Engineer's Guide to Career Growth and Professional Awareness*, 1996.

CHAPTER 11

PHASES OF RETIREMENT PLANNING
BEGINNING WITH THE YOUNG ENGINEERS

Retirement planning is a lifelong activity that is best started with your first job after college and continuing throughout your career and even after retirement. An overview of the different phases of retirement planning which will happen during your lifetime is shown in Figure 11-1.

The first phase is early career planning followed by mid-career and then late career. In early and mid-careers, your primary retirement planning focus is on your ability to save and invest. The late-career phase covers the last five years of your work. This is the time frame when you will have to decide when you are going to retire, where you are going to live and how to best draw on your investments to support your retirement. The end of your career marks the start of the next major phases of early retirement, mid-retirement, and late retirement. In early retirement you may travel. Generally you are highly social and may even work part-time. In mid-retirement your activities start to slow down. Finally, in late retirement you may require assisted living, be concerned with estate planning, and making your final arrangements.

As you move through all these phases there are four major focus areas one needs to consider when retirement planning and these areas are:

Focus Area	Phases
1. Financial (income)	All phases
2. Benefits (primary is medical)	Late career and all through retirement
3. Residence (where to live)	Late career and all through retirement
4. Activities (your time)	Late career and all through retirement

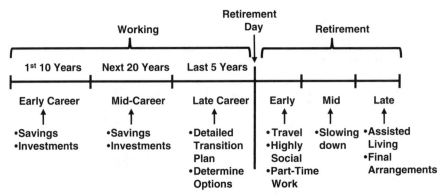

FIGURE 11-1   Overview of retirement planning activities during your career.

The financial focus area covers all phases of your career from the very beginning to the very end. While you are still working, the focus of the financial area is how to best save and develop your investments for retirement. After retirement, the financial area focus is on how to best convert your investments to income and avoid taxes.

The benefits area comes in play after you retire. After retirement one of the major expenses is medical benefits. With the cost of health care and prescription drugs skyrocketing, this becomes the major concern of many retirees. Having a good medical plan so that your retirement savings are not quickly depleted in the event you become ill is primary.

Your residence is the next focus area and this comes into play when people make a decision on where they would like to live when retired. Do they want to stay in the same home, move to a condo for less maintenance, or even move to a warm climate to get out of the cold winters? How is your tax liability affected by changing residence? Do taxes go up or down? How are your medical benefits affected by a change in residence? Do you want to leave family members, relatives, and friends? These are some of the major concerns you will have to consider when selecting your residence for retirement.

The last focus area is activities which deals with what type of activities do you want to be doing once you are retired. Do you want to work part-time? Or maybe do volunteer work, start a hobby, or travel?

Each of these focus areas is dependent upon the other focus areas and decisions in one area may affect the others. For example, moving your residence out of state may affect your financial situation and your activities. Medical policies may require you to live in the same area as your company after retirement. Your travel activities may quickly deplete your savings. For these reasons, you will need to make retirement planning a life-long activity. Good retirement planning should result in a well-balanced plan for each of the focus areas and hopefully allow you to retire in the manner that you would like.

EARLY CAREER RETIREMENT PLANNING

For this phase the most important area of retirement planning is financial. When you're young and just starting your career it's an optimal time to start retirement planning as Jeffrey Wuorio points out in his book, *Retirement Planning* [1]. The earlier you start, the better off you will be when you retire and more options will be available to you. The longer you wait to start planning for retirement, the more pressure you have to save for lost time.

Financial advisors recommend starting early when planning your retirement. The following is an example of the importance of starting young.

Let's take the example of Joe versus John. Joe starts investing in his 401(k) at age 25 and contributes $10,000 a year for 10 years, then stops contributing. If he earned a 7% rate of return on his money, at age 60, his account would be worth approximately $630,000. Now John doesn't start saving until age 35. He saves the same amount of $10,000 per year, but does this for 25 years, until he is 60 years old. At the same 7% rate of return, John's account would be worth approximately $630,000. Joe contributed only $100,000 while John contributed $250,000 and their accounts where worth the same at age 60. If Joe would have kept saving that same amount of $10,000 a year until age 60, his account would've been worth over $1.3 million.

The optimum is setting up your finances such that you pay yourself first 15–20% of your salary (see Chapter 49 "Finances"). Following this, one of the best investments you can make is participating in your company-sponsored retirement plan. Most 401(k)s offer some company match. A common company match is 50 cents on the dollar up to 6%, similar to getting a 50% return on your money. Most companies offer a wide variety of mutual funds to invest this money in. Target Date Retirement Funds and Asset Allocation Funds are a wise place to invest if these are offered in your plan. They are managed according to your age and risk tolerance. For example, a 2035 Target Date Fund would be suitable for an individual who plans to retire at or near year 2035; the fund is a blend of stocks, bonds, and cash, and will automatically get more conservative as you get closer to retirement.

Another tip to help you keep on track over the years is not specifying an amount for your contributions but specifying a percentage of your salary. This way your contributions will automatically increase with your raises and salary increases.

If your company does not have a qualified retirement plan or you are self-employed, then seek out the help of a financial advisor to assist you in setting up your retirement plan immediately.

Another retirement option to check into is the pension benefits of your company. If your company has a pension plan, make sure you qualify for it. Most companies who have a pension plan require you to work for them for at least 10 years before you qualify for a pension. This is something important to consider when you are thinking about leaving the company for another opportunity. If you leave before you become vested or qualify for a pension you could be giving up a large amount of money, especially if you are only one or two years from qualifying for a pension. Thoroughly understanding your retirement benefits could save you thousands of dollars later in life.

MID-CAREER RETIREMENT PLANNING

For this phase the most important area of retirement planning is financial. If you started investing for retirement early in your career, then by the time you reach mid-career (40–55 years old), you should have a substantial portfolio. During your mid-career years you should periodically sit down with all your recent statements from all your investments and see if they are performing to your expectations. Hopefully, as you made adjustments and changes over the years they have resulted in realizing the earnings you wanted. Measure the specific performance of each investment and determine if they are on track with your plan. If not, you are going to have to make some adjustments.

Here are some factors that you should consider as you do a mid-career assessment of your progress.

1. Have there been any significant events impacting your investments and what actions should you take?
   a. College education for children
   b. Wedding
   c. Serious illness
   d. Market downturns
2. Are you putting away the maximum you can?
   a. Have contributions increased with salary increases?
   b. Has the company match changed?
3. Is your spouse also contributing to a retirement plan?
   a. Is it possible to save more?
4. How have your tax liabilities changed? Should you be investing more as a tax shelter?
5. Finally, are you ahead or behind your plan? What adjustments should you make?

These are all important things to consider when reviewing your investment portfolio. Seek out the advice of your financial planner when making any adjustments. There are many excellent books written on financial planning and retirement investing [2–4].

LATE CAREER PLANNING

As you mature and come to the final years of your career, the tendency is to coast into retirement. When asked "What are your plans for your transition to retirement?" The answer often comes back, nothing really, just going to retire when the time comes. These engineers suddenly go into panic mode when their retirement is less than a few years away. They are totally unprepared for making the transition and often end up returning to work since they did not plan appropriately. Detailed retirement planning should start as early as 5 years before retirement. At this phase all the four major areas need to have a detailed plan (finances, benefits, residence, and activities).

There are many factors to consider when you take the final curve in the career road leading to retirement. Career retirement counselors tell us this will be one of the toughest decisions a person has to make. Many choices and options have to be decided upon.

Finances. At the top of the list of things to be done before you retire is to determine how you are going to use your investments, pensions, Social Security, 401(k), and other assets, to live on during retirement. This is no simple task and is going to involve your company, spouse, and financial advisor. Your objective is to lay out a plan that will give you the income you need as long as you live and hopefully not run out.

Identify the sources of money you plan to draw on, the options for each and then compute a potential yearly income for each year. This will give you a projected income that you can count on during retirement for planning purposes.

The biggest problem most people face is what age is best to retire at? The answer to this question is not simple. It depends on many factors. First and foremost is your health. Being healthy offers you the option to work longer and retire later. Correspondingly, you will be able to receive more benefits. If health is an issue you may want to retire early. When is your spouse retiring is another consideration. Ask yourself, "when can I afford to retire?" The answer to this question depends on your debt level going into retirement, your standard of living, and how much you were able to save and invest over your lifetime. In engineering terms, this is a very complex computation that involves many variables. There are many solutions and only through taking your time and working through all the variables with your spouse, financial advisor, and company human resources, will you come to a satisfactory solution. This is why you want to start analyzing options and considering potential outcomes at least 5 years before you retire. To wait until a couple of weeks before retiring leaves you no time and few options. Let's illustrate this with a simple example. An income projection sheet for a couple planning on retiring is shown in Table 11-1. The husband is 4 years older than the wife and both are working and have retirement investments. This is only one possible scenario and the purpose of showing it is to identify some of the options and the many choices that need to be made.

TABLE 11-1 Project Income Worksheet for Couple Retiring

	Age									Remarks
Husband's age	65	66	67	68	69	70	71	72	73	Retire after 65 years
Wife's age	61	62	63	64	65	66	67	68	69	Retire after 65 years
Husband's work income	75	15								
Husband's pension		18	18	18	18	18	18	18	0	Run out after 7 years
Husband's 401 (k)		20	20	20	20	20	20	20	20	Good until 82 years old
Husband's Social Security		25	25	25	25	25	25	25	25	Reduce benefits for surviving spouse
Wife's work income	55	55	55	55	55					
Wife's pension						12	12	12	12	Good until 72 years old
Wife's 401 (k)						10	10	10	10	Good until 78 years old
Wife's Social Security						8	8	8	8	Full benefits
Total yearly income ($K)	130	133	118	118	118	93	93	93	75	
% of normal yearly income	100%	102%	91%	91%	91%	72%	72%	72%	58%	

Both have elected to retire at age 65, which means the wife will be working four years longer than the husband. The husband plans on drawing from his company pension, 401(k), and Social Security. The first year after retiring he plans on working part-time to help with expenses. He has elected Social Security benefits based on surviving spouse and not the level or single person options. His pension is good only for seven years and runs out when he is 73 years old. He projects his 401(k) is good until he is 82 years old. The wife is planning on working longer to help with benefits and will retire 4 years later. She plans on selecting her Social Security benefits based on the single person option. For this example, the couple's total yearly income is shown at the bottom. Up to his age 70 they are living on 91% of their original working income. Between ages 70 and 72 the income drops to 72% and down to 58% thereafter. This is only one of many possible scenarios for the couple due to all the options that are available with retirement benefits and retiring age. If the couple decides to retire early or later or she decides to retire at the same time, the numbers all change. Trying to decide which option is best is going to take time and you should't wait until the last two weeks before retirement. Making decisions in haste that you have to live with the rest of your life is not a good plan.

Benefits. What about medical benefits? Did you save enough income to pay those high premiums? What is your plan to transition from company-paid monthly medical premiums to you paying the monthly premiums? For some people the plan is to return part-time to work simply for the medical benefits. Most companies offer retirees the capability to convert their medical benefits through plans that offer lower premiums that they can get by themselves. Look into other groups and organizations such as engineering societies and other professional associations for economical benefits. Check with people you know who have recently retired and ask about the medical benefits and coverage. Is it what they expected? How much does it cost and what are the monthly premiums?

Residence. Where do you plan on living when you retire? Are you going to stay in the same house or downsize to something more economical? Are you counting on retirement income from the sale of your home? Will the market support a quick sale of your home? Many people plan on retiring and moving to a retirement community in another state. They put their home up for sale and move to Florida or Arizona. The problem that may arise is medical benefits often do not allow you to live in another state. Many medical plans require you to reside in the state you retired in. On multiple occasions, I have had engineers tell me they ended up selling their retirement home only to return back to the city where they retired because of medical benefits.

Activities. What are you going to do with your free time after you retire? Many people just say relax and once they retire, become so bored with no planned activities. You will still need activities in your life and stopping cold turkey is not recommended. To ease the transition to retirement, counselors recommend continuing some type of part-time activity to keep you busy. Here are some great activities you could consider.

1. Volunteer work for a charity organization
2. Starting a new hobby
3. Working part-time
4. Taking a class at the university
5. Teaching a class
6. Writing a book
7. Consulting

An important aspect to consider in your transition to retirement career plan is how the spouse is going to handle you being around 24 hours a day, 7 days a week. The spouse is not ready for the invasion and this often leads to conflict. If you plan on retiring soon, get your spouse involved in the planning and even see a counselor if you can. Have a strategy of how you are going to cohabit the same home "24-7" once you retire.

There is excellent help available and many resources you can draw upon for assistance once you enter the last 5 years of working. Key among these is the US government Social Security website with lots of excellent information (www.ssa.gov). Another good organization to join is the American Association of Retired People (AARP). Their website is (www.aarp.org) and it is loaded with helpful tips on nearly aspect of retirement. The AARP is a national organization with many local chapters. Check for a local chapter and its meeting places in your area.

BEST AGE TO RETIRE

What is the best age to retire? And the answer is, "that depends." The answer to this question is specific to each individual. However, the following information from a study conducted by the Boeing Company on the longevity of retired engineers from the company, gives us some food for thought. The study identified the years to death for engineers on the basis of age at retirement. This data is shown in Table 11-2.

The data identified the average engineer who retired at age 55 lived to 78 or enjoyed 23 years of retirement. Correspondingly, the average engineer who waited and retired at age 65, only lived for 10 months. Wow! The data clearly indicates a correlation between retiring early and living longer versus retiring later and dying sooner.

How is your health? What are you doing to ensure you are in good health when you retire and you will live to enjoy your retirement? Based on this data, a good transition to your retirement career plan should contain provisions for your health and regular exercise.

TABLE 11-2   Life Expectancy for Age at Retirement

Age of Retirement	Age at Death	Years to Death from Retirement
55	78	23
56	76	20
57	74	17
58	73	15
59	71	12
60	69	9
61	68	7
62	68	6
63	67	4
64	66	2
65	66	10 months

(From Boeing Study 10-2-84).

You will be eligible for full Social Security benefits if you "retire" at age 65 as it stands today. The standard retirement age of 65 is scheduled to increase gradually from 65 to 67 by the year 2027. For each year you wait beyond age 65 to collect, the benefits will increase slightly. Currently, if you retire at age 62 you can receive 80% of the benefits had you waited until age 65. Does that mean you should wait until age 65? Not necessarily as Ernst & Young point out in their book, *Ernst & Young Personal Financial Planning Guide* [2]. The amount you receive between age 62 and 65 is 35 months of benefits. This is a sizable amount and if you waited until age 65 to start collecting the higher benefit, it would take more than 12 years of higher benefits to make up the difference.

So the answer to this question, "what is the best age to retire" clearly is an individual choice and is based on the individual health and financial situation. No one answer fits all.

BEST TIME OF YEAR TO RETIRE

When is the best time of the year to retire is another factor to consider. The formula your company utilizes to compute retirement benefits may affect the time of the year. Generally, companies will compute a retirement benefit based on the salary you received in the last full 3 years you worked. Therefore, any promotion or salary increase you receive in the last 3 years actually represents a huge raise for the rest of your life! Why not strive for one last promotion as you near retirement. Also waiting to retiring just after your last work anniversary or full year could mean a more significant amount of money. Another factor to consider is when your company awards vacation. If, for example, you earn 4 weeks of vacation each year and the company grants your 4 weeks every April, then it is best to wait until after April to retire since you will be able to collect an additional 4 weeks of vacation pay.

You should not select your retirement date until you have identified a good plan and one you feel confident with. Remember that you are going to live with the decisions you are going to make for the rest of your life; therefore, you need high confidence in the choices which are right for you. You do not want to make long shot guesses in hopes you will get lucky. You want decisions that are solid and you should feel very confident of your plan.

Once you get to the point of selecting a date to retire, your first actions are meeting with your human resources and financial advisors. It may take months for human resources to put together the paperwork and get it approved for your retirement. They do not like surprises, especially when people show up unexpectantly and announce they are retiring that day and think human resources can get everything arranged that quickly. The same holds true for your financial advisors. It takes weeks to process paperwork and setup payments. In either case, it maybe 1 or 2 months before you see your first check. I have heard stories from engineers where they planned exotic travel immediately after retirement as a kick-off. They never realized it would take several months before they would see their first retirement check and as a result had to delay their plans. Make sure you have sufficient liquid funds to cover your expenses between the time you retire and you get your first retirement check.

TRANSITIONING INTO RETIREMENT

How you transition from working into retirement is strictly an individual's choice. In this section, I share some of the strategies to help you in your planning. One thing to consider for your transition strategy is the amount of change and stress you are placing on you and your spouse as part of the transition. The more changes you make, the harder and more stressful it is going to be. According to retirement counselors, the most difficult transition is one where your life completely changes over almost overnight. Let me describe this scenario: the husband decides to retire in a couple of months, wants the wife to quit her job also, puts the home up for sale and buys a new retirement residence in another state and several days after retiring moves to the new home leaving all their family, relatives, and friends behind to start a new life in a new place. Clearly, this would be stretching the limits of any person's ability to handle change and making all the major stress factors (loss of job, change of residence, and loss of family and friends) come into play all at once. Your transition plan should control the amount of change at any given time and subsequently minimize the stress. Here are some transition strategies for retiring.

One strategy is to retire and then return to the company to work part-time in special assignments. The company views the engineer as such a valuable asset they are willing to hire back the engineer after retirement. The engineer receives full retirement benefits and then returns to work part-time for a modest salary and medical benefits. The benefits to the company are the engineer can be used as a resource and help the company. This frees up more time for the engineer but yet provides a small income and medical benefits.

Another retirement strategy is to become an independent consultant and work part-time helping and giving advice to other companies. Or possibly become a teacher and instruct students at a local university. There are also many seminar companies that hire older engineers to teach 1 or 2 day classes and the pay is excellent. This allows the engineer to control the amount of time and days to work offering more freedom while providing a little extra income.

Some people simply leave work and never return to their profession again. For these people, they usually have something lined up that they always wanted to do and now that they have the time, they are going to do something else. They have a plan for the transition and often are working at it years in advance, so there is not a sudden and stressful life change.

One final strategy is stay working beyond 65 years old and hope the company offers you a special retirement package. People select this option mainly because they either have not planned and prepared for retirement and therefore try to work for as long as they can or they really enjoy work and prefer to remain active. Their career strategy is to have the company offer a special incentive bonus to take retirement and in the meantime continue to save for retirement and plan for the day.

The key point here is you must have a strategy for transitioning to retirement to be successful. Controlling the amount and velocity of change while minimizing stress is the goal. Waiting until the last minute and making decisions in haste or waiting until you are retired to figure out what to do next is not a good transition plan.

Three Phases of Retirement

Once you retire, your planning is not done by any means. Due to the constantly changing economy, your health, changing Social Security benefits, increasing medical costs, and changing tax laws, you will still need to assess and monitor your situation regularly. Now let's look at the three phases of retirement and how your planning is affected by each.

Early Retirement. In the early retirement phase you are probably going to do the most travel and be highly social. Your ability to get around and driving/mobility will still be good allowing you to enjoy many social events. For this reason, your expenditures will normally be the highest during this phase. Some financial advisors recommend that you should plan for withdrawing more income from your investment savings during this time and reducing the income to live upon in later years since you will not travel and partake in as many social events. Careful financial planning in this stage is required to make sure you do not overspend and end up running short later in life. If you have not already prepared a will you should absolutely do this during this time. In addition, working with a financial advisor and tax expert, you should start your estate planning to ensure all your hard-earned money would go to the people you want without putting a heavy tax burden on them.

Mid-Retirement. At this phase your mobility reduces and possibly your health issues arise; subsequently couples tend to be more home-bound. The exotic vacations stop and life settles down to just being around the house with family and friends. Your spending declines and moving to maintenance-free low cost living is the norm. You need less income to live on. Health is the primary concern and medical treatment is the primary concern. Making sure you have good medical benefits is key at this point.

Late Retirement. Although not easy to plan for, late retirement is when the planning for assisted living, making arrangements for your estate as well as your final arrangements. Most people are concerned at this point about not being a burden to their children and the quality of life. Living wills, writing the terms and conditions of your final days are very important. Appointing a power of attorney to oversee your estate is a great way to reduce the burden on yourself or loved ones in the final days.

SUMMARY

This chapter provided an overview of the key elements involved in retirement planning, but you will need to dig deeper into each of the areas. The best time to start retirement planning is when you are young, the sooner you get started the better. As you move through all the phases of your career, there are four major focus areas one needs to consider when retirement planning and these areas are: (1) financial (income); (2) benefits (primary is medical); (3) residence (where to live); and (4) activities (your time).

What is the best age to retire? The answer is dependent on how you did your retirement planning, your health, and your personal desires. It is specific to each individual. No one answer fits all. When the time comes, you will need the assistance of your company's Human Resource department, financial advisors, and tax experts. There will be many decisions, research, and educating yourself, to ensure you select the best choices for you. In most cases, you have to live with the choices you made for the rest of your life. Making all these choices is not done in matter of hours, days, or weeks; you should start this type of planning when you are about 5 years from retirement.

How you transition from working to retirement is strictly an individual's choice. The key point here is you must have a strategy for transitioning to retirement to be successful. Controlling the amount and velocity of change while minimizing stress is the goal. Waiting until the last minute and making decisions in haste or waiting until you are retired to figure out what to do next is not a good transition plan.

Once retired, your planning does not stop due to the constantly changing economy, your health, changing Social Security benefits, increasing medical costs, and changing tax laws. You will still need to assess and monitor your situation regularly. As you age in retirement, your lifestyle will change significantly and being prepared for this financially and medically is the best plan. Develop a support structure of family, friends, and advisors, to help in your retirement with people who you can trust and call upon.

ASSIGNMENTS AND DISCUSSION TOPICS

1 What are the four primary focus areas you consider when retirement planning?
2 Why is it so important to start young?
3 What is the best age to retire?
4 Name two transitions to retirement strategies?
5 What are the money sources for your retirement?
6 Why is having activities so important after retiring?
7 Describe the phases of retirement.

REFERENCES

1. Wuorio, Jeffery J., *Retirement Planning, The Complete Idiot's Guide*, Alpha Books, 2007.
2. Ernst & Young, *Personal Financial Planning Guide*, John Wiley & Sons, 2004.
3. Hebeler, Henry, *Your Winning Retirement Planning*, John Wiley & Sons, 2001.
4. O'Shaughnessy, Lynn, *Retirement Bible*, John Wiley & Sons, 2001.

PART 2

COMPANY STRUCTURES, ORGANIZATIONS, AND BARRIERS IMPACTING YOUR CAREER

CHAPTER 12

HOW COMPANY SIZE AFFECTS YOUR CAREER

You can have a very successful career working in either small companies or large corporations. Understanding the advantages and disadvantages of each and knowing the appropriate career actions is key to having a successful career. In this chapter, we will highlight how your career actions change with the size of the company and provide helpful tips to be successful regardless of the company size.

▶ *Career Tip.* The size of your company will have a significant impact on your career planning. If you work in a small company your career actions will be different than if you work in a large corporation.

SMALL COMPANY STRUCTURE AND THE IMPACT ON YOUR CAREER

A generalized small company structure is shown in Figure 12-1. By small company, we are referring to organizations with typically less than 150 people. Generally, the organization is run by the company owner. The department manager reports directly to the owner along with the company technical specialist.

The department manager is responsible for running the entire organization and the technical specialist is the lead person for technical design and performance of the product.

The lead engineer directs the work of the engineering team and generally handles daily tasks associated with design, build, and testing of the products. The lead engineer may work alongside of the marketing and accounting personnel. Reporting to the lead engineer is the product design team. This team typically consists of the people responsible for producing the product

The Engineer's Career Guide. By John A. Hoschette
Copyright © 2010 John Wiley & Sons, Inc.

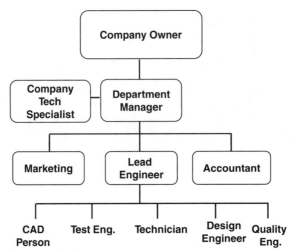

FIGURE 12-1   Generalized small company structure.

and may include computer-aided design (CAD) engineer, test engineers, technicians, design engineers, and quality engineers.

Working for a small company has its own unique advantages and disadvantages as well as special challenges. These are summarized in Figure 12-2.

In a small company you can easily have visibility with upper management. It is common for everyone to know everyone else and if you want to

Advantages
• Small size allows you to know everyone
• Easy visibility with upper management
• Broader job assignments
• Fast moving, no large bureaucracies
• Poised for rapid growth
• In on the ground floor

Disadvantages
• Easily go out of business
• Usually smaller number of contracts
• May be lower job security
• Typically lower pay and reduced benefits

Special Challenges
• Limited resources—must develop
  support network outside the company
• Ability to work with little resources and
  independently
• Oftentimes must do it all—research,
  design, build, test, marketing

FIGURE 12-2   Advantages, disadvantages, and special challenges when working in small companies.

talk to the owner, you simply walk down the hall and knock on their door. Your job assignments are broader in scope and you often do it all. You design, build, test, and market. Small companies offer the advantage of being fast-moving with no large bureaucracies to deal with. Decisions are rapidly made and you are in on the ground floor. You may be betting your career that your company is soon to become the next Microsoft or Apple Computer and that's okay if it fits your style. Your company may have phenomenal growth and you will be there to grow with it.

In a small company environment, you must be able to deal with rapid change and the fast moving pace. You must be flexible and readily adapt to change; this requires a special type of person. You may not want to work for a small company if you do not like an environment with rapid changes and operating uncertainties. Do you have this type of personality? Can you handle change and do you enjoy learning about other areas of engineering outside of your technical expertise? Would you be willing to take on assignments, say in chemical, mechanical, electrical, or even software, if you had to? If this type of assignment excites you, then working in a small company is a good match for you. If, on the other hand, you do not want to broaden your technical background and would prefer working only in your area of expertise, then working for a small company may not be a good match for you.

Working for a small company also has some disadvantages for career growth. It is very easy for a small company to quickly go out of business. I have seen cases where employees show up to work one morning only to find a sign posted on the locked doors informing them: "Closed, No Longer in Business." Therefore, working in a small company requires that you keep an eye on the profitability of your company on a quarterly basis. Does your company have the financial reserves to sustain payroll and jobs through several bad quarters? Or is your company operating on a shoestring budget and susceptible to layoffs at the first sign of poor financial performance?

Small companies usually have only one or two products or contracts. As long as the products sell or the contract continues, the company is solid financially. Having only one or two major products does not leave one with many career options. Your work is confined to a single product or contract without many career growth options. Typically, there is less job security in a small company and the benefits are not as good as the larger companies.

Working for a small company presents special challenges. In a small company there are usually very limited resources to call upon. You must be able to handle a variety of tasks outside your engineering discipline. To be successful at your assignments, you must develop a network outside the company; a network of people and other companies you can quickly call upon for help and guidance when you are faced with a problem you cannot solve yourself. In addition, you must be able to work independently and with little supervision or guidance. You must be able to do it all: research, design, build, test, marketing, sales, and customer interfacing.

▶ *Career Tip.* To be successful in a small company you need to have the skills to work in a highly dynamic environment which may rely on limited resources.

Have you considered your career growth paths in a small company? If you are the department manager, you have several options. These may include: buy out the owner, start a business on your own, or grow the business so another department, similar to yours, will be created and you will become the manager of two departments.

If you are the lead engineer, unless the department manager retires or leaves, your only hope is to grow the business and start another department and get appointed department manager. Your career path should show that you are the best candidate to take on the department manager role at the first opportunity that the business expands and requires another department.

If you are an engineer, your career path is getting appointed to lead engineer position. In order for you to do this, the lead engineer needs to move or the business needs to grow so another department is added. Your career path should show that you are capable of handling the lead engineering position when the opportunity arises.

If you consider all the options for career growth in a small company, they all lead through the action of growing the business and developing new and more profitable products. To grow the business requires a different skill set than you learned in the school of engineering. It requires a skill set of handling customers, recognizing the need for new products, and coming up with new products that meet the customers' needs. It requires a deeply ingrained entrepreneurial spirit. Are you open to learning this new skill set?

I have observed over the years that some of the leaders in large corporations actually started out in small businesses. It is interesting to note that many large corporations like to hire people from smaller companies. The reason is, people from a small company are generally more rounded individuals and have had exposure to all aspects of the business. They come with a broad experience base and are generally better leaders because of this experience.

▶ *Career Tip.* Working in a small company provides the opportunity to take on more diversified tasks and gives you a broader experience base. These are highly sought after characteristics in large corporations.

Another interesting fact that engineers coming from small companies have shared with me is that large corporations appear to be made up of groups of small businesses. When you look into it, there are many similarities between small companies and groups or product lines in major corporations. Both often operate with similar business principles.

Over the past 30 years, I have seen many engineers switch careers between large corporations and small companies and greatly enjoy the change. I have seen engineers who start out in large corporations and leave the large bureaucracy for a small company to be very happy. I have also seen engineers start in small companies and leave for the more stable environment of large corporations.

Correspondingly, I have met engineers who have left large corporations to work for small companies, only to find out their personality does not match that of a small company, and happily return to the large corporation. I have also seen engineers who start in a small company, change to a large corporation and then realize their personality does not match that required for large corporations, and return to small companies.

▶ *Career Tip.* You are going to be more successful if your personality style matches the style of your company.

The key to a successful career for you is to honestly look at your personality and determine if you are best suited for working for the size of company where you are presently employed.

LARGE CORPORATION STRUCTURES AND THE IMPACT ON YOUR CAREER

A generalized reporting structure that is typical of most large engineering corporations and companies is shown in Figure 12-3.

At the top are the CEO and corporate board members. Reporting to the board is the next level down which is usually comprised of vice presidents responsible for the corporation's various lines of business. The vice president is usually responsible for a large business area that could contain many product lines. A typical vice president will normally have between 1,000 and 2,000 people reporting to them. Reporting to the vice presidents are the business area directors. Directors generally have between 500 and 700 people reporting to them. Reporting to the directors are the managers which have 100–150 people they are responsible for. Below the manager are the lead engineers who are generally responsible for the development of specific parts of a product. And reporting to the lead engineers are the engineers designing, building, and testing the products.

Working for a large corporation has its own unique advantages and disadvantages as well as special challenges. These are summarized in Figure 12-4.

In a large corporation, you generally have more job security since there are a large number of contracts and products in the corporation portfolio. This

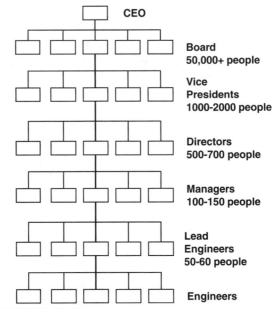

FIGURE 12-3   Generalized large company structure.

Advantages
- Better job security (many products/contracts)
- Better benefits and salary
  - Medical, educational, higher salaries
- Abundance of resources
  - Technical experts, laboratories, and facilities
  - Mentors
- More opportunity for advancement
- Larger career support structure
  - HR, training

Disadvantages
- Little or no upper management visibility
- Competition for raises and promotions
- Small scope in job assignments—focus on one specific task
- Large bureaucracies / slow moving

Special Challenges
- Ability to work in team environment
- Ability to work with constant supervision
- Oftentimes work on one small task, contribution may seem insignificant

FIGURE 12-4   Advantages, disadvantages, and special challenges when working in large corporations.

offers the advantage to the engineer of being able to transfer from one product line to another, as jobs come and go, without having to leave the company.

▶ *Career Tip.* Large corporations tend to offer more job security and better benefits compared with smaller companies.

Another advantage is that corporations usually have better medical benefits, better chances of obtaining an advanced degree on company funding, and generally the salaries are higher. A large corporation generally has a larger abundance of resources the engineer can call upon when needed, as compared to small companies with very limited resources. In addition, there are generally more opportunities for advancement and more of a career support structure in terms of HR and training offered.

Working in large corporations may also present some disadvantages for career growth. In a large corporation there is often little visibility, or none at all, with upper management for the average engineer. When I worked at Honeywell, I had 5 years of service before I met the CEO. In a large corporation, you have to work at getting upper management visibility. It is not simply walking down the hall to the owner's office like in a small company.

There is more competition for raises and promotions in a large corporation. You generally have assignments smaller in scope and directed toward one specific task. Finally, you have to deal with bureaucracies in large corporations. In a small company, decisions are made quickly and usually involve only one or two people. In a corporation, the decisions often have to be approved "Up Chain" and may take several days, weeks, or even months to obtain.

▶ *Career Tip.* Large corporations may have many people competing for a single position resulting in more competition for promotions.

There are special career challenges associated with employment in large corporations. One career challenge is working on large teams or doing your job in a team environment. You must be able to work effectively with people of all levels and different backgrounds. Team meetings are common and getting a team approval of your design can take a significant amount of effort. You must also be able to work under constant supervision. With all the levels of management in a major corporation, there are constant reviews to track progress. Finally, your tasks may seem insignificant in the grand scheme of things and many engineers find this very discouraging.

You often work on one single small task, or your tasks may use very little of your formal university training and capabilities, leaving the junior engineer wondering why he or she had to take all those years of advanced calculus. Only after you have proven you can handle it, is the scope of your workload increased.

SUMMARY

Whether you are working for a small company or a large corporation you can have a successful career. Both have advantages and disadvantages, as well as special career challenges. Working in a small company requires you to be able to work with limited resources and oftentimes do more than just engineering. Your job assignments will have a broader scope than in corporations. In general, you have less security and benefits, but you are positioned right if your company business is successful.

Working in a large corporation requires you to be able to handle the bureaucracy and large team environment. Your job assignments will be more focused and narrow in scope. In general, you have better benefits and security, but there will be more competition for raises and promotions.

The key is recognizing your type of personality and if it fits the personality required by the size of your company. Are you a small company person or a large corporation person, or something in-between? If you have a match, this is great. If you feel you have a mismatch, it is time to start career planning and make some changes.

Have you identified any career actions you want to take as a result of reading this chapter? Please make sure to capture them now before you forget by recording them in the notes section at the back of the book.

ASSIGNMENTS AND DISCUSSION TOPICS

1 Which is better to work for: a large corporation or small company?
2 Name two benefits of working for a small company. What are the disadvantages?
3 Name two benefits of working for a large corporation. What are the disadvantages?
4 What should you do if your personal style does not match your company? Retrain or move on?

CHAPTER 13

THE CORPORATE LADDER
CAREERS OPEN TO YOU

How high up the corporate ladder do you want to climb: to the middle or the very top? The first step to climbing the corporate ladder is to determine the steps in the ladder and where they lead to. Corporations have a hierarchical engineering ladder with a definition for every job level on the ladder. If you are not aware of the engineering hierarchy in your corporation, then meet with your supervisor or someone senior in the company and map it out. Another good place to learn about the engineering hierarchy is your Human Resources department. Oftentimes, companies will post this information on the company website. Make sure you check out all three sources since this hierarchy is often highly dynamic and usually changing on a yearly basis.

▶ *Career Tip.* To successfully climb the corporate ladder, first determine its structure and where the steps lead.

CAREER PATHS FOR ENGINEERS WITHIN A CORPORATION

A typical ladder structure for larger engineering corporations is shown in Figure 13-1. At the bottom are the nonsupervisory engineering levels. Most companies have between four and five levels of engineering before one reaches the staff or supervisory level. I have identified these levels as E1 through E5 in the figure.

In the lower levels (E1–E2) of engineering you are expected to be a good team player and learn from the senior engineers, who are considered the pros. Your assignments are often one or two days long and usually accomplished by yourself, or with the help of another person or mentor.

As you move up the engineering ranks, your responsibility greatly increases. In the middle levels (E3–E4) you start to lead small teams to

The Engineer's Career Guide. By John A. Hoschette
Copyright © 2010 John Wiley & Sons, Inc.

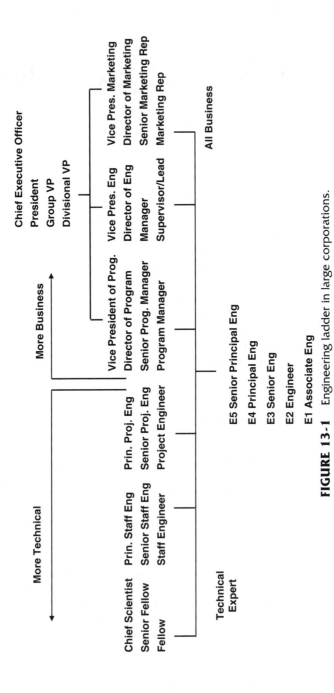

FIGURE 13-1 Engineering ladder in large corporations.

accomplish specific objectives. The objectives are usually well-defined. The teams are two to three people and assignments may extend over several weeks or months. You are primarily responsible for technical work. You slowly transition from being an individual contributor to team leader and guiding the work of others to accomplish your objectives.

In the upper levels (E4–E5) of engineering, you are responsible for directing large teams of engineers with various backgrounds. Often the objectives are not well-defined and it is up to you to plan things out. Besides giving the team technical direction you are also responsible for maintaining schedule and cost control. You must have good interpersonal skills to ensure work is being accomplished by other people rather than doing it yourself.

Once you have reached the staff level, several career choices are available. The typical choices are shown in Figure 13-1. The career ladder typically splits into five or six different paths. The engineer has the opportunity to pursue several different technical career paths and several different management career paths. They range from the very technical on the left to mostly business on the far right. The three technical ladders on the left are comprised of people who are not interested in becoming business leaders and are rewarded by finding new technology.

Staying technical means your career interests follow along the fellow and system engineering paths where your primary focus is technical advancement. Having a successful career does not only mean becoming president of a company. You can have a very successful career and remain technical. Your success is not necessarily measured in terms of titles, power, and profits. It is measured in terms of breakthroughs, papers published, patents, technical awards, and what is right for you.

ENGINEERING FELLOWS' LADDER

At the far left is the fellows' ladder, which is the most technically-oriented ladder. These people are usually PhDs who have dedicated a great deal of time to furthering science and technology. They are usually the research scientists of the company. To get these positions, they have spent their life focusing their career on solving technical issues. They are often recognized as technical experts in the company. One example might be if a company is building a car, they could be the expert in spark plugs, or another example might be the world's leading authority in tires. Fellows are typically recognized technical experts within the corporation and also known for their expertise throughout industry. Stereotypically they are thought of as having silver hair, a messy office with patents, plaques, and awards.

If your career aspirations are to remain purely technical then this ladder is for you. If you already have an advanced degree in engineering, then you have a good start. However, with the rapid pace of changing technology you will have to return for updating in new technologies and make lifelong learning part of your career plans. If you do not have an advanced degree in engineering, then returning for a Master's or PhD should be in your career plans.

Usually fellows are most interested in the pure science of the problem. Fellows mainly deal with technical issues and avoid management or cost issues whenever possible. Learn who they are in your company. They are the best technical resources you have in the company. Fellows are the technical lifeblood of the company. Usually, they are easy to approach and often enjoy helping younger engineers work toward the best technical solutions. Their many years of experience, and typically large network of technical contacts, can be extremely valuable in helping you solve technical issues.

Engineering fellows have most likely seen nearly everything and they are very thorough with technical details. When interfacing with fellows, be very precise, pay attention to detail, and do not guess about the technical subject you are discussing. Make sure to double-check your numbers before presenting any data. You should think of fellows like your university professors— they want to see the technical details and analysis of your work. When you interface with technical fellows, make sure you show them your best technical work.

▶ *Career Tip.* Engineering fellows are the technical experts in the corporation. They make great coworkers to have as a friend when you encounter difficult technical issues.

Associating with fellows can accelerate your career or dead-end it quickly. If they like your work and believe in you, they can sponsor you to upper management. The senior fellows often help management assess who should be on the team and who should not. If your work is top notch and they like it, more than likely you will receive a favorable nod for the more technically challenging assignments. Correspondingly, if your work is sloppy and full of mistakes, they will quickly label you as technically inferior and not a good choice for leading technical assignments.

STAFF/SYSTEMS ENGINEER LADDER

The next ladder to the right in Figure 13-1 is the staff or systems engineering ladder. These people usually have a Master's degree and deal with putting systems together. The staff or systems engineers generally work on the big picture technical issues, such as the overall performance of the products. If, for example, the staff or systems engineer were building a car, he or she might work on the technical problems associated with the overall mileage performance of a car. This may include multiple technical areas that need optimizing, such as aerodynamic shaping, motor efficiency, weight, and size. Their assignments require them to be technically proficient in multiple areas.

If your career aspirations are to remain technical but you like to work the "Big Picture" or overall system aspects, then this ladder is for you. If you already have an advanced degree in engineering, then you have a good start, but you will need to broaden your areas of technical knowledge. For instance, if you have an electrical degree, this will mean you should take more technical classes in mechanical, software, or chemical engineering, or other engineering disciplines your company considers critical to its survival. The value the staff engineer brings to the corporation is to be knowledgeable in multiple engineering fields and able to technically direct the work of engineers of all types of disciplines. With the large teams and diverse backgrounds of the team members these days, the staff engineer plays a vital role.

If you do not have an advanced degree, then returning for a Master's or PhD should be in your career plans. However, with the rapid pace of changing technology you will have to return for updating in new technologies and make lifelong learning part of your career plans.

Staff engineers, like fellows, concentrate mainly on technical issues. However, with their broad background they often also serve in advisory roles to management, on technical as well as schedule and cost issues, of the project. They are very analytical and usually responsible for the big picture on how things are coming together technically. You can usually recognize them by the fact that they typically have the best computers in the group and their office walls sport collections of flow charts and graphs showing trade studies.

Staff engineers primarily work to determine the best overall product configuration. They are primarily interested in the product requirements or specifications. When conflict arises or performance compromises have to be made, they are the ones doing the trade analysis and determining the optimum approach. They are highly analytical and often talk about product performance requirements, budgets, and performance margin. If you try to get an answer out of them, they will usually qualify it with the phrase "That depends upon."

▶ *Career Tip.* Staff/system engineers think in terms of the big picture, so make sure you understand the big picture when working with them.

Interfacing to staff or system engineers can also help accelerate or dead-end your career. System engineers analyze the big picture; they want to know what all the causes and effects are. If you have the opportunity to interface with system engineers make sure to present the big picture and how your part relates to the overall plan. Know your requirements forward and backward and exactly how your design is performing against requirements. If you go into their office with only the small picture, they might quickly label you as not understanding the problem. Being labeled incompetent is exactly what you don't want your supervisor to hear. Having your supervisor hear from the system engineer about how you have a great understanding of the requirements, and how your design performance is superior, is exactly what you need for career advancement.

PROJECT ENGINEERING LADDER

The next ladder to the right in Figure 13-1 is the project engineering ladder. These people have an engineering background and generally possess an advanced degree. Project engineers deal primarily with running the technical aspect of projects. They are usually responsible for determining the project plan that defines the engineering tasks to be accomplished, generating a plan, assigning work, and tracking progress. They are responsible for ensuring the technical vitality of the project as well as ensuring that the staffing, budgets, and cost planning are met. They have responsibility for both technical and limited cost performance. Their primary focus is on organizing and executing the engineering tasks of the program, this requires them to have good solid background and knowledge in multiple engineering disciplines.

The project engineer supports the program manager in accomplishing the project goals. The project engineer must have business skills such as team leadership, project planning, financial budgeting, and a broad technical background. The project engineer must be able to define the work scope, plan the effort, establish engineering budgets, determine the master program plan, showing all tasks and track progress. When problems are encountered, the project engineer's role becomes coming up with a recovery plan. This may include getting special engineering talent assigned and organizing special teams.

If your career aspirations are to some day be the technical lead, then this ladder is for you. If you already have an advanced degree in engineering, then you have a good start. However, having an engineering degree is only the prerequisite for this ladder. You must also have some business training. The project engineer must have business skills and this is a new skill set for the engineer. The engineer should return for business training and even obtain a business degree. Some of the specific skills you will need are dealing with difficult people, team leadership, motivation, and project planning. You will be responsible for generating proposals and then executing the programs. If you do not have an advanced degree in business, then returning for a Master's degree should be in your career plans.

▶ *Career Tip.* Make sure you know what the project plan is, what your tasks are, and how you are doing against the plan when you interface with project engineers.

If you interface with project engineers, make sure you know what the project plan is, your tasks and how you are doing against the plan. They expect engineers to understand the critical tasks they need to accomplish, have a good technical approach, and be able to report how they are doing against the plan. They often provide feedback to the supervisors on engineer's performance, so let them see you at your best every time you interface with them.

PROGRAM MANAGER LADDER

The next ladder to the right in Figure 13-1 is the program manager ladder. These are people who typically run programs or projects and have an engineering degree or background. They are responsible for cost and schedule performance and have little to do with performing the actual technical work. They organize teams of engineers to accomplish the tasks on their programs and they are generally the primary interface to the customer. Program managers are easy to recognize because of the huge schedules hung up in their offices. The schedules usually show every task planned and the progress made.

The program managers control the funding for programs and have little to do with personnel and salary administration (raises). The program managers are focused on getting the job done on time at the cost they quoted. They are primarily responsible for organizing the team to build and ship products, so the company generates revenue and makes a profit to pay salaries.

If your career aspirations are to someday control the program financials and deal directly with customers then this ladder is for you. The program manager must have business skills that may be a new skill set for the engineer. This means the engineer must return for business training and even obtain a business degree. Many universities recognize this fact and now offer a Master of Business Administration degree program specifically geared toward the returning engineer. If you do not have an advanced degree in business, then returning for a Master's degree should be in your career plans.

Program managers play the vital role of organizing the team, determining the tasks to be worked, and establishing a budget and plan to accomplish their goals. Program managers not only work with engineering, they also work with nontechnical departments in the company. Program managers direct the work of personnel from such departments as contracts, finance, legal, and marketing. In this capacity, they must have a broad business background and understand the financial as well as legal implications of the companies' work.

▶ *Career Tip.* Program managers are primarily interested in getting the job done right, on time, and within cost.

Program managers expect you, as the engineer, to understand your assignment and have the technical knowledge to complete it on schedule.

They expect you to be able to discuss your progress against the plan, control spending, and present alternative ways to meet the development schedule. If you run into roadblocks or technical difficulties, program managers are not interested in excuses. Their bottom line is getting the job done. You may find they can be abrasive at times, because their main focus is the job (work to be done). They can do a lot to help you get a promotion or prevent it. Your supervisor is normally required to obtain feedback on your performance from the program manager on your projects. What kind of feedback would you expect the program manager on your project to provide your supervisor? Have you considered this? You can be doing a great technical job, but if you are far behind schedule and causing cost overruns, this is not the type of work a program manager wants on their project. Balancing technical work with cost and schedule consideration is a primary concern of the program manager; please remember this the next time you interface with a program manager.

MANAGEMENT LADDER

Engineering managers are responsible for personnel and profits and usually possess a technical background. The management ladder is not only responsible for profits but they also control hiring, raises, promotions, and demotions. Managers typically do not deal in the day-to-day technical work, but generally rely on their technical staff for this. They focus their time and energy on making sure the company has the appropriate staff, work to resolve personnel-related issues, help staff and de-staff teams, and help the engineers in career development.

The engineering manager also plays a vital role in the company of handling the everyday issues of simply running the business and remaining profitable. It is this ladder that determines the products to be worked on, the people to be assigned to projects, which products and projects are continued, and which are stopped. The managers must handle all the salary administration of the company, ensure policies and procedures are being followed, and provide rating and ranking of the engineer's performance. They must have great people skills that allow them to tactfully resolve people problems in the best interest of the company and the employee, and they often work with the Human Resources department to accomplish this.

If your career aspirations are to become an engineering manager some day, then this ladder is for you. If you already have an advanced degree in engineering you have a good start. The engineering manager must have the people and business skills and this is an entirely new skill set for the engineer. The engineering managers spend most of their time helping engineers with their career and staffing program. Their primary job is to provide the best engineering personnel they can for a program. In addition, they are responsible for conducting performance reviews and handling the administrative tasks of the company like training, benefits, vacations, and salaries.

This is the ladder your supervisor may be in. Their primary concern is that you are a contributing member of the organization, you are performing your job tasks as needed and accomplishing them in a timely manner, you are providing sound technical solutions to problems, and you are growing in technical knowledge and background. They are concerned about your ability to successfully interface with other team members and senior members of the organization.

They deal mostly in making sure the company is getting the best human resources possible and people are performing their jobs. They control people, salaries, and costs. They often function like coaches for the organization. They assemble teams, make the work assignments, and help motivate the team. They often do all the hiring and employee performance appraisals. Supervisors are responsible for ensuring the policies and procedures of the company are followed.

In addition, they have to deal with the unpleasant side of the people problems, namely poor work performance, legal issues, and even potential lawsuits. These are all new skills the engineer must acquire if hoping to become a manager some day. If your career aspirations are to some day become an engineering manager, then I highly recommend you read the literature and books dedicated to transforming an engineer into a manager.

MARKETING OR BUSINESS DEVELOPMENT LADDER

The career ladder furthest to the right is the marketing or business development ladder. This ladder is responsible for developing new business through marketing of the products. They often have an engineering degree as well as advanced training in business marketing. Their primary function for the company is to sell products and help with the customer interfacing. Generally, this type of position draws upon a technical engineering background since many of the products today are quite complex and highly technical in nature. Marketing people spend the majority of their time traveling and meeting with customers. They are the company's lifeline to the customer. They work on order forecasting, predicting market trends, and public relations for the company. They require a special skill set on how to sell and market products and work through customer problems with engineering to get quick resolution.

If your career aspirations are to become a marketing manager someday, then this ladder is for you. If you already have a degree in engineering, then you have a good start. However, having an engineering degree is usually only the prerequisite for this ladder. The marketing manager must have the people and business skills and this is an entirely new skill set for the engineer. The engineer must return for business training and even obtain a marketing degree. If you do not have an advanced degree in marketing or business, then returning for a Master's degree should be in your career plans.

Career Ladders Leading To Executive Levels

The corporate engineering ladders are discussed first to let you know they exist, to make sure you are aware of it, and to get you thinking about your career path. Second, as your career develops and you move up the ranks, you will have to decide which ladder fits your plans, whether you want to be technically or business-oriented. As you move up you will need additional education and training. You can shorten the time it takes to move up the ladder by getting the appropriate training before you advance. If you like technical work, you should prepare yourself by taking more technical training as you grow. If you desire a more management-oriented career, then you should take more business training. There is no right or wrong career path. There is only what is right for you.

As shown in Figure 13-1, the right three career ladders of Program Management, Engineering Management, and Marketing generally lead to the top positions of the corporation. Oftentimes, to become CEO of a corporation, you must have spent time in at least two of the three business ladders. This is often one of the key selection criteria required since you need experience in multiple facets of the business to successfully run the corporation. For this reason, most executives will have worked on several different career ladders before they make it to the top. Therefore, taking assignments in different career ladders can be a very good career move. In many corporations, it is the program management ladder that most often leads to the vice presidential level.

Many engineers aspire to become managers and are willing to take the training required to successfully carry out the assignments. However, a fair portion of these engineers leave the management ranks after they discover they are not cut out for this position. So before you take the leap into management, meet with your mentors and other senior level people to find out all that is involved in being a manager.

Many engineers go into management with the hopes of quickly climbing the corporate ladder to become CEO. To this thought, I would like to point out a piece of hard evidence to the contrary. The CEO of a major corporation is 1 in 100,000 employees. That means your odds are approximately 1 in 100,000. And the changing of the CEO only occurs about once every five years. To put it bluntly, these are not very good odds and if you check some state lotteries, you have more of a chance of winning the lottery than you do of becoming CEO. However, I do not want to discourage you. All I want for you to realize is that without a good career plan, any advancement is highly unlikely and if you aspire to become CEO, you should have a great career plan.

Moving Up the Ladder Means More Responsibility

As you move up the ladder, each level becomes more demanding. You must decide for yourself what level is right for you. If you have a family and that is important to you, you must decide if becoming Vice President is right for you.

Usually, to become Vice President the family experiences many compromises. Some people stop at the lower levels, others at the middle levels. For these people, this is comfortable and a good choice. You have to decide what is good for you. How high you decide to climb on the corporate ladder is strictly a personal decision everyone must make for themselves. Beware the higher you go the more responsibility you accept.

▶ *Career Tip.* The higher you go up the ladder, the more responsibility you accept.

The bottom line here is that you need to define what success is for you. If it is getting to level E3 and you make it, then this is success for you. If becoming a supervisor is what you aspire to and you make it, then this is success for you. Success for you may not be measured in terms of titles, raises, or promotions; it may be as simple as doing the best job you can. Not everyone can become president of the company, nor do they want to. However, everyone should reach their own definition of success.

Now that you have had time to review and understand all the levels, which do you think is the best one? Many male engineers quickly respond "CEO, he is on the top." When I asked the women engineers they mostly respond, "The one you are most happy with!" If you are at that level where you thoroughly enjoy the work, then don't beat yourself up because you are not the CEO, and enjoy the position you presently have.

▶ *Career Tip.* The best position on the corporate ladder is the one that makes you most fulfilled and happy.

If you are unhappy at the level you are at, your work will show it. Make a change and move to the level you are happy with. This could mean moving up as well as down or even lateral.

There is another very difficult position for most engineers. It is the first-line supervisor. Everything in the company comes together at this level. All the great ideas generated by the president of the company on how to run things, must be implemented by the first-line supervisor. All the policies geared to running the company must be enforced by the first-line supervisor: vacations, health care, drug testing, expense reports, employee training, security, raises and promotions, project successes and failures, personnel problems, and department budgeting to name just a few.

▶ *Career Tip.* One of the toughest positions on the ladder is the first-line supervisor. Remember that when you are talking with them.

Put simply, all the direction from management above falls on them, and all the headaches from employees below rise up to meet them. They are the single point in the company where everything comes together. They are required to do everything and most often it all has to be done now. They are usually overloaded with work.

Take it easy on your supervisor. They're trying to do the best job they can, so give them a break. Often, first-level supervisors quit and return to the technical ladder rather than continuing on the management ladder, because it is a very demanding level. Learning to deal with the stress and handle multiple priorities at once is key to career advancement at this level. The first-line supervisor is like a filter function in math. Those engineers who do not develop great time management skills and people skills are quickly filtered out of the management ranks.

Make your supervisor your friend, not your enemy. The last thing they need is you running into their office every time something goes wrong. Remember, when you go in to talk to your supervisor, he or she probably has a thousand things on their mind.

Let me make several interesting observations about the engineering ladder structure. Between the lowest level and the top there can be as many as 14 1evels. Not all organizations have this many 1evels in the corporate structure. In any case, let's look at how long it would take you to get to the top if there were 12 levels. If you could get a promotion every 3 years (and this is considered very fast rising), it would take you 36 years. Therefore, you had better make every one count.

▶ *Career Tip.* Only with great career planning is advancement possible, and if you aspire to become CEO you need a career plan that yields promotions every 3 years.

An easier way is to know someone at the top and have them help you rise to the top. This is called sponsorship and believe me, it happens. The alternative to coping with all this structure would be to start your own company, but that's another story for another book.

SUMMARY

The corporate ladder has many steps and multiple paths. Some ladders lead to purely technical careers where rewards are in terms of technology break-throughs, patents, and publishing papers. The other ladders lead to business careers where the rewards are in terms of management positions, profits, and people management. The key to being successful is knowing the different ladder paths and which ones you are best-suited for. In addition, all paths to the upper levels require significantly more training and greatly expanding one's skill base. As a result, lifelong learning should be part of your career plan.

Have you identified any career actions you want to take as a result of reading this chapter? If so, please make sure to capture these ideas before you forget by recording them in the notes section at the back of the book.

ASSIGNMENTS AND DISCUSSION TOPICS

1   The first homework assignment you must complete is to map out the engineering ladder in your company. A good place to start is with your boss. Inquire about the levels within the company; next ask to see the organization chart and where you fit in. Another good place to check is the company website. How about your program manager on the project you work on? Does he or she have an organization chart?

2   If you can, obtain a description for each of the levels. Usually, every company at one time or another has written job descriptions for each level. It is interesting to compare the descriptions as you move up the ladder. If you do not understand the descriptions, a good place to get clarification is your supervisor or Human Resources department.

3   Obtain a copy of the job descriptions for your level and the next level up. First study the description for your current level and see how you compare to it. Remember, if you want to get promoted, make sure you are fulfilling every requirement. Next, study the level above yours and see how the levels compare.

4   The final homework assignment is to determine what your career path will be. How high up the ladder do you want to go and how much time will it take? What actions must you take to reach your goal?

CHAPTER 14

INVISIBLE COMPANY BARRIERS STOPPING YOUR CAREER GROWTH

Invisible barriers exist all over the corporate ladder. In this chapter we will expose some of these invisible barriers, so you know what you are up against. Are these barriers insurmountable? Absolutely not. However, it is going to take hard work, a keen sense to detect when you are up against one, and the knowledge on how to break through these barriers to be successful.

ADVANCED DEGREE BARRIER

One barrier that stretches across all corporate ladders is the advanced degree barrier. The typical pyramid structure of the company with the Chief Executive Officer (CEO) at the top is shown in Figure 14-1. I have taken a general survey of the types of degrees that people have at various positions in the company in several companies. Below the first-line supervisor, about 80% of the people have bachelor degrees, about 15% have Master's degrees, and about 5% have PhD degrees. In between the first-line supervisor and the director level, about 50% of the people have PhD degrees, 40% have Master's degrees, and 10% have Bachelor degrees. Between the director level and the CEO, about 60% have PhD degrees and 40% have Master's degrees. Often at this level a director or VP will have multiple degrees: a doctorate in engineering along with a Master's degree in business. In my survey, I found there was no one at these upper levels with only a Bachelor degree.

So the message comes out loud and clear. If you want to become an upper level manager someday, you will need an advanced degree.

The Engineer's Career Guide. By John A. Hoschette
Copyright © 2010 John Wiley & Sons, Inc.

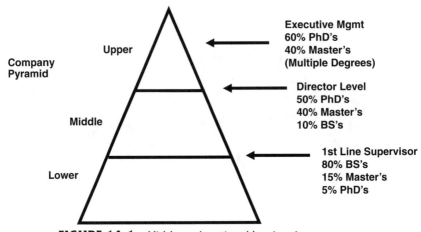

FIGURE 14-1 Hidden educational barriers in a company.

▶ *Career Tip.* An advanced degree and sometimes even multiple degrees are required to climb high up the corporate ladder.

The only way you are going to find this out is by asking people about the degrees they have obtained. This should reveal two interesting things. First, how prevalent advanced degrees are in your organization. Second, the type of advanced degrees that virtually ensure getting ahead. This leads to another invisible barrier, which I refer to as the technical background barrier.

TECHNICAL BACKGROUND BARRIER

One common practice in corporations is only to promote engineers to the upper levels if only they have the right technical background. This is because corporations have built their business based on being the technology leader in specific areas. Therefore, the people with the right technical background are readily promoted while others with technical backgrounds perceived as not appropriate are passed over. Let's look at this example as illustrated on the left-hand side of Figure 14-2. If you are an electrical engineer working for a heavily oriented mechanical company and all the managers above you are mechanical engineers, that tells you something about the promotion policies being followed. Similarly, if you are a software programmer in a chemical company and all the managers have advanced degrees in chemical engineering, it tells you something about how far you can go.

Investigate and try to determine what types of degrees people have above you. If you have the opportunity to interface with a vice president, simply ask about their background. They are usually most willing to share their experiences and most significant career accomplishments. This opens your

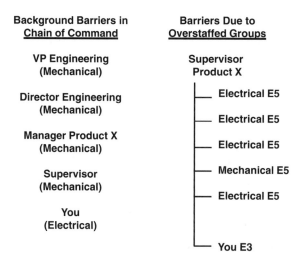

FIGURE 14-2 Technical background barriers and overstaffing barriers.

door to opportunity. Ask what they would recommend for a younger employee. Hopefully, at that point, you are hearing the shortcuts that you need to know and their hidden criteria for promoting.

OVERSTAFFING BARRIERS

In the organization on the right-hand side in Figure 14-2, you are assigned to a group that has all level 5s in the group. You are at level 3. Do you think you stand a good chance of promotion in this group when the group is overstaffed with level 5s already? I seriously doubt it; in fact, my experience has shown you are the only one doing real work. So if they promote you, the group has to find someone else to accomplish the assignments. This condition is called overstaffing and seriously eliminates any chance for advancement of junior engineers. Have you determined the grade levels of people in your group? If you are in an overstaffed group or top-heavy group as it is sometimes referred to, your lack of career growth may have nothing to do with performance and everything to do with being top-heavy. Your supervisor cannot afford to promote you since they already have too many senior-level people. This condition occurs quite often and I have experienced it myself. In fact, when I was appointed a supervisor for the first time, the group I was put in charge of had three level E5 engineers. This was against company policy; each engineering group should only have one. So, my first job was to find jobs for two of the E5s in other parts of the company.

▶ *Career Tip.* Knowing the engineering levels for all the people in your group will quickly identify if you are up against an overstaffing barrier.

Correspondingly if your group lacks any senior-level people, this is your opportunity to move up quickly and fill the senior-level position. It is important to know the level for each person in your group and determine if you are in an overstaffed or understaffed situation. Then take the appropriate career action.

▶ *Career Tip.* Working in a group that has few or no senior-level people is being in the right spot at the right time.

DEPARTMENT CHARTER BARRIER

Another hidden career barrier that is often not discussed is the department charter barrier. This career barrier is the fact that according to the department's charter, all work in the department can be performed, for example, by level E3 engineers or lower. Another way to think of it is the most challenging assignment in the department only requires a level E3 and does require an upper level of engineer (E4–E5). Therefore, if you are a level E3 looking to get promoted to E4, it will not occur since the department cannot justify using these upper level workers to get work done. The department is incapable of promoting an engineer from level E3 to level E4 since, by definition, they do not have work in the department for an E4.

If you are a level E3 and hoping to get promoted to E4 in a department like this, you do not stand a chance in this department. The department does not have the charter to do level E4 work and therefore, by definition, there will be no E4 engineers working in the department. Have you checked your department to determine if they have a department charter like this one? Much to my surprise, I found myself in a department that had a charter like this and quickly realized that I had to change departments to get my next promotion.

▶ *Career Tip.* Hidden career barriers exist in most engineering groups. To overcome these barriers, you have to do research to find out if you are up against one of them.

SUMMARY

Do you think managers have a bias on the type of people they promote? Absolutely. Have you ever asked managers above you what their backgrounds are? This knowledge can be very beneficial in determining the type of career barrier you are up against.

Are there other barriers in the corporate ladder? Yes. Another common invisible career barrier or promotion selection criteria is the engineering school or area of the United States that you attended school. If all the managers are from one engineering school or one region and you are not from this school or region, then you have an invisible career barrier you must overcome.

If you are a woman, check to see how many women are at each level of the ladder in the organization. If you find the percentage does not stay the same and it drastically drops off as women move up the ladder, you have gender barriers.

All the barriers discussed in this chapter can be overcome with hard work and knowledge of what the next best career move is. The challenge is to quickly discover when you are up against one of these invisible career barriers and take the appropriate actions to overcome them.

Have you identified any career actions you want to take as a result of reading this chapter? If so, please make sure to capture these ideas before you forget by recording them in the notes section at the back of the book.

ASSIGNMENTS AND DISCUSSION TOPICS

1 Inquire and find out the backgrounds of the managers in your chain of command. What advanced degrees do they possess and what are their technical backgrounds as well as the engineering schools they graduated from. How do your degrees and engineering background compare to theirs?

2 Find out the engineering levels of everyone in your peer group. Are you in an overstaffed or understaffed situation?

CHAPTER 15

GETTING AHEAD IN PRODUCT-ORIENTED ORGANIZATIONS

Most companies utilize teams of engineers organized into either product-oriented or functional-matrix structures. A typical product-oriented organization is shown in Figure 15-1. In the product-oriented organization everyone works on the same product. Often the entire department is responsible for getting the product out the door. Everyone in the product-oriented organization reports to a single upper-level manager at the top of the organization. The product-oriented organization in Figure 15-1 illustrates the reporting organization for the ZX50 car. Reporting to the departmental manager are the individual subsystem departments, in this case, four different departments. For the ZX50 car, there is the engine department responsible for designing and testing engines, the body-design department responsible for the exterior body, and so on. Each of these departments performs a necessary job within the organization.

Reporting to the department manager is often the program manager and usually one or two staff engineers. The program manager determines what is to be worked on and the engineering department determines the best technical approach. One very positive advantage to this type of organization is that your work direction and salary review often come through your supervisor. One person, your supervisor, provides the work direction and hands out the raises.

There are several advantages to working in a product-oriented organization. First, the organization gets to build the entire product. It is very rewarding to see the entire product come together in your department. It gives you a real sense of accomplishment. When the product is a huge success, management knows whom to reward. Second, everyone reports through the same chain and decisions are more easily made in this type of organization.

The Engineer's Career Guide. By John A. Hoschette
Copyright © 2010 John Wiley & Sons, Inc.

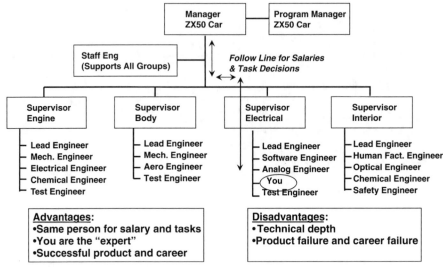

FIGURE 15-1   Typical product-oriented organization found in large corporations.

Third, you often work with people who have very different backgrounds and gaining new perspectives from them greatly broadens yours. Since staffing of organizations tends to be fairly lean, you might be the only one in the group with your type of background; an excellent and advantageous position, since there is no competition. Another advantage is that if you have the same background as the supervisor or manager, they can more easily appreciate and understand the work you are doing. Managers tend to promote people with backgrounds similar to their own.

There are some disadvantages to working in a product-oriented organization. When the product development stops, in most cases so does your career advancement. Also, if the product runs into trouble (due to poor planning or performance), you may have that cloud hanging over you or be tagged as coming from a troubled or failed program.

Another disadvantage of the product-oriented organization is the fact that your boss may not have the same technical background as you. If you are a chemical engineer and they are electrical, it is often hard for them to appreciate the great job you are doing. And finally, when the product comes to the end of its development, there may be no place for you to go unless a new product is being developed. Working on a single product often creates engineers who know everything about one little thing. A difficult position to be in if you plan on running the entire organization some day.

SUMMARY

The organizational structure of your engineering or product teams will drastically affect your career actions. If you are in a product organization your career advancement actions should focus on producing successful products and the managers you directly report to. Getting ahead in this type of organization will require that you develop many skills outside of your engineering major and also have excellent people skills. The more rounded and big picture person you are the better, since to become a leader in this organization, you must be able to manage a team of very diverse backgrounds and talents who make your product successful.

Have you identified any career actions you want to take as a result of reading this chapter? If so, please make sure to capture these ideas before you forget by recording them in the notes section at the back of the book.

ASSIGNMENTS AND DISCUSSION TOPICS

1   Determine how your company is organized. Is it a product-oriented organization? Who reports to whom? Can you identify all the levels and the person at each level between you and the vice president? If you can, you're in great shape! If you don't know who they are, how can you expect them to know who you are and promote you?

CHAPTER 16

GETTING AHEAD IN THE FUNCTIONAL-MATRIX ORGANIZATION

A typical functional-matrix organization is shown in Figure 16-1. In this structure, the company is organized around functional groups. Everyone in the functional group usually has the same background but may not work on the same product. The left-hand column of the functional-matrix structure identifies the engineering departments in the company according to function. The top row of the matrix identifies the projects within the company. For the example shown, three types of vehicles are produced: the ZX50 and ZX90 cars and the S5 truck. The top row of the matrix identifies the program manager(s) responsible for getting work done.

In this type of organization, the functional manager is usually responsible for making the salary and promotion decisions, while the program manager usually determines the work to be accomplished. Companies utilize the matrix organization since it is a very efficient way to run the business of large organizations. Highly skilled engineers of a particular technical background are grouped together. When a program manager needs an engineer with a particular expertise they contact the functional managers of the group to get an engineer assigned to the program. The functional manager determines the best person for the job on the basis of the skills and experience needed and then assigns that person to support the program. This type of structure allows the company to quickly move people on and off programs and hopefully to assign the best person to the job.

Working in a functional-matrix organization offers several advantages for your career. First, you are working with people who have a background similar to yours. This is advantageous when you get stuck on a problem. Usually, someone in your department has faced a similar problem and is readily available to help you. Second, you may work on more than one product at a

The Engineer's Career Guide. By John A. Hoschette
Copyright © 2010 John Wiley & Sons, Inc.

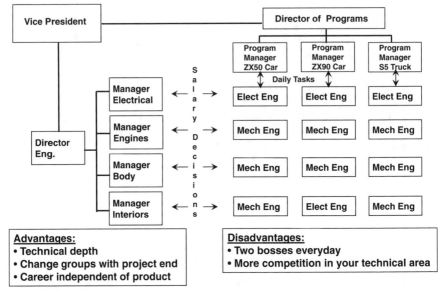

FIGURE 16-1   Typical functional organization found in large corporations.

time, providing more job security. This arrangement also provides more protection when the product fails since career is not directly tied to the success of only this product. Third, in a functional-matrix organization you can quickly move from one product to another as they come and go.

Working in the functional-matrix organization also offers some disadvantages. Most of the time, work direction comes from your program manager and technical direction comes from your supervisor. This places you in a very tricky situation when the program manager and your supervisor disagree. If you follow your supervisor's direction, it may make the program manager upset and they could very possibly move you off their program. If you follow your program manager's direction, it may upset your supervisor and affect your raise, since they directly control your salary.

In a functional-matrix organization you really have two supervisors to please on any given day: your functional manager and the program manager. You have to find a way to keep both happy, if you are going to get that promotion or raise. Often the program manager and department supervisor sit down and discuss your performance to determine the raise you will be getting. The ultimate decision is up to your department supervisor, but you had better make sure the program manager agrees with his assessment of your worth.

It doesn't hurt to simply ask your functional manager and program manager how you're doing. It's better to find out well in advance of job review time that they think your performance needs significant improvement. Asking in advance provides you with time to improve your performance and subsequently change their perceptions before your performance reviews.

Another disadvantage to the functional-matrix organization is that you are placed with a group of people who all have similar background to yours. You may find it is tough to shine or stand out in the crowd when everyone there is doing the same thing. You have to look for ways to be different.

SUMMARY

In a functional organization you really have two bosses to keep happy. This means a lot of extra work for you. These two are your functional manger and the program manger. To get your next promotion both these people must agree you are ready for it. To ignore one and focus your career actions only on the other is taking a big chance. A very common mistake by engineers in a functional-matrix organization is to ignore their functional manager who they rarely see, and instead, focus on the program manager who they are in daily contact with. The program manager gives you the opportunity to prove yourself; however, it is your functional manager who ultimately has to recommend and support your promotion to upper management. Functional managers rarely promotes engineers who they know little about or seldom interface with.

Most companies are not strictly organized in either a product-oriented or functional-matrix fashion but rather are a mixture of both. In order to move ahead in either organization you must know how it works and what the pitfalls are. No one organization type is better for your career than the other. The important thing is to recognize there are differences, and that what works in one, may not work in the other. Find out how your company is organized and who determines your raises. You could be working hard to impress someone in the organization who has no involvement with your raise. Make sure you report your progress on a continuing basis to the person who conducts your performance appraisals. In this way, they will always know what a great job you are doing.

Have you identified any career actions you want to take as a result of reading this chapter? If so, please make sure to capture these ideas before you forget by recording them in the notes section at the back of the book.

ASSIGNMENTS AND DISCUSSION TOPICS

1 Determine how your company is organized. Is it a functional organization? Who reports to whom? Can you identify all the levels and the person at each level between you and the vice president engineering? If you can, you're in great shape! If you don't know who they are, how can you expect them to know who you are and promote you?
2 Who is your program manager? Can you name the people in the program management organization?

PART 3

DETERMINING THE CRITERIA BY WHICH YOU ARE MEASURED AND WHAT IT TAKES TO GET BETTER PERFORMANCE RATINGS AS WELL AS PROMOTED

CHAPTER 17

DETERMINING HOW THE COMPANY MEASURES YOUR PERFORMANCE

Do you realize your manager and other upper-level superiors are evaluating your performance every day? Do you know the criteria they are using to rate your performance? Do you know what they consider as minimum acceptable performance, average performance, and outstanding performance? Do you know what they consider as important and not important? If you have not taken the time to discuss and find out what they consider important or the performance required for a promotion, you could be wasting a lot of time and energy.

The analogy is one of a person driving to a new destination they have never been before and do not bother to get directions. They rely on their gut feelings and quickly jump into the car and take off. They know the destination is located near a freeway exit and they will recognize it when they see it. They do not even know if they are headed in the correct direction on the freeway, but since they are traveling at a speed of 65 miles per hour, it doesn't matter because they arrive sooner. You can waste a lot of time driving around and around hoping the exit is just around the next turn. Just like you can waste a lot of time working toward a promotion if you do not know the formal criteria you will be assessed by.

If you don't know the formal criteria by which you are being judged, you are acting like this person. Blindly driving around on the freeway. You rapidly perform task after task hoping the next one will get you promoted. You assume the quicker you get the task done, the quicker you will be promoted. You hope everything you do will result in getting the big raise or promotion. Just remember, it is very difficult to hit a target that cannot be seen. It is almost like you are operating with a blindfold on as far as your career is concerned.

The Engineer's Career Guide. By John A. Hoschette
Copyright © 2010 John Wiley & Sons, Inc.

▶ *Career Tip.* Don't leave your next promotion up to luck. Take the blind-fold off.

In this chapter, we will discuss the methods you can use to find out the formal criteria being used to make judgments on your performance. In Chapter 18, we will discuss the informal criteria. Knowing the formal and informal criteria will, in effect, remove the blindfold. Once you have clearly identified the criteria by which you are assessed against, you will be able to see exactly what you must do to get promoted. With the criteria clearly identified, it is easier to obtain the raise or promotion. It should also come sooner since you can use every work task to your benefit and to help fulfill the criteria.

FORMAL CRITERIA: WHAT ARE THEY COMPRISED OF?

In every company, there are formal criteria by which you are assessed. You must understand the formal criteria for successful career development. The formal criteria are usually well-defined and documented.

The formal criteria are manifested in three ways. The first is through the job performance review process. The second is through the job performance criteria or guidelines. These guidelines summarize the performance expected of the employee at each level of the engineering ladder. The third way the formal criteria are manifested is through the promotion review process. Each company has its own promotion review process that all managers must follow to ensure employee promotion approval. This process is usually a very formal one that is well-defined and strictly adhered to. We will now explore how you can clearly identify and deal with each of the three formal criteria.

UNDERSTANDING THE JOB PERFORMANCE REVIEW PROCESS

The first formal criterion with which you must become thoroughly familiar is the job performance review process or job appraisal method utilized by your company. On a periodic basis your manager must formally document your performance on the job. Some companies review employee job performance once a year or once every other year. You must find out everything involved in the job performance review process for your company. How often do they conduct job reviews? When is your next review? What criteria are involved? A good way to clearly identify the criteria is through the paperwork processed during your job review.

In preparation for a job performance review, a manager will fill out some type of form which formally documents your performance to date. Some companies use a standardized form and other companies simply document your progress in a memo.

Ask your manager for a copy of the form used if you have not yet had a job review or performance appraisal. Study the form and make sure you are familiar with everything on the form. A sample employee job performance review form is shown in Figure 17-1. Keep in mind as we review this form, the criteria utilized by your company.

PERFORMANCE COMMUNICATION PROGRAM

RATING FORM

EMPLOYEE NAME & JOB TITLE	EMPLOYEE SS NUMBER	GRADE	YEARS/MONTHS IN GRADE

DIVISION/DEPARTMENT	APPRAISAL DATE	PERFORMANCE PERIOD FROM TO

WHEN COMPLETING THIS FORM REFERENCE THE BASELINE AGREEMENT AND DISCUSSION QUESTIONS

PERFORMANCE CHANGE RATING

☐ 1 GROWING RAPIDLY IN GRADE - EMPLOYEE IS DEMONSTRATING RAPID IMPROVEMENT IN PERFORMANCE & CAPABILITIES TO HIS/HER PAST PERFORMANCE & OTHERS IN GRADE.

☐ 2 PROGRESSING WITH/GROWING IN GRADE - EMPLOYEE IS DEMONSTRATING IMPROVEMENT IN PERFORMANCE COMPARED TO PAST AND OTHERS IN GRADE.

☐ 3 PROGRESS LESS/ DECLINING IN GRADE - EMPLOYEE PERFORMANCE IS UNCHANGED WITH NO GROWTH OR HAS DECLINED COMPARED TO PAST & OTHERS IN GRADE.

JOB REQUIREMENTS RATING

☐ 1 CONSISTENTLY EXCEEDS JOB REQUIREMENTS

☐ 2 MEETS/OCCASIONALLY EXCEEDS JOB REQUIREMENTS

☐ 3 DOES NOT MEET JOB REQUIRMENTS

OVERALL RATING *

____

* (PERFORMANCE CHANGE + JOB REQUIREMENTS) / 2

CAREER DEVELOPMENT

I WANT A CAREER DEVELOPMENT PROGRAM DISCUSSION WITH MY SUPERVISOR ☐ YES ☐ NO

REMARKS, SUMMARY, STATEMENTS, ETC.

SIGNATURE INDICATES A PERFORMANCE COMMUNICATION DISCUSSION BETWEEN EMPLOYEE & SUPERIOR HAS OCCURED

EMPLOYEE	SUPERVISOR	MANAGER

DISTRIBUTION: WHITE - HUMAN RESOURCES YELLOW - EMPLOYEE PINK - DEPARTMENT FILE

FIGURE 17-1 Sample employee job performance review form.

Across the top of the form is usually the employee data. This data includes the employee name, job title, grade level, years in grade or job, department, and some reference to the last appraisal date. Review your form and see what data the company considers important. Why do they consider it important? For example, some companies have a standing policy that an employee must be in a grade for a minimum of 3 years before they can be promoted. This is the reason for the time-in-grade block on the form. If your company has a similar policy, what do you think your chances are for a promotion if you have only been in the grade 1 year? Does your company have any other hidden prerequisites for promotion of which you need to be aware? For this reason, you must understand the importance of all the data on the form.

Below the general information employee data is the performance summary section. This is the bottom line. All forms have one. This section is where, in a matter of two or three lines, your entire performance is summarized. Most people do not realize the importance of this section. This section usually summarizes your performance and identifies some type of overall rating. Your raise is usually computed on the basis of these ratings; another way you can think of this section is in terms of promotability or dollars.

For the example shown, a three rating results in a demotion or salary reduction. In other words, it may cost you thousands of dollars and years of setback. For example, a two rating means status quo. Exhibit just average performance and you'll receive the average raise and years to the next promotion. Finally, a one rating means outstanding performance, you've earned a large raise and, by all means, keep up the good work. You will soon be experiencing that promotion. After identifying the hidden meanings behind these formal performance rating criteria, most people immediately start to pay more attention to this section.

Do you understand the section of the job performance form that summarizes your rating? How do the ratings get translated into dollars? What do they mean in terms of promotions? If you don't know, you are operating with a blindfold on. Ask your manager and they will tell you how the rating translates to raises and promotions.

Below the performance summary section is the career development section. This section identifies your desire to discuss a career development program with your manager. As far as I'm concerned, there are only two reasons why you would not want to check yes in this section. One reason is that you are making more money than you think you should and therefore you don't feel you need career development. The other reason is that you plan on retiring next year and therefore career development is not on your mind. If you do not fit into either of these categories then you should be requesting a career development discussion with your manager.

If your company's form does not have a section similar to this, ask your manager for extra time to discuss your career plans. You may want to schedule this talk when you both have more time to discuss your future plans. Do not try to get through your performance review and have a career development meeting at the same time. It is too much to try to cover in one meeting.

The next section on the form is the remarks and summary section. Your manager usually fills out this section. In this section, your manager will try to summarize your performance since the last performance appraisal. Study the remarks carefully. On what matters have comments been made? What has not been noted? Do you know why things were missed? What do they consider important? What do they consider unimportant? After you answer these questions, you may come to realize your manager's opinion of what is important differed from yours.

HANDLING NEGATIVE FEEDBACK AND CRITICISM

At some point in the performance review you will get to the "Needs Improvement" part. In this part, the manager informs you of things you need to improve upon. Make sure you spend enough time on this subject. It may be hard to sit there and listen to your manager spell out all the things you need to improve upon, but it is a must for career development. By identifying your shortcomings, your manager is telling you the very things you need to improve upon to get the raise or promotion.

Whatever you do, please, don't take the feedback too personally. Try not to find an opening in the commentary wherein you rebut anything you interpret as detrimental. You are only going to lose. Your best bet is to listen patiently to the feedback and make sure you understand it. From this criticism you can learn what they consider important that you are not doing. If the manager lowers both guns at you, you need to consider whether you really want to continue working for them or if, perhaps, it's time you move on. Most performance appraisals do not go this poorly even though they may seem that way. Listening to criticism is tough but here are some tips on how to turn this difficult time to your advantage.

▶ *Career Tip.* "Tag-on" statements turn improvement areas into strengths during appraisals.

One way to turn this difficult time to your advantage is to use a tag-on statement after each criticism. Typical tag-on statements go something like this:

> I understand now. If I improve my performance doing ... (state the item your manager just identified) ... then do I stand a better chance for a raise?

> Let me make sure I understand, one of the reasons I did not get the promotion was because ... (name the item your manager just

identified) ... and if I correct or improve myself there should be no reason why I would not get the promotion next time?

Let's summarize, the areas I need to show improvement on in order to get the promotion by the next appraisal time are ... (name your improvement areas) ... and if I improve these things I stand a good chance for advancement.

Are these the only things I need to improve upon to qualify for the promotion by the next appraisal time?

By adding these tag-on statements to the criticism you are doing three things. First, you are clearly identifying what has caused you to miss the big raise or the promotion this appraisal time (removing the blindfold). Second, you are hopefully getting your manager to identify everything you need to do to get the promotion next time (identifying any hidden agendas your manager may have). And third, you are sending the message that you expect the promotion by the next appraisal time since you will be improving your performance in each of the identified areas (setting the deadline for the raise or promotion). You have clearly sent the message to your manager that you understand your weaknesses. You will correct the problems and after doing so you expect the raise or promotion next time. Remember, people rarely get promotions they do not ask for.

▶ *Career Tip.* People rarely get promotions they do not ask for.

Another tip is to think of each criticism as a step closer to the next big raise or promotion. This should help make acceptance of the criticism much easier. Every criticism that your manager identifies becomes another reason why you deserve the promotion after you have proven to your manager that you corrected the problem. The more reasons your manager identifies, the more reasons you will have in your defense once you have overcome the problems. When your manager does not identify areas for improvement, then it is time to worry. If this happens, you need to get your manager to open up more.

▶ *Career Tip.* Criticisms, improvement areas, and lack of skills are the areas to concentrate on for most improvement.

My own experience has shown the quickest and best promotions came only after I got my manager to really open up and clearly identify what I had to improve. It was only after we got everything out in the open and I started to improve my performance that they soon realized there were no reasons not to promote me.

At the end of your performance review you should have a clear understanding of what you need to do from what is recorded on the appraisal form. From this point on, getting the raise or promotion becomes much easier.

This brings us to the last section on the appraisal form — the signature block. Who has to sign it? The signature block tells you immediately who controls your raises and promotions. Do you know who will sign your appraisal and approve it? If you do not know them, then they surely do not know you. And most managers do not promote people they do not know. Some people do not even know who will see their appraisal form. Meeting with, and getting to know, the people who sign your appraisal form is the key to career development.

The sample form in Figure 17-1 may not look at all like your performance appraisal form. I have used it as a guide to help you study your form(s). Some companies use appraisal forms that are several pages long. Some forms require the manager to rate you in each area of your job that the company considers important. Whatever type of form your company uses, make sure you are familiar with everything on the form. Understand the hidden meanings behind each block. Only after you have taken the time to understand the formal criteria defined on the performance appraisal form can you expect to use it as a guide for your career development.

Determining How Job Performance Is Measured in Your Company

The second type of formal criteria you must be aware of is the criteria your company requires you to meet at each level of the engineering ladder. The starting point to discovering these criteria is your manager. Your manager has written guidelines that define the performance expected at each level of engineering. Meet with your manager and ask them for a copy of the criteria. Usually, these criteria have been developed over time and with the help of the personnel department. If your manager does not have a copy, stop by the personnel department and request a copy. Make sure you obtain the formal description of your level and the level you hope to reach when promoted.

Study the guidelines. Try to identify all the things you have accomplished at your level and the things you need to work on. Next, look at the level above yours and study it to find out what you have to do to reach the next level. In studying the guidelines you will probably generate more questions than answers, which is good.

Since every company will have its own guidelines, Figure 17-2 is a sample guideline based on several companies. Studying this guideline should help you in understanding yours. This sample guideline is organized in a matrix fashion with the job grade levels defined in the first column and the criteria categories defined across the top. For this example, five levels of engineering

Grade Level	Technical Requirements	Technical Judgment	Technical Challenge	Leadership and Work Direction	Management of Cost and Schedule	Team and Interaction
I **Junior Level**	Supportive of Project 4 to 5 Years Training	Evaluate and Recommend Technical Solutions	Applies Known Techniques	Can Explain and Coordinate Work with Members of Equal Grade	Performs Assigned tasks within specified cost & schedule	Customer Contact Not Normal Coordinates with own group
II	Perform basic Engr. Tasks Perform analytical prediction of results	Evaluate and Recommend Effective Solutions Justify solutions based on facts	Technical path usually defined Define tasks to be performed	Can Explain and Coordinate individuals of same or lower grade	Performs Required tasks within cost and schedule Estimates supporting goods and services required	Coordinates with Depts. and Projects Infrequent Customer Contact
III	Experienced Performer in a specialty Capable of performing in broad range of assignments	Evaluate Alternatives and Select technical approach Justify Alternative selected	Can identify Tech Cost and Sched. constraints Can apply original approaches based on established methods	Can direct small or specialized team in pursuit of a task objective	Accomplish Task/Team objectives within cost & schedule Anticipates Problems and initiates Action	Regular Supporting Role in customer contacts Usually limited Technical Exchange
IV	Experienced leader in her/his field Has depth of knowledge in related fields Demonstrated ability	Select and Implement state-of-the-art solutions Precedence may not exist	Technical method not established Original or creative approach may be required Makes Tech., Cost, Sched. Judgments	Can organize and direct a small or specialized team in pursuit of project objectives	Accomplish Team objectives within cost and schedule Cost and schedule implications typically have significant project impact	Regular Tech External Contact with customers Normally has a limited role for tech interface Communication and Judgment key to success
V **Senior Level**	Recognized Authority Grad Work or Adv. Work in Field Advice respected by Mgmt. and Customers	Technical Options are respected internally and by customer Technical decisions reviewed by results only	Technical solutions beyond industry precedents Assimilates complex problems and develops tech solutions	Can define the need for and direct the work of a group concerned with a variety of engineering disciplines	Accomplish Team objectives within cost and schedule Select Alternatives involving Cost, Sched., and Tech Trade-offs	Regular Ext. Customer Contact Lead role in Customer interface and Tech solutions Key individual to ensure project success
	Technical			**People, Cost, and Schedule**		

FIGURE 17-2 Typical engineering grade level performance guidelines.

are defined and six performance categories are identified for each job level. The job levels are ranked from entry level Grade Level I engineering to the most senior Grade Level V. The six performance categories contain two groups of skills. The left three columns identify the technical skills needed by the engineer and the right three columns identify the interpersonal or team leadership skills needed.

First, let me try to show you the differences between each engineering level with an example of how the job responsibilities vary for each level. At engineering level I, you might receive an assignment that requires you to analyze some data. The data has all been collected, the analysis has been defined and programmed into a computer, and the program output is on a graphics plotter. Your job is to enter the data and generate the plots with the existing program and computer.

At grade II, your assignment may be to collect the data from the test, enter it into the computer, modify the program if necessary, and plot the results.

At level III, your assignment would be to collect the data. However, first you have to assemble the hardware, plan tests, and obtain the help of a technician to collect data, organize the results for input into the computer, and plot them out with the help of junior engineers.

At level IV, your assignment will be to collect data and analyze it. However you must first organize a team, get a time and cost estimate to accomplish the tasks, schedule tasks and make assignments to collect the data, write a program to analyze the data, choose a computer to complete the task, identify type of language, and define the plots to be generated. You will do this by organizing a team of engineers, computer programmers, and technicians.

At level V, your assignment will be to figure out what has to be done. What tests are to be done? What data is to be collected? What does theory predict? First you must organize a team, get cost estimates established, brief management on plans, and get approval. Next, schedule tasks and make assignments to accomplish your experiments and collect data. You must oversee the team's choice of computers and language as well as plotters. Once the data has been collected, you must get the team to write a final report and you must present the results of your effort to management. If technical problems arise you are expected to determine the best methods to modify the experiments and resolve the issues.

This simple example shows how to identify the differences between grade levels. The example also shows that as you move up the chain interpersonal skills and team leadership skills become more important. Managers are not really expecting much in the way of team leadership skills from junior engineers. However, you should be developing them as you go because management is expecting good leadership skills from the upper grade level engineers. Therefore, the criteria on the right of the chart become more and more important the higher you advance.

All managers interpret these generalized guidelines differently. Some managers firmly believe that all you have to do is be good on the left-hand side of the chart while other managers believe the right-hand side of the chart is most important. All managers think differently. On your company's guidelines, do you know what your manager considers the most important? You could be working to impress your manager with your technical judgment skills and they may be thinking that leadership is most important.

Another common mistake engineers make when looking at these guidelines is interpreting the level one should be rated at. Your job assignment may allow you to perform at many levels. Let's assume an engineer has the following ratings for each performance category:

> Technical Requirements Level III
> Technical Judgment Level III
> Technical Challenge Level III
> Leadership and Work Direction Level IV
> Cost and Schedule Level II
> Interaction Level II

What level would you rate this engineer? Most people respond with level III. However, from a management point of view this is not an appropriate rating. Engineers are rated on their lowest level of performance.

To be rated a level III, the engineer must demonstrate performance at level III in all categories before they will be considered ready for promotion to that level. This level II engineer has a good start on promotion to level III but some areas must be worked on before promotion.

Why have I put these guidelines into this book? Not for you to study them and learn about engineering levels, or for you to determine how you are performing. I did it because it's a cheat sheet! Copy down these guidelines on a single sheet of paper, make sure they are readable, and take them to your manager. Explain that you got them out of a career development book and sit down with your manager and ask for an opinion. How important these criteria are and what does your manager see as the real differences between grade levels? From that point on, don't talk but only listen. The guidelines are too general for anyone to really determine the criteria for each grade level. Therefore, anything said defines the criteria as your manager sees it.

Everything they will tell you will be exactly what you will need to know about their criteria for development. Have them explain what they consider the most important criteria for your next level. How do they rate you for each category and where do you need to improve? Remember, you should be all ears at this point because they will be telling you everything they consider important. Try and absorb as much as you can, you should be like a sponge soaking it all up. As they are talking, make notes all over the guidelines. Mark them up together, cross out, and change things to their satisfaction. When you are done, you will now have everything on one sheet of paper that identifies exactly what your manager thinks and what you must do to earn the next promotion. In effect, you will have a piece of paper that has most of the answers on it, a perfect cheat sheet to help you shorten the time to your next promotion.

▶ *Career Tip.* Make an appointment with your supervisor and jointly go over the grade level performance guidelines!

If the sample job level guidelines are too different from your company guidelines, then use yours. Meet with your manager on an informal basis and review the company's job grade guidelines together. I recommend the best place to do this is at your desk or office, where it is not intimidating for you or your manager. Your manager's office can be too intimidating. You also stand a better chance of not being interrupted.

Ask your manager to explain the performance expected by the company at your level. Then ask them to explain the performance it would take to be at a level up from yours. Once you get them talking, don't interrupt. Make notes in the side margins for everything you can. If you do this right, at the end of the conversation you will have identified the formal criteria you must meet for your level and the level above yours.

During your talk try to get your manager to identify some specific things you have to demonstrate in your work that will show that you are meeting the criteria for your job, and possibly, for the next level. For instance, you might ask, "If I complete my assignments on time and within cost, does that signify performance at my level or the one above my level? How might I perform my tasks such that I will have demonstrated I meet the criteria for the next level? What exactly do you see as the difference between my level and the one above? How do I get broader assignments so I can demonstrate performance above my level? What exactly do I need to do on my present assignment to increase my chances for a raise or promotion?"

Your manager's response to your questions will help lift off the blinders and provide guidance for your career advancement. You now have a clear vision of the formal criteria you must meet for career advancement.

The guidelines are usually written in such general terms that anything defined specifically is really what your manager thinks you need to do. Ask as many open-ended questions about the guidelines as you can. The reason for this is that any answer given you will define the criteria. Hopefully they will be describing exactly what they believe you have to do to get the raise or promotion.

Most managers will not feel apprehensive about discussing this with you in a relaxed, nonformal setting. If you wait until job review to have this conversation, it will be too late and too formal. Your manager may be as nervous as you are in a formal job review and may feel you are trying to second-guess them. You must have this conversation well in advance of your job review. This way your manager will not feel like they are being put on the spot and you will have enough time to demonstrate the performance needed before the formal job review. In any case, make sure you listen and ask them to clarify anything you may not understand. The more they talk, the clearer the picture you have of the formal criteria you must meet. Take as many notes as you can and be sure to review them often.

Most engineers believe that the minute they are performing at a level above theirs the company will immediately promote them. It is very discouraging for them to find out that this is not the case. The engineer must go through a formal promotion review process and be assessed "ready for the promotion." In larger corporations this formal promotion review process often takes months and may involve a multitude of other people. Therefore, the third group of formal criteria you must be aware of in your company is the promotion review process.

UNDERSTANDING THE PROMOTION REVIEW PROCESS AND USING IT TO YOUR ADVANTAGE

The formal promotion process identifies the steps your manager must go through to get approval. For the lower levels on the engineering ladder, this process may just involve your manager and their superior. As you move up the engineering ladder, the promotion process becomes more complex and may include promotion review boards made of several people as well as the use of company totems. For career advancement, you must have a clear understanding of the promotion process for your company.

▶ *Career Tip.* Learn everything you can about the promotion process for your company.

To help you better understand the steps companies go through to promote someone, I will describe two examples that represent what most companies follow for promoting people. These processes are generalized in nature but show the dynamics that may be occurring in your company; this should help you to identify and understand your company's process.

To aid in this discussion, I have diagrammed the spheres of influence often involved in the promotion review process. Figure 17-3 shows the spheres of influence you need to be aware of in the promotion review process. The sphere at the bottom is you, the next highest on the ladder is your immediate lead engineer, and above the lead engineer is your manager. Above your manager are their superior or manager as well as the company totem and promotion review board.

I have drawn the spheres in relation to how you most likely perceive them. Your lead engineer has the responsibility for assigning the daily tasks and doing most of the interfacing with you; their responsibility is to take care of most of the problems. Your supervisor simply has too many people reporting to them to spend the amount of time that they should with everyone. Therefore, your supervisor calls on the help of the lead engineers to hand out assignments and make sure everything is being accomplished.

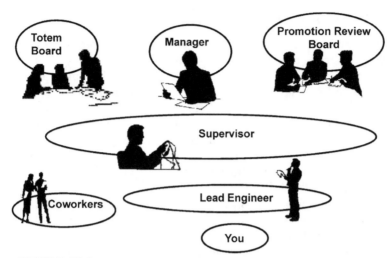

FIGURE 17-3 Promotion review process and spheres of influence.

Behind your supervisor are three spheres you need to be aware of, usually these spheres are not very visible to the engineer. However, knowledge of their existence is a must for career advancement. The first sphere is your supervisor's superior or manager. The manager must approve all the raises and promotions your supervisor requests. In addition to your manager, there are two other spheres: the totem board and the promotion review board.

The totem board is usually comprised of second level managers and possibly directors. The purpose of the totem board is to compare and rank the performance of all the engineers in the company. The managers meet and rank the performance of all engineers relative to one another. You can think of this as similar to class ranking. Totems are usually done once a year. Usually, the top-ranked individuals in the totem are the first to be promoted. The bottom ranked employees are put on improvement plans or laid off.

In addition to learning about the ranking process, do you know what your rank is for the grade level you are in? One simple way to know, is to ask your supervisor. While most supervisors will not tell you your exact rank, you can ask which quartile are you in? First, second, third, or fourth?

The promotion review board is usually made up of a cross section of people in the company. The board may contain senior staff engineers and upper level managers and Human Resources personnel. The purpose of the promotion review board is to review each candidate for promotion and determine if they fulfill the criteria established. The promotion review board is a nonbiased third-party group whose function is to ensure that each candidate meets the criteria. Usually, your supervisor and their manager present to the review board on your behalf the reasons why you deserve the promotion. The review board evaluates the merits of your case and makes a decision based on the evidence presented.

Lower Level Engineering Promotions

To understand all the dynamics involved in getting a promotion, let's first review an example of a lower level promotion where only your supervisor and manager are involved. Generally, for a lower level promotion only your supervisor and manager need to decide that you are ready. The first thing your supervisor does is to check with the lead engineer about your performance. How do you get along with your lead engineer? Do you support the lead engineer and follow their direction? Or are you on poor terms with the lead engineer and constantly in disagreement? What do you think the lead engineer's recommendation will be about you? Correspondingly, do you have enough visibility so that your supervisor can see that you are ready for a promotion or does the lead engineer do all the interfacing? This is a delicate situation you must keep in balance. You must follow the direction provided by your lead engineer and get visibility with your supervisor. Going around your lead engineer and always directly interfacing with your supervisor can cause problems; you can hurt your chances for a promotion by doing this.

Try to maintain a balance in this situation. Make sure your lead engineer is always aware of what you are doing. If you have to discuss things directly with your supervisor, make sure the lead engineer knows why you are doing this. On the other hand, if your lead engineer is not giving you the chance to report your progress to your supervisor, discuss the situation with the lead. Try and get them to allow you to report progress at least once in a while. Every lead engineer and supervisor differ in how much interfacing they allow the junior engineers to have.

▶ *Career Tip.* Make sure your lead engineer and supervisor know what you are accomplishing on a weekly basis.

Once the supervisor and the lead engineer decide you are ready for a promotion, then comes the job of convincing your manager. The supervisor usually carries forward a recommendation for your promotion. You must keep in mind that the manager may have anywhere from 30 to 50 engineers reporting to them and at any given time, four or five of those engineers may also be up for promotion. What will make the manager approve your promotion over others? This is the question, or problem, you must address.

The dynamics involved in obtaining the approval for your promotion from the manager will depend upon several things. First, does your performance warrant a promotion? Second, what is the relation between your supervisor and manager? If they have a poor relationship, it could be years before you are promoted. Third, who else is up for promotion? Often, a limited number of promotions are allowed at any given time due to budget constraints. Is this the first time you have been nominated? Are there others who have been waiting longer? And, finally, how well does your manager know you? If you have never met, you can rest assured they do not know you and, consequently, unless your supervisor does a wonderful selling effort, you will most likely not get promoted.

Higher Level Engineering Promotions

Now let's look at the dynamics involved in a higher level promotion where your manager must utilize the totem board results and present your case to the promotion review board. What happens here is that the manager, with the help of your supervisor, prepares a small brief or summary as to why you deserve a promotion. The brief contains a description of your accomplishments and shows how you meet the criteria for promotion. In addition to the brief, your ranking on the totem is reviewed. This is a check and balance process. If you meet the criteria then you should be performing at the top of your grade and ranked accordingly. If you are ranked at the top of your grade then your performance should normally meet the criteria for promotion as defined by the promotion review board. If you're not at the top of your rank or do not meet all criteria then your chances for getting the promotion approved are not good.

Once you understand this process, you will quickly realize you must know two things. The first is where you stand on the totem and the second is what the board's criteria are for promotion. Determining what both of these are may seem impossible, but it is doable. Finding out where you rank in the totem, or in relation to others, takes tact and patience. Your manager knows where you rank in the totem and does not like to tell anyone. Therefore, you may have to do a little fishing. Find out when the last totem ranking occurred and when the next is scheduled. If your company does not use a totem then you must explore how the company ranks engineers. Your supervisor should have this information. Most supervisors will share with you how and when ranking occurs. The next thing you must find out is where you stand on the totem ranking or in relation to others being considered for promotion. The best approach is simply to ask. Most supervisors will inform you that this information cannot be shared. You might find out something by asking what quarter you are ranked in: the top, upper middle, lower middle, or bottom. Your manager may share this piece of information with you since they are not telling you the exact rank. If you are in the top quarter, you're in great shape. If you're below the top quarter, you have your work cut out for you.

The second action is to find out the criteria used by the promotion review board to determine if someone is ready for a promotion. Sometimes this information is closely guarded and sometimes it is readily available. The first place to start asking is your supervisor. If your supervisor has the information and shares it with you, you are in great shape. Study it and determine where you exceed, meet, or fail to meet the criteria.

▶ *Career Tip.* Find someone in the company who was recently promoted to the level you want to be and network with them to discover their experiences!

The next place to check is with someone who was just promoted to the level, which you are looking to get promoted to. Chances are they did not make it on the first try and needed to do some additional things to meet the criteria. By talking to others who recently received a promotion you can get a pretty good idea of what the promotion review board criteria are or what is involved in the promotion review process. If you do not know of anyone, check with personnel and find out who in the company was just promoted. People who were just promoted are usually glad to share this information with you. They know how hard it can be to get promoted.

The important point here is to check all the available sources you can about the promotion process. The promotion process will vary from company to company and manager to manager. It is a waste of energy and time to go after the promotion blindfolded. You can literally remove part of the blindfold by doing research on the promotion process used in your company. Your challenge is to find out all you can about the promotion process in your company, and what the formal criteria are that you will be judged against. The information is all there. It's up to you to do the legwork and find it.

SUMMARY

Your manager and other upper level superiors are evaluating your performance every day. In every company, there are formal criteria by which you are judged. You must understand the formal criteria for successful career development. The formal criteria are usually well-defined and documented.

The formal criteria are manifested in three ways. The first is through the job performance review process. This is usually a yearly assessment and review of your performance as determined by your supervisor in meeting the company expectations. Understanding this process and the form used is a key element for career advancement.

The second is through the job performance criteria, which is the performance expected of engineers at each level in the organization. The third way is through the promotion review process. The promotion review process and actions your supervisor has to follow to get you a promotion are usually well-documented and strictly adhered to large corporations. In smaller companies this formal criteria may not be as well documented. Your career advancement will depend on you understanding all three of these and knowing what actions you need to take to show that you are ready for the next promotion based on this formal criteria.

If you are in doubt about any of the formal criteria the best person to seek help from is your supervisor. Sit down with your supervisor, clarify the formal criteria by which you are being measured. Use the tools suggested in the chapter to help you identify the formal criteria and any potential improvement areas. Next visit with your mentor, they should be able to help you also.

Have you identified any career actions you want to take as a result of reading this chapter? If so, please make sure to capture these ideas before you forget by recording them in the notes section at the back of the book.

ASSIGNMENTS AND DISCUSSION TOPICS

1 Study the forms your company uses for job appraisals or job reviews. Do you understand each block of information and what it means in terms of your performance?
2 Who signs off on your job appraisal? Do they know you?
3 Make a list of all the things you were critiqued on during your last job appraisal and identify something specific you can do for each one to show improvement.
4 Does your company store job definitions and performance criteria on a website? Review the criteria for the level you are and the next level you aspire to achieve.
5 Meet with your manager and discuss the criteria you are expected to meet for your present job level as well as the performance expected from you for the next level. (Remember the cheat sheet.)
6 What is involved in the promotion review process for your company? Are totems or rankings used? If so when do they occur and what is your rank? Is a promotion review board used? Who are the members? What criteria are they using?
7 How do you find out who was recently promoted? (Hint: Human Resources has all this data.)

CHAPTER 18

DETERMINING THE INFORMAL CRITERIA BY WHICH YOU ARE ASSESSED

Most engineers do not realize that informal criteria often play a bigger role than imagined in evaluating one's performance. Informal criteria are very hard to define, and therefore, you must be aware of and recognize the subtle hints that are available to you. The informal criteria are those intangible things that your manager and superiors use to judge your performance. Informal criteria are usually based on personal biases or beliefs and styles as to how to best perform the job. These beliefs are often a result of your manager's work experience as well as their background and training. The informal criteria are usually never documented and sometimes very hard to determine.

UNDERSTANDING THE DANGERS AND BENEFITS OF THE INFORMAL CRITERIA

Since most of the informal criteria are style-related, you have to be acutely aware that people have different styles or methods when working and solving problems. We all select a style or method of performing work on the basis of our experiences, training, and personal beliefs. Since we all have different training and experiences there will always be differences in opinion on the best way to complete the task. These differences in opinions or styles can cause serious career problems when your performance is rated solely on whether or not a person likes or dislikes your style. My objective in this chapter is to identify some common styles and the dangers associated with these styles.

The real problem for your career occurs when you and your supervisor have directly opposite styles for accomplishing work. Either style will achieve

The Engineer's Career Guide. By John A. Hoschette
Copyright © 2010 John Wiley & Sons, Inc.

the high-quality result, but are totally different in their approach. This is a very dangerous situation because even when you are meeting or exceeding work expectations. Simply having opposite styles from your supervisor's may result in getting rated as a "Poor Performer."

Let's identify some common examples of how supervisors are using informal criteria to judge you and without being conscious they are doing so. We will work on helping you to identify and recognize when you and your supervisor have different work styles and how to successfully overcome this.

INFORMAL CRITERIA—HOW TO IDENTIFY THEM

For the remainder of this chapter, I will identify common informal criteria many supervisors use to judge people. Along with this I will discuss how you can use this knowledge to improve your performance ratings, and hopefully, significantly improve your supervisor's opinion of you and your style.

PICKING THE ONE AND ONLY SOLUTION

Over the years, I have NEVER had a lack of methods to solve a particular problem. When seeking other's advice I found that the lead engineer had one method, my supervisor had a different method, my fellow engineers had another method, the Program Manager had still another solution, and I sometimes even had a different method to solve the problem.

My quandary was one of having too many choices to solve the problem rather than having none. I quickly learned who I should listen to when put in this situation; it was my supervisor. They were the most brilliant men and women I had ever met. I usually tried to highlight how I implemented their brilliant approach in my work.

▶ *Career Tip.* When faced with many different equal options on how to solve the problem, always pick your supervisor's approach.

I learned early on in my career that when all things are fairly equal in science, the obvious solution was the one most politically correct. This was tough for me to accept in my early career since engineering classes always taught there is only one correct answer to a problem. If you do something wrong in deriving the answer to the problem, you are marked off points. Subconsciously, this trains one to believe there is only one correct method of solving any problems (your way). When on the job, I tried applying this opinion that there is one possible method of solving a problem; it ended in disaster for me.

What finally did work for me was selecting the supervisor's method and letting my coworkers know how brilliant a solution it was. If the solution worked, I was complimented for following my supervisor's advice and labeled a great employee. If the solution, per chance, did not work, I was not blamed since I was only following the boss' direction.

Obviously if the supervisor did not have a good technical solution, I made sure to ask if they had considered the benefits of other solutions. Then we jointly discovered my solution was the proper way to go. Once again my supervisor was brilliant for recognizing the problems with their approach and discovering a better approach.

NEATNESS OR APPEARANCE

Take a good look at your managers and their superiors. How do they dress and what is their style of hair? How do you dress and wear your hair for instance? Some managers like to come to work everyday wearing a suit with neat, short hair. Or in the case of women, some wear full makeup and jewelry. If you are a man and prefer to dress very casually, wear your hair long, or in the case of women wear neither makeup nor jewelry, how do you think your superiors relate to you? Might he or she be thinking you are a sloppy person or that you need more discipline and character in your life?

Or conversely, if you like to dress up and wear a good business shirt with a tie and maybe even a sport coat, and your supervisor likes the very casual dress of jeans and tee shirt, what do you think they are thinking of you— you're uptight and too stuffy? In the case of a women supervisor who wears neither makeup nor jewelry, she may be wondering why you, as a woman, are wearing makeup and jewelry. You obviously have your priorities in the wrong spot.

In either case, just because you dress differently, it has nothing to do with your performance. You are being judged simply by your difference in appearance. The personal beliefs of the supervisor (style) influence their opinion about you.

▶ *Career Tip.* You can dress the way you want outside of work, but the style of your dress will impact your relationships with others at work and ultimately your career advancement! Dress for success at work.

The important thing to remember is to observe your style of dressing compared to your supervisors. Neither style (neat versus casual) is right or wrong. They are simply different. If your style of dress is similar to your supervisor then you might be able to relate to one another better. When they are opposite, problems may arise. Who should change, you or your supervisor? I have found that trying to change your supervisor to fit your style is NOT a good career move. If anyone is going to have to change their style, it more than likely is you. You are going to have to change your style. Take a few minutes and observe your supervisor, your lead engineers, and other key individuals in the organization, and how they dress. A good place to do this is at your next staff meeting. It is a good idea to mimic your manager in dress and office appearance unless your manager is clearly a maverick in the company.

NEATNESS OF OFFICE

This is a classic example of style difference. Let me share a personal example with you. I have a neat office style, I like a clean and clear desk with everything filed away. I dust my office weekly and make sure it is cleaned up every time my boss comes to visit. I like to give the appearance I am in control of things and I am, so I want my office to reflect it. My personal feeling is that people with messy offices reflect an image of not as much in control and generally not good managers.

However, I worked for a great manager who was just the opposite. He filed by piles on his desk and table. He could not even get his filing cabinets shut because he stuffed drawers until they were overflowing. In his office he had boxes stacked from floor to the ceiling because he had never unpacked from the last two moves. He could find anything he needed as long as you did not disturb his piles. He had hieroglyphic sketching on his white board from the last century because he never erased anything.

We were working on a particularly tough project and things were not going well. I had been putting in overtime to help, but things were not improving at the rate my supervisor wished. He called a meeting in my office to discuss what was going on. He came to my office and we proceeded to discuss the problems and potential solutions. After about 10 minutes, he suddenly announced he discovered the problem. My office was too clean and I obviously had not been putting in the time needed on the project.

I was furious at his remark, I had spent overtime working on the problem and even staying extra late at night and on weekends to clean up my office. I kept my cool and realized what had happened. He had judged me on my office style which was directly opposite his. I realized then what I needed to do. We finished the conversation and he left with me agreeing to put more time in and we planned for a follow-up meeting in a week.

A week went by and this time I was ready for him. I took out several dozen folders and threw them across the top of my table and desk. I made sure every square inch of my desk and table was covered. I even managed to create a pile of folders on the floor. It was completely ugly, the place was a mess, and I could barely stand it. I had paper and hardware all over the office.

He arrived promptly on-time for the meeting and I quickly started giving status on all I had done during the week. I apologized for the mess, told him I was too busy to clean up. We had only talked for about 10 minutes when he suddenly announced there was no need to continue. He informed me I had obviously totally engaged in the project and had enough to do since I did not have enough time to clean up my office. He indicated further discussion would only slow my progress and he was confident I would solve the problem shortly. He stood up and left before I even had a chance to complete giving him a progress report. We had overcome a very difficult problem and expected to complete the project on plan as a result. His leaving before I could finish made me realize how my work was being judged by the condition of my office. Making a mess out of my office before he arrived was somewhat the right thing to do.

The moral of the story is your supervisor may be judging you on your style compared to theirs. If they are similar, you have less to worry about. If they are different, you better have a plan on how to deal with their totally opposite style.

Take a few minutes and observe your manager's office. Some managers keep a very messy office and feel it is a sign of being busy. Other managers keep a neat, well-organized office and claim they are in control. How do you keep your office compared to your manager? If you are neat when they are messy or vice versa, then they may not be able to relate to you. Both styles produce results. The danger is when you and your boss' style are opposite of each other.

VERBAL VERSUS GRAPHICALLY REPORTING

There are generally two styles of presenting information. One style is verbally. For this method, people tend not to write instructions or reports but communicate the results verbally. These people often say, "Tell me what happened, I'd like to hear everything." Also, verbal people use very little paperwork to support their discussions, the best way to communicate is to simply sit down and talk things through.

The second style is graphically. For this style, people present results in written form or transmit ideas by presenting charts or drawing pictures and diagrams. They usually have large amounts of graphs and pictures for support. A good way to find out which style a person prefers is by noting how they give directions. If someone gives you directions to a place by writing words (turn right, go north, etc.), he or she is usually a verbal person. If a person draws you a map, then he or she is a graphics person. What is your manager's style? What is yours?

Now let's look at an example of a verbal supervisor getting a status report from a graphic engineer. The verbal supervisor will ask the engineer to simply inform them of the progress. The graphic engineer will start out talking but quickly start pulling out graph after graph to show all the support material. At the fifth or sixth graph the supervisor is totally lost in all the data and simply informs the engineer to put it all down and tell him what is going on. The engineer has been trying to do this all along and is totally taken back by the remark. The supervisor does not relate to all the graphs and wants a simple verbal explanation. Both are frustrated with each other at this point and both blame the other person for the disconnect. Neither party realized the disconnect and the supervisor informally judges the engineer as a bad communicator and not a good engineer.

Now let's reverse the styles. Let's assume the supervisor is a graphics person and the reporting engineering is a verbal person. The verbal engineer is convinced the best way to brief the boss is by sitting down and verbally reporting status. The verbal engineer shows up at the graphic supervisor's office with nothing in hand and starts reporting progress verbally. After about 10 minutes the supervisor is lost and wants to see the data and graphs from which the engineer drew the conclusions. The engineer does not have time to generate all the data but has a good grasp of the meaning and is giving the boss a great verbal report. The boss is frustrated at this point and starts asking for the charts and graphs. What happened? Difference of styles! The supervisor informally believes the engineer is a poor communicator because of the apparent lack of necessary graphic support material. The engineer is frustrated because they feel the verbal report is the best way to communicate the information. Neither supervisor nor engineer relates very well to one another's communication style, yet neither style is better than the other.

▶ *Career Tip.* Learn to present information in both verbal and graphic style. Use the appropriate style for each person you are presenting to.

If you are a verbal person trying to report results to your manager and they are a picture person, they may have a tough time relating to you. They may be constantly asking you for something graphic to look at in your report. On the other hand, if you are a picture person and your supervisor is a verbal person, you will have a different problem. They will not be interested in reading reports or looking at graphs—they want you to tell the results.

ARE YOU A DETAIL PERSON OR BIG PICTURE PERSON?

Some people like to talk details and not worry about the big picture. They believe the devil is in the details. Detail people will ask for all the supporting data and go through each piece to ensure its integrity and applicability.

The opposite style of a detail-oriented person is the big picture person. The big picture person does not want to get into all the details, but to understand the progress, issues, and status at a more global level.

Let's take a look at an example of these two types of people trying to discuss the progress on a project. Let's assume the supervisor is a big picture person and the engineer is a detail person. The supervisor asks the engineer to report progress on the program. The supervisor is expecting to hear an overall assessment of the status with the bottom line of: "Are we on track to complete the project on time?" The detail engineer assumes the supervisor wants to hear all the details and immediately launches into a detailed status of the project/program with little or no indication of how they relate to the big picture. The supervisor stops the engineer after 10 minutes and indicates this is way too much detail. What is the big picture? At the same time, the supervisor is thinking informally, this engineer has no idea of what's going on. The detail engineer feels he or she knows exactly what's going on and cannot believe the supervisor is not interested in the details. Once again a difference in style has led to a total disconnect. Neither style is right or wrong; it is when the styles are directly opposite that conflict occurs.

What type of a person is your manager and what type are you? If you are both the same you will have more in common. If you have different styles you may have to consider altering your style when working together.

ARE YOU PEOPLE-ORIENTED OR RESULTS-ORIENTED?

Some managers are very people-oriented, which means they are really interested in you, your family, and how you are doing. When you interface with these people, they want to hear about you just as much as they want to hear about work. You can recognize these people by the type of office they keep. Generally, they will have pictures of their family, trophies they have won, and many personal items. These people believe socializing at work is part of conducting business. So you want to remember to talk about yourself, as well as your work, when interfacing with other engineers and supervisors who are people-oriented.

The directly opposite style is results-oriented. These managers are only interested in hearing about and obtaining the final results. They are not interested in discussing personal matters; they are in it for business only. They are not interested in socializing and consider you to be wasting company time when you socialize at work.

Now let's take a look at when a people-oriented manager interfaces with a results-oriented engineer. The manager tries to start the meeting out socializing and asking the engineer about his or her weekend. The results-oriented engineer, feeling very uncomfortable about small talk and socializing, ignores the question and immediately starts reporting progress. The manager thinks how rude the engineer is and informally identifies the engineer as having very poor people skills. This engineer is perceived as not a good leader.

Now let's reverse the roles. This time the manager is results-oriented and the engineer is people-oriented. The engineer enters the manager's office and immediately starts to socialize and ask about the manager's weekend. The people-oriented engineer continues to socialize until the manager can no longer stand it and abruptly announces that this is work time and we should not waste so much time discussing nonwork items. What happened again? Total style differences! The manager is informally convinced that the engineer will never make a good leader since he or she wastes too much time socializing. In both examples, neither style is better than the other; they are simply different. It is when you and your supervisor have different styles that trouble arises.

▶ *Career Tip.* The upper levels of management are generally results-oriented people. When interfacing with these upper-level managers, get to the point quickly—no more than two charts and start with the bottom line!

How does your style relate to your managers'? If they are the same then you and your manager should be able to communicate openly and easily. If they are opposite, then you may want to communicate more in the style of your manager.

PROCRASTINATOR OR PLANNER?

How people handle business and deadlines generally fall into different styles. One style is what is referred to as a procrastinator style where the individual waits until the last minute for everything, and is usually in a panic mode and seems to always be fighting fires. Sometimes this style is referred to a fire putter-outer. In this style, the person usually waits to the last possible minute and often works late into the night to meet impossible deadlines. A good example of this mode of operation is how people handle term papers and finals in college. The procrastinator person was the engineer who waited until the last night possible to start writing the term paper or lab report, and ended up working all through the night to complete the assignment just in time. He or she turned the paper in after putting in a humongous amount of work in a very short period of time and received an "A" on the paper.

If your manager is a procrastinator person he or she will most likely expect the same from you, instant response and last minute panics to be taken care of. If you do not work well under these circumstances, you will have a hard time impressing your manager.

The opposite style is the planner style. For this style, the person has to have everything planned out well in advance and always strives to avoid last minute panics. In college, this type of engineer always had their term paper completed days in advance of when it was due and handed in extra credit work to make sure it shined above the rest. This engineer also got an "A" on the term paper. It is interesting to note that the engineers used a totally different style of work to complete the assignment and both got an equivalent grade "A."

Now let's observe these two styles at work and how complex working together can be. For the first example let's assume the manager has the style of planner and the engineer has the style of procrastinator. Now the real world problem. The manager gets a call from the president of the company and the engineering team is expected to demonstrate the new product in 3 weeks. The manager immediately calls a meeting of the team and announces the plan to demonstrate the new product to the president in 3 weeks. To be prepared in advance for the demonstration, the manager asks a procrastinator style engineer to lead the effort. The manager describes how the first week the engineer will call a team meeting to generate a complete plan of everything that has to be done. The second week the team will start working the details for everything, as well as start assembling the hardware and software and decide what will be shown. Three days before the demonstration the team will do a practice run to ensure everything goes perfectly. And finally, the entire team will review everything the day before to be absolutely ready. The day of the demonstration, they will assemble early in the morning and quickly run through everything. The team should be all set by the time the president arrives.

The procrastinator style engineer immediately wonders what all the fuss is about and realizes they could pull the whole thing off in just 3 days. So the lead engineer never calls the team meeting because they believe there will be plenty of time. One week goes by and the manager does not hear about the team meeting nor any progress. After checking in to why the team meeting never took place, the manager becomes angry when the procrastinator engineer explains there is no reason to panic because there is plenty of time and the team had other more important things to work on. Another week goes by and nothing else happens. The procrastinating engineer has delayed everything until the last week, and by this time the manager is in complete panic, a state that a planner style considers a failure and never wants to get into. Finally, with just 2 days to go, the procrastinating engineer calls a team meeting, gets everyone organized, but they are going to have to work late the night before. A deadline situation the manager has wanted to avoid at all costs.

The manager goes home the night before totally upset since they do not even know if the demonstration is going to happen. The engineer and team work late into the night and get the entire thing working perfectly.

The next day the engineer and team put on a successful demonstration for the president and the president congratulates the team. Excellent work! The engineer feels wonderful; they made it happen and the president was totally impressed because of the engineer's work.

And what is the manager thinking? Informally, fire the engineer! They jeopardized the entire demonstration by waiting until the last minute and that is no way to run a business. This procrastinating engineer is viewed by the manager as a poor performer even though the demonstration was a huge success. The engineer is not a good member for the team or the department. Totally different styles led to the problem.

▶ *Career Tip.* Working for a procrastinator is extremely stressful. If your supervisor is one, be prepared to be constantly in panic mode. Get yourself a stress relief plan.

Now let's look at the same example only reverse the roles. This time the manager is the procrastinator style and the engineer is the planner style. In this case, even though the manager received a call 3 weeks in advance, the manager procrastinated and did nothing to prepare the week of the demonstration. Then in a panic, the manager calls the team together and announces the demonstration must come off this week and the planning style engineer is in charge. Immediately the planner engineer instinct kicks in and determines there is not enough time to pull this off successfully. So the engineer panics and argues with the manager to get the demonstration delayed—there is not enough time to prepare! The manager counters with yes, there is plenty of time. Three whole days and the engineer will just have to deal with it.

The engineer panics and starts working late into the night to make things happen, something a planner usually despises and is uncomfortable with— remember it is not their style. The engineer complains to the manager daily about the prospects of not making the deadline and things have to be postponed. After 4 days of working late and most of the night before the day of the demonstration, everything is ready to go. The engineer is totally freaked out and a nervous wreck.

The demonstration to the president is a huge success and the president congratulates the team. Excellent work! The engineer believes they deserve an award for meeting an impossible deadline. The engineer is stressed out and never wants to work under these conditions again. Poor planning on the manager's part caused all the unnecessary extra work and panic. It was all totally avoidable.

And what is the manager thinking? Informally, the engineer cannot work under tight deadlines. The engineer will never make it because impossible deadlines are the norm in business. This engineer is a "poor" performer even though the demonstration was a huge success. The engineer is not a good leader for the team or the department. Totally different styles led to the problem.

These styles are totally different and will often cause conflict among team members. Each style appears to be out of control to the other style. Neither style is right or wrong; they are simply different.

Now the big question: how does your style compare to your managers'? If they are different, who is going to change?

How do you find out what style your manager is? Simply observe them in their office, meetings, and hallway conversations. Once you realize their style and how this relates to your style, you can start to deal with the conflict since you will know what you are up against. Realizing this informal criteria, and its impact on career, make work and career advancement quite a bit simpler.

These are some of the more common informal criteria that managers use to judge people. You need to be aware of these informal criteria and constantly checking for others that your manager may have. The informal criteria are often not obvious and they may not be easy to recognize, but they are there. With a little bit of research and acute observation, you will begin to recognize other informal criteria. Your task is to start looking for and recognizing these informal criteria, and then use them to your advantage. In other words, take off the remaining part of the blindfold and see what you have to do to get the promotion.

SUMMARY

Most engineers do not realize that informal criteria often play a bigger role than imagined in evaluating one's performance. Informal criteria are very hard to define and therefore you must be aware of the subtle hints that are available to you in identifying them. Informal criteria are usually based on personal biases or beliefs and styles as to how to best perform the job.

We all select a style or method of performing work on the basis of our experiences, training, and personal beliefs. Since we all have different training and experiences there will always be differences in opinion on the best way to complete the task. These differences in opinions or styles can cause serious career problems when our performance is rated on whether or not a person likes or dislikes our style.

The real problem for your career occurs when you and your supervisor have directly opposite styles for accomplishing work. This is a very dangerous situation because even when you are meeting or exceeding work expectations, simply having opposite styles from your supervisor's, may result in getting rated as a "Poor Performer."

Once you realize you have a difference in style from your supervisor, you can start to deal with the conflict since you will know what you are up against. Realizing this informal criteria, and its impact on career, make work and career advancement quite a bit simpler.

Have you identified any career actions you want to take as a result of reading this chapter? If so, please make sure to capture these ideas before you forget by recording them in the notes section at the back of the book.

ASSIGNMENTS AND DISCUSSION TOPICS

1   Which is more important: the formal company criteria or the informal criteria supervisors use to assess your performance?
2   Is selecting your supervisor's approach always the best way to go?
3   Observe your supervisor's office. What style does it indicate he or she is?

4 When your supervisor's style is directly opposite yours, who should change? Should the supervisor be adaptive and change for your benefit?

5 When you are in charge of a team and the members of the team have different styles, what should you do?

6 Identify your styles and make note of it.

7 Do these styles exist in other nonengineering departments of the company?

8 Are these the only styles with directly opposing approaches to work? Can you identify others?

9 If you are a manager, do you feel you should have to change to match others' styles?

CHAPTER 19

THE LEADING REASONS ENGINEERS ARE NOT SUCCESSFUL

Most engineers are and were rated in the top quarter of their high school graduating class as evident from surveys conducted on the admission criteria for engineering universities [1–4]. Admission to engineering schools requires above average grades in math and science [5]. In order to graduate, the average engineering student must spend four years studying a variety of extremely complex and abstract subjects. Therefore, one would expect these highly trained and intelligent engineers would be highly successful once they leave school.

However, this is not the case. Some engineers are not successful on the job principally because they do not develop the broad outlook and basic human relations skills that are so important to achieving in a team environment. In this chapter, we will explore some of the most common reasons why engineers are not successful and how you can easily avoid making these career blunders.

DEFINITION OF BEING UNSUCCESSFUL

The definition of being unsuccessful takes on many different meanings for people. First, it is necessary to describe what this is. If a person is reassigned or removed from a project against their will because they are too disruptive to the team or fail to get the work done, this should be considered as being unsuccessful. The most unsuccessful action is termination from the company.

Being unsuccessful may come in more subtle forms. For example, success is thought of as moving up the corporate ladder. Therefore, stagnation at the same level year after year rather than advancing can also be unsuccessful.

The Engineer's Career Guide. By John A. Hoschette
Copyright © 2010 John Wiley & Sons, Inc.

Receiving minimal raises, being assigned trivial duties, constant reassignment, or continual transfer, are all forms of being unsuccessful. Being pigeon-holed into one job or only one type of assignment can also be. The engineer can also be technically unsuccessful as well. If the engineer fails to resolve technical problems or the company loses a significant amount of money because of a poor technical approach, these might also be considered as unsuccessful.

The definition of being unsuccessful is different for every person. What is common to all, is the engineer does not reach the goals they intended to reach and there is no career growth. Let's explore some of the common causes of being unsuccessful for engineers.

LEADING CAUSES FOR BEING UNSUCCESSFUL

Studies have shown that there are a wide variety of root causes for engineers being unsuccessful [6]. Probably, the most common and noteworthy among these are:

1. Inept or poor communication skills
2. Poor relations with the supervisor
3. Inflexibility
4. Poor and lax work habits
5. Too much independence
6. Technical incompetence

These reasons for being unsuccessful are highlighted since they cross all technical fields and are probably the most prevalent ones cited on job appraisals. They are the problems that are likely to continue to haunt you throughout your career unless you aggressively do something to correct them. A special item to note is that the top four reasons are related to people skills and not engineering skills.

INEPT OR POOR COMMUNICATION SKILLS

Good communication skills are absolutely necessary to move up in the company. The inability to effectively communicate is what often keeps an engineer from advancing.

Good communication skills are required in a variety of areas. For example, good communication skills are required for reporting progress to management, giving specific direction to subordinates, dealing with customers, describing complex problems over the phone, presiding at meetings, writing specifications or reports, and even interfacing with people over computer networks and in video conferences.

Engineers get paid to resolve complex problems using communication skills to bring together the resources, people, and technical knowledge necessary for success. Lack of good communication skills could limit the success of the project and your career.

By nature and training, engineers tend to focus on technicalities rather than people and this often tends to make engineers poor communicators. Many believe technical skills are all that count and they will be rewarded accordingly. Technical skills do make up a portion of the criteria for advancement but communication skills are also another important part of the criteria.

An engineer who is able to communicate clearly (in writing and verbally) and has a good technical understanding is usually recognized by superiors as someone with high potential. To illustrate how important good communication skills can be, let's look at some examples.

Clear, concise, and easily understandable writing is a must for engineers who often write specifications and technical reports. These specifications usually define requirements for a product or process. Poorly written specifications for these products or processes can cause a multitude of problems. First, poorly written specifications need to be interpreted and rewritten so that the true meaning becomes apparent. This can cause delays in the start of the project because it appears the product is being designed to unsafe or potentially life-threatening specifications. All of this may result in cost and schedule overruns, potential lawsuits, or even death.

▶ *Career Tip.* Take writing classes and improve your capabilities.

Poor technical writing skills can be improved upon, but it takes time, practice, and more practice. One way to improve your writing skills is by taking technical writing courses at your local college or university. Another way I discovered to be helpful was look for guidance in old reports, specifications, or other documents previously generated by other people in the organization. If they are notable, adapt their outlines, forms, and style of writing as much as you can. Getting your hands on an old report that was well received can save you hours of rewriting and editing.

Clear and concise verbal skills are a must for career advancement. You need to be able to do this during your meetings with management as well as with fellow employees. You must be able to verbally communicate the important points contained in the technical charts, graphs, and reports that you have written. You must also be able to verbally give clear and concise instructions to people so they know what must be done. This is especially true when working on potentially dangerous projects such as nuclear reactors, high-voltage equipment, or with corrosive and poisonous chemicals.

Developing good verbal skills takes time, and you can take advantage of shortcuts by listening and studying how engineers in the upper echelon of the organization report verbally. On what do they put the most emphasis—cost, schedule, or technical aspects? What is the standard form for presenting verbal results? Do they use projectors, handouts, drawings, computer plots, or some other type of graphics to support verbal reports? Seeking out opportunities to speak before groups or perhaps present technical papers is excellent training.

There are organizations that help to improve speaking skills. Probably the most common of these is the Toastmasters [7]. This is a national organization with local chapters dedicated to helping people learn how to make speeches and presentations in front of a group. By studying the speaking styles within your company, and through practice, you can significantly improve your verbal communication skills.

▶ *Career Tip.* Practice giving your verbal reports before you go into a meeting. Enroll in classes on making presentations. Refer to my website for information on making Great Technical Presentations (www.careerdevelopmentcoaches.com)

Poor verbal and written skills can be overcome, but this will take time. Once an engineer has made a poor impression with written reports or verbal presentations, they may not be asked to assist in certain key activities because of the initial impression. For this reason, communication skills are extremely important for the new engineer during the first year. Communication skills may even rank higher than technical skills since new engineers are not given the more technically challenging tasks.

POOR RELATIONS WITH THE SUPERVISOR

The personal relationship an engineer has with their supervisor is probably the dominant factor in determining success. Many engineers, and especially recent graduates, do not understand this and greatly underestimate the importance of their supervisor. They naively believe that if they do a good job, their supervisor will always recognize it and they will be successful. Doing a good job is not enough. What counts is doing the right job and having your supervisor recognize this. Previously, it was highlighted that in the business world, there may be many solutions to a problem. The challenge is to pick the right solution. This is where the engineer/supervisor relationship comes in.

In order to advance, you must have a good relationship with your supervisor. Engineers must be able to discuss problems, report progress or lack of it, identify solutions, and finally get the supervisor's approval. The engineer must understand their role on the team and relationship with the supervisor as one of cooperation and providing assistance.

The quality of your work will greatly suffer if the relationship with your supervisor is not good. Your supervisor is not your enemy; they have been in your position before and have a pretty good understanding of what needs to be done and how to do it. He needs your cooperation and support, not a strained and troublesome relationship. This relationship with your supervisor should be similar to that of a hand and a glove; the hand and glove go together. You and your supervisor work tightly together and depend upon each other to get the work done.

How would you rate your relationship with your supervisor? Would you classify it as good or would you classify it as poor? Or have you never really thought about it? If things are going well between you and your supervisor you should be able to openly and candidly discuss work and problems. You should feel you can ask for guidance and it should be given willingly. Your supervisor should be defining a course of action and you should be implementing it. Not all relationships run that smoothly and there will be disagreements. The important thing is that you can air your differences, agree to disagree and move on. The optimal is to have things run smoothly.

There are danger signs to watch for in your relationship with your supervisor. The first is that the two of you are usually in disagreement. Do they always seem to be saying white when you're saying black? Are they nagging and berating you and discounting your work? Does your supervisor continually redo all your work with no explanation or continually give you a sense that you are incompetent? If you answer yes to any of these questions, then it's time to realize you have a very poor relationship with your supervisor, which can be career limiting for you. It's time to sit down with your supervisor, and discuss your observations and feelings.

▶ *Career Tip.* Don't ignore poor relationships with your supervisor; they will only get worse unless you work to improve them.

Try to find a common ground. Identify what you can agree upon and concentrate on this rather than on how much you disagree. Identify what you can do to help the situation. Ask for inputs. Do not ignore the situation; it will only get worse.

If the poor relations are caused by different technical opinions, the best thing to do is agree to disagree. This is where you both agree that there is an honest difference of opinion. You are each declaring that you respect the other's opinion but you disagree. By doing this, you are allowing each other to feel their opinion is valuable, just not your choice or way of thinking. Neither of you is right or wrong.

If you disagree with your supervisor on your performance, it is a different matter. To change an opinion of your performance, you will have to start performing the way they want you to and not the way you want to. You will have to clearly demonstrate actions that meet any criteria. Changing your supervisor's opinion will be hard and it will take time. It will also take a large amount of effort on your part. If, after trying all this, you feel you are not making any progress, it is time to move on and look for a new supervisor/department to work in.

INFLEXIBILITY

There are a couple of sayings that characterize inflexible behavior. The first is "There are two ways to do the job, the wrong way, and my way." The second is "If you want things done right, you have to do them yourself."

If you find yourself agreeing with these approaches to problems, chances are you are very inflexible. It may even be a problem that could be limiting your career advancement.

The objective of most classes when you are in college is to learn how to apply well-defined rules and formulas to come up with the one right answer. All other methods lose points and your grade drops. This leaves the engineer with the impression that there is only one right answer and all others are incorrect. On the job, however, problems are not well-defined and there exists a multitude of correct answers. If you apply this attitude of only one right answer exists, and it is your's, you more than likely will come across as inflexible.

Solving problems on the job requires a team effort, with inputs and solutions suggested or derived from many people. Often cost and schedules do not allow you to score a perfect 100. In fact, the final solution may be far from optimal. All this requires the engineer to maintain a balance in his work. Remain as flexible as possible and, with time and experience, you will learn to find the optimum solution.

Being inflexible can cause the team and you a multitude of problems. Most people will pull away from someone who is too inflexible. If you are inflexible, others will feel you will only implement your own solution to the problem and their suggestions do not really count. This is being un-successful for a team leader. Clearly, working on a team requires that you consider inputs from everyone and choose the best solution regardless of whose idea it may be.

Engineers have different backgrounds, styles, and formal schooling. Therefore, you must be able to work with each of these different styles. Flexibility is the key word. You must change your style in relation to those around you to stimulate them, as well as yourself, in order to get the best performance. An example I like to use is that of the coach of a successful football team who displays a different style when talking to the defensive line than he does when talking to the quarterback. He must be flexible and change his style in order to get the most from his players. Similarly, you must be flexible to get the most out of your career.

There are several questions you can ask yourself to help determine if you are flexible or inflexible. If you are presented new information, do you take the time to evaluate it or simply discount it since you already have the best solution? Do you tend to deal with all people and problems in the same way, or do you try to tailor your approach to that particular problem or circumstance to attain the best result? Do you ever change your mind?

▶ *Career Tip.* Ask your coworkers if they consider you flexible. The answers may surprise you!

Do you strictly follow every company policy, procedure, and regulation, with no exceptions? Is there, and has there always been, only one correct way to get the task accomplished?

Flexibility is key to career growth. Inflexibility is only good when it comes to compromising personal, moral, or ethical standards.

POOR OR LAX WORK HABITS

Poor or lax work habits as well as work schedules are other reasons why engineers fail. Poor or lax work habits and schedules can take many different avenues. Are you easily distracted as you go about your workday? Do people come and go continually in your office, leaving you with the feeling you just never seem to have the time to complete the work at hand? Are you unable to concentrate for very long on a problem before you find yourself wandering off to another problem? In other words, do you jump to the next problem or task before you have completed the present one? Do you spend too much time visiting with other people and having friendly non-work-related conversations? Do you get caught up in all the details and never seem to find a solution? Does it seem there is always some little detail coming up at the last minute to nullify your work? These are all signs of poor or lax work habits.

▶ *Career Tip.* Poor work habits left unchecked will cause your career to be unsuccessful.

The reason these habits are so career limiting is the dramatic effect they have on the quality and quantity of work you accomplish. Being easily distracted can turn an easy, short task into a major one. This type of performance causes cost overruns and missed schedules. Jumping from problem to problem leaves you with nothing ever being completed. You have worked on everything as the supervisor wanted you to, but you have accomplished nothing. Again, this is being in an unsuccessful mode for the engineer.

Getting caught up in the details and wandering aimlessly is very career limiting. The engineers of today are in the middle of an information explosion; they must be able to sort through massive amounts of information to determine what is important and what is not. In order to do this, they must be well-disciplined and organized. This means having methods of sorting and storing information, keeping good records, and summarizing whenever possible. It means you have the ability to organize and run complex experiments and keep track of all variables.

Lack of discipline and being disorganized are two major reasons why most projects fail or end in cost overruns. I cannot believe the number of times experiments or tests had to be rerun simply because the engineer was not organized, failed to pay attention to details, or never bothered to record test conditions or results. Poor and lax work habits cost the company time and profits and could even lead to injury or death in extreme cases.

TOO MUCH INDEPENDENCE

Most engineers are members of a design team. Team members must work closely with each other and draw from each other in order to accomplish their goals. If an engineer becomes too independent, the team can suffer. Some engineers believe they can do it all and try to take on bigger assignments than they are capable of handling. Others want to work alone with no one overseeing their work. In either case, these engineers often go off into a corner and get stuck because they are too timid or have too much ego to ask other team members for help. The net result is that their inability to solve the problems is often discovered too late by the team. Everyone is affected and the outcome is poor performance.

Don't be too independent; it is too career limiting. Try as best you can to be a team player. Take the time to share results with other team members, and do not be afraid to ask questions and get clarification if you need it. You may not have all the solutions but neither does anyone else on the team. By sharing ideas and discussing problems you may find that someone else may have the answer you are looking for. Better yet, you may have the answer to someone else's problem. You will find the solutions through sharing and working together as a team. Remember, any team regardless of the sport (football, baseball, volleyball) needs all of its team members working together to score. Your work team is no different.

If you are a loner, it will take some time to make the transition to team efforts. Take things slowly at first and at a minimum, at least show up for team meetings. If you are afraid to share your work in front of everyone, then schedule special time to review your work with only the team leader. Try to find at least one other person on the team with whom you feel comfortable sharing results. The point is that you must adapt to survive, and only you can do it.

If you are a team person but are experiencing problems working on the team, try to discover why. Is it personalities, different backgrounds, or the way the team is organized? Try to identify specific things causing you problems. Ask if others are having similar problems. You may find that you are not alone. Some teams gel and everything clicks; other teams develop a multitude of problems and try your patience. Stick with it. Teams do not last forever and there will be another team and other tasks. The best way to move up is to do your best on the present task, and prove you are ready for a more challenging one. If you're not up to the challenge of the present task, what makes you think you will do any better on a more difficult one?

TECHNICAL INCOMPETENCE

Last on the list of reasons why engineers fail is technical incompetence [6]. This comes as a surprise to most people. An engineer must learn to face technical challenges to accomplish two things. The first is to broaden his technical knowledge and experience base. The second is to technically update himself periodically throughout his career. To neglect either results in being unsuccessful in the long run.

The products of today are extremely complex, requiring a multitude of engineering disciplines working together to successfully get them out the door. Take, for example, a copy machine. This machine requires that electrical, mechanical, optical, and chemical processes must all work together. If you expect some day to advance to the position of team leader, you had better expand or broaden your background to be able to deal with each of these disciplines. To be a really good team leader, you must understand the customer. This may mean looking into sales and marketing and broadening your knowledge of these fields also. The point here is that to continue career development you must broaden the scope of your knowledge.

If you prefer to remain competent in one area only, then you must periodically update your knowledge in that field. This updating may take the form of attending seminars or returning for classes at a university. Why is this necessary? Because the world of engineering is rapidly changing and not updating yourself could result in being unsuccessful in the long run.

YOU DON'T KNOW WHAT YOU DON'T KNOW

The first time my lead engineer said to me "You don't know what you don't know and these are the things that often stop you from being successful," I was totally dumbfounded. "What do you mean by that," I asked? He proceeded to explain it in very simple terms. He said the things you learned in school and understand, you already know about and you can usually come up with a solution or manage the problems to find a solution. However, if you do not know about something that is capable of causing your project to be unsuccessful, then you cannot prepare for this in advance and are most often doomed to suffer a setback. Therefore, in engineering, it is not the things you know that will hurt you, it is the things that you do not know that are the most likely to cause you to fail. To illustrate this, here are some simple examples. In electronics class, the professor lectures on how to build a great amplifier. However, when it comes time in the lab to build up the amplifier circuit, what you don't know is that all good amplifier circuits need to have great grounding and shielding otherwise they will oscillate and become useless. So, if you are unaware of this, you will most likely build a useless oscillator rather than a superior amplifier.

Another example you might consider is chemistry class. You study chemical reactions and plan a lab experiment. However, no one mentions to you that the test tubes need to be completely clean and the water has to be filtered ultrapure before you start. Not knowing this, you march into the lab and start mixing chemicals, only to get completely unexpected results since you used contaminated test tubes and unclean water.

▶ *Career Tip.* Life is too short to make unnecessary mistakes. Learn from others what you don't know.

So I asked how I would find out about the things I don't know. My lead engineer replied that this is where having great relationships with your coworkers and great networking skills come in. He pointed out that the chances of you doing something so totally unique that no one in the entire company has ever done before are very, very slim. Knowing this is the case, it is your task and challenge to find the people who have done similar tasks, and spend some time discussing your approach and things to watch out for. This is a lesson I have high regard for and has saved me from setbacks many times.

BULL IN A CHINA SHOP

Engineers usually become very excited when a new project is launched and want to get things moving quickly. They rush into the assignment and do not take the time to come up with a well thought-out plan. They feel the need to just get going. The important thing is to show activity and progress quickly.

In the engineer's rush for progress, they end up stepping on peoples' toes, wasting effort, and usually end up breaking things. The perfect analogy is a bull in a china shop. Fine china shops are usually very small with many displays around the shop. One has to move carefully and make sure not to knock over or break anything. Well, a bull is not a graceful animal and a china shop is no place for one. The bull will move about with no regard for the china knocking down displays and breaking china.

Some engineers have the mentality of a "bull in a china shop" when it comes to working in a lab or dealing with people. Their method of dealing with equipment is beat it when it doesn't work, throw things around, and in general break equipment. When it comes to dealing with people, they are often abrupt and abrasive resulting in alienation of team members. They end up causing more problems than they solve.

Just remember, the best thing you can do with a bull is get them out of the shop and isolate them in a field far, far away from anything they can destroy. Don't be a bull in a china shop when it comes to engineering.

TAKING ON MORE THAN YOU CAN HANDLE

One setback experienced by some engineers is simply taking on way too much; they greatly overestimate the amount of work they can handle. Rather than doing an excellent job on small easy obtainable tasks, they try to take on more tasks and accept assignments they are incapable of completing. They simply do not have the time and resources to complete the project. This results in the manager perceiving the engineer as incapable of completing the assignment. An analogy here is the student who takes an extra credit load to graduate early, only to do poorly in several classes resulting in lower grades and a lower GPA.

Contrast this with an engineer who knows their limitations and selects easily manageable assignments and gets them done in advance of the deadline. The manager perceives this engineer as ready to take on more and a good candidate for promotion. The analogy here is the student who reduces the credit load and graduates a quarter later, but was able to raise the GPA since they had more time to concentrate on their classes and get better grades.

TRYING TO DO TOO MUCH TOO FAST

The analogy here is that your ultimate goal is to run a marathon. But you are at the infant stage. You must first learn to sit up, crawl, stand up, walk, jog, run, sprint, and then run a marathon. This is the successful way to get from the infant stage to running in a marathon; you build up to it. This is the same for engineering projects.

There are some managers who believe going through all these stages is too costly and if the team is really going to accomplish their goal on time, they need to start out running 10 miles a day when they have not learned to walk. This is usually not going to work.

A perfect example I have personally experienced was the build and test of multiple computer boards for a computer. The design team built all the computer boards at once since we were behind schedule and rather than testing each board individually in the box, the decision was made to just plug all the boards into the box and turn on the power to see if it worked. If it worked, we would instantly be on track. Why wait to test the boards individually? This will take too much time. Let's just start running marathons from day one.

When the power was turned on, there was an immediate flash and smoke came rising out of the box setting off the fire detection systems. After the smoke cleared, it was discovered that two of the boards had their power supply lines mistakenly tied to the ground, which shorted out the power supply in the box, burned up traces in the backplane and completely destroyed the two boards as well as damaging other boards in the box. It was a disaster.

It took us six months to get back to the point of test where the box burned up. This time the team made sure everything worked at each step along the way before we moved on. We learned to sit up, crawl, stand up, walk, jog, run, sprint, and finally run a marathon.

▶ *Career Tip.* Have a development plan that is achievable with intermediate steps that will lead you to winning a marathon.

THINGS TAKE LONGER THAN PLANNED

Some engineers are extremely optimistic about the time it takes to accomplish tasks. They assume everything is going to go perfectly the first time and plan their tasks accordingly, leaving no time for rework, errors correction, or unforeseen delays. As a result, they are forever reporting to management they are behind.

When planning out your tasks, plan in extra time that will allow you to recover from setbacks and unforeseen delays. Some of the most common examples for engineers are: allowing for the time to get parts ordered and received, time to rework products for unexpected failures, and the amount of time it takes to get things done when several people are involved. As the number of people that are required to touch the product or help in some way goes up, the chances for delay increase dramatically.

Some projects I have worked on have come to a screaming halt when the procurement person left for a 2-week vacation in the middle of parts ordering. Or the quality department shut down the operations while equipment was recalibrated. Or the assembly materials like glue and lubricants suddenly went out-of-date, and the production line was shut down until new material was ordered and put into the line.

A classic example that happened to me was the need for a special oscilloscope to take measurements on a circuit to complete the testing. The lab did not have one but I was able to locate the correct model of the oscilloscope on the other side of the plant. I quickly ran over to the lab and grabbed the oscilloscope and started walking back to the lab thinking I only lost about two hours. Not realizing the plant was a union shop and to move any equipment in the plant, a union person had to move it. About halfway back to the lab, I was stopped in my tracks by the union steward who quickly pointed out the violation and quickly relieved me of the oscilloscope. To teach the engineer a lesson not to move any equipment, the union steward made sure the oscilloscope arrived at the lab about two days later. So what started out as an unexpected delay of an hour turned into a 2-day delay. It was all totally unplanned and definitely put me way behind my overly optimistic schedule. After that instance, I made sure to always plan a little padding in my schedule in case I ran into unforeseen delays.

▶ *Career Tip.* Plan a little extra time in your schedule to help with unforeseen problems and you will complete your work on schedule most of the time.

Most people would realize this is the case in life and adjust accordingly; especially after going to college and pursuing a degree. Very few students today complete their degree in a 4-year plan, even though they have a plan to do so. The norm is more like 5 years. This gets to the next example of being overly optimistic when being a team leader and developing plans.

OVERLY OPTIMISTIC PROGRESS SCHEDULING

Industry rewards those who can get things done quicker than planned and under cost. As a result, when engineers become team leaders, they will generate a project plan that assumes everything will go perfectly as planned or they generate a plan that shows completion of the project ahead of the actual due date. For some reason, many team leaders often conveniently forget how long things really take and come up with a schedule that has progress being made in terms of hours and days, when these tasks normally take days and weeks to accomplish. The schedule is designed to make the leader appear as a hero.

The reality of the situation is that the team leader has generated an overly optimistic project schedule with a very low probability of successfully completing it as planned. Then just several days into the plan, the first setback occurs putting the team behind schedule for the rest of the project with no hope of recovery. For the remainder of the project, because of overly optimistic planning, the team is reporting to management as being behind schedule. Not a good situation to be in, especially if the completion date for the project is several years away.

The key here is developing a project plan or schedule that is based on realistic task duration with planned recovery times if setbacks should occur. If you are assigned to a project that has an overly optimistic plan or schedule, you are going to have to discuss this with your supervisor so that they are aware of the situation.

▶ *Career Tip.* Schedule realistic progress.

NOT WILLING TO PUT IN THE EXTRA EFFORT

Most managers realize people can accomplish about 80% of what they plan to do. Many distractions and delays come into play during the normal work day that impede or slow down an employee's progress. To counter this, a manager may assign each employee about 120% of the normal daily work, hoping to get a full 100% daily output, if they are 80% efficient. Simply raising the expected output is a manager's only hope to accomplish work as planned.

Another method of making up for employee inefficiencies is by putting in extra effort. Engineering companies may have the unwritten rule that engineers do what it takes to get the job done, which is what they are paid for. Many engineering companies expect employees to put in the extra effort if it is required to stay on track or make up for errors. The combination of overly optimistic scheduling, running at 80% efficiency, and the belief that most engineers are willing to put in the extra effort generally results in management expecting employees to put in extra time. This usually translates to late nights and weekend work schedules. Oftentimes, you donate your time for the good of the company.

If you are not willing to put in the extra effort, it will be evident to your managers and coworkers. If you are looking for career advancement, this is not the image you want to portray to management.

SUMMARY

Being successful at the university does not guarantee success in the business world of engineering, and even very successful engineers can fail once they leave the university. The most common causes of engineers being unsuccessful are poor communication skills, poor relations with the supervisor, inflexibility, poor work habits, too much independence, and technical incompetence.

Being unsuccessful can take many forms. Some of these forms are stagnation at one level in the company, minimal raises, constant reassignment, poor assignments, and ultimately, being dismissed. Setbacks need not be final, and with effort you can turn your performance around. To do so, you must constantly work at improving yourself on a daily basis.

Have you identified any career actions you want to take as a result of reading this chapter? If so, please make sure to capture these ideas before you forget by recording them in the notes section at the back of the book.

ASSIGNMENTS AND DISCUSSION TOPICS

1 What is the definition of being unsuccessful for you?
2 Why are communication skills so important for an engineer?
3 Is having good people skills really a requirement for an engineer?
4 How can you avoid having too much independence?

REFERENCES

1. Burkin, Kurt, *US College Admissions Criteria*, University of Wisconsin-Madison, http://www.education.umd.edu/EDPA/faculty/cabrera/College%20Admission%20Criteria.pdf.

2. Barreto, David, and Quevedo, Antonio, "Student Profile of the Incoming First Year Class of the College of Engineering at UPRM and Their Academic Performance After Their First Year," presented at 2005 ASEE Annual Conference, Portland, OR, June 13, 2005.

3. University of Texas Tech Admissions website, http://www.admissions.ttu.edu/freshman/requirements/default.asp.

4. Purdue University admissions website, http://www.ipfw.edu/admissions/requirements/beginning.shtml.

5. College Admissions Info website, http://www.collegeadmissions.ws/college2.php.

6. McAlister, J., "Why Engineers Fail," *Machine Design Magazine*, February 23, 1984.

7. http://www.toastmasters.org/.

PART 4

IMPROVING YOUR PERFORMANCE ON THE JOB AND STANDING OUT FROM YOUR PEERS

CHAPTER 20

THE ENGINEERING PROCESS IN YOUR COMPANY
LEARNING THE BUSINESS

The engineering process is simply the steps that companies go through to design, build, test, and deliver products to their customers. This engineering process can be subdivided into two processes that your company utilizes and that you must learn to be successful. The first is the product design and development process. This process defines the engineering steps involved in creating products. It is the sequence of tasks engineers follow to design products, build early units, and eventually turn over to manufacturing for production.

The second process is the department flow process. This process identifies the typical departments of the company that contribute or help build and test the product. Each department provides a special service or capability to support in the product development.

Therefore, the engineer must know (1) the sequence of steps in product development and (2) which departments perform these steps. Understanding both of these processes for your company is an essential prerequisite to career advancement. Once you have learned these processes as defined by your company, you can use this information in a multitude of ways to benefit your career. Learning these processes is often referred to as "learning the business."

PRODUCT DESIGN AND DEVELOPMENT PROCESS

The design process varies from company to company and from product to product. A generic product design and development process is shown in Figure 20-1. This process or sequence of steps is shown in a timeline diagram.

The Engineer's Career Guide. By John A. Hoschette
Copyright © 2010 John Wiley & Sons, Inc.

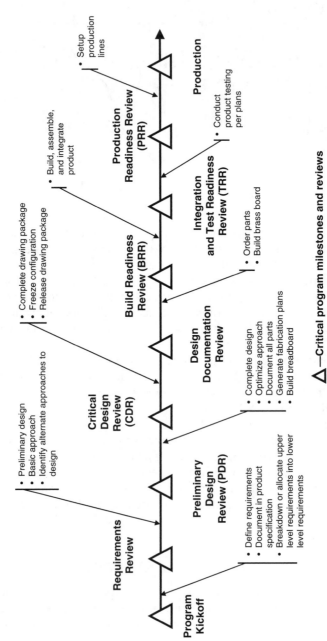

FIGURE 20-1 Product design and development process.

△ —Critical program milestones and reviews

The figure shows a timeline with the following milestones (left to right):

Program Kickoff

Requirements Review
- Define requirements
- Document in product specification
- Breakdown or allocate upper level requirements into lower level requirements

Preliminary Design Review (PDR)
- Preliminary design
- Basic approach
- Identify alternate approaches to design

Critical Design Review (CDR)
- Complete design
- Optimize approach
- Document all parts
- Generate fabrication plans
- Build breadboard

Design Documentation Review
- Complete drawing package
- Freeze configuration
- Release drawing package

Build Readiness Review (BRR)
- Order parts
- Build brass board

Integration and Test Readiness Review (TRR)
- Build, assemble, and integrate product

Production Readiness Review (PRR)
- Conduct product testing per plans

Production
- Setup production lines

Major program events are planned to provide a system of checks and balances as the product progresses from design to production.

The timeline starts with the first major event, which is usually the program kickoff meeting. The program kickoff meeting signifies the official start of the effort. The purpose of the meeting is to get all the team members together, in the same spot at the same time, to review the plans for the program. Generally, the overall program schedule is reviewed with significant events identified. The team organization is identified and roles and responsibilities are agreed upon. And probably the most important is that the technical objectives are reviewed and agreed upon.

After the program kickoff, the next step is "requirements" development. At this stage in the program, the engineer defines all the requirements the new product will be required to meet. The standard practice is to document these requirements in a product specification. Oftentimes, only a few or top-level requirements are known. Therefore, the engineer is required to work with the customer and other engineers to further develop all the requirements. In addition to finding new requirements, the engineer will break down or allocate down the top-level requirements to lower-level requirements.

A good example of breaking down a top-level requirement into subrequirements might be a products power source. The top-level requirement is for the product to be powered by 120 V AC. The engineer knows the electronics cannot run directly on 120 V AC, therefore lower-level requirements are identified for the product to contain a power supply generating $+15$ V DC, GND, -15 V DC, and $+5$ V DC. And even these requirements can be broken down further into requirements such as nominal current required for each DC voltage, ripple voltage, turn on surge current, and maximum rated current. This is a simple example to illustrate the point. Actual product specifications can contain hundreds of pages and may include thousands of requirements.

Have you ever written a specification before? Does your company have special requirements on the form and content of specifications? One significant time-saving step might be to ask other engineers for a copy of specifications they have written on previous designs. It is much easier to modify an already existing specification than it is to create one from scratch. Maybe your company has an online process that describes how to write a specification. If so, this is a great starting point.

▶ *Career Tip.* Check with other engineers for copies of specifications they have written or go online in your company to see what guidelines and specifications might already exist.

Most specifications contain language that not only has technical meaning but legal implications also. For example, the industry standard interpretation for use of the terms "will" and "shall." If the term "will" is used, it is taken to mean the requirement is a goal and it is not absolutely necessary the product meets the requirement. If the term "shall" is used, it is interpreted to mean the product must absolutely meet the requirement. A good example of the difference can be shown with a simple example. If the requirement is written, for example: the product will weigh less than 5 pounds. This is considered a goal and if the final weight comes in slightly over 5 pounds, it is not a problem. If the requirement is written, the product shall weigh 5 pounds or less. This is considered a hard and fast requirement. If the product comes in slightly over at 5.02 pounds, it means the product does not conform to the requirements and redesign is necessary. The difference between the use of "will" and "shall" is significant and can end up costing the company thousands of dollars.

Another significant and often overlooked skill the engineer must acquire is the ability to write a requirement that has a tolerance band and is easily measurable and verifiable. This may sound easy, but oftentimes it can be extremely difficult to do and can result in very expensive testing. A good example of a well-written requirement is the following: the product shall weigh 5.0 ± 0.2 pounds. The tolerance band of ± -0.2 pounds allows variance in the product weight with the hope all the products final weight will be within the tolerance band. Also, most scales can easily measure to an accuracy of 0.01 pound. Therefore no special or expensive weight scales will be required to verify performance.

A good example of an easily understood requirement that can be difficult to verify is automobiles' gas consumption or miles per gallon performance. It is easy to specify a requirement of so many miles per gallon (MPG) performance but testing to verify it meets the requirement can be very difficult and expensive. Many conditions must be considered when testing this requirement, including size of the engine, horsepower, size and weight of the car, number of passengers, road conditions, gasoline grade, accelerating and decelerating methods, idle time at stop light, and so on.

▶ *Career Tip.* Make sure the requirements you place on a product can be measured and easily verified through the simplest testing possible.

At the end of the requirements definition stage comes the requirement review or systems requirements review (SRR). This meeting is primarily run by the systems engineers of the program. Its main purpose is to systematically go through all the requirements the product will be built to, and get an agreement from the customer and/or management on these requirements. Generally, each engineer is asked to present a list of their requirements and the reasons behind selecting these requirements. This meeting may last only a few hours or could take as long as a week. It may be a very formal meeting with the customer or informal with just internal staff. The depth and style of the meeting is generally left up to management and the customer to decide. The SRR signifies the end of requirements phase and the beginning of the product design.

The next step in the process is to create a preliminary design of the product or subsystem. This is only a preliminary design where you identify alternative means for accomplishing or meeting your design requirements. During the preliminary design stage, basic analysis of options is completed and a preferred approach is selected for further development and refinement. At this stage, the engineer has a good idea of the approach and completed some preliminary design with analysis and modeling to support the design. The design is usually documented in draft or preliminary sketches with basic modeling completed to support the conclusions presented.

Remember those lab notebooks the university tried so hard to get you to develop and maintain during the lab classes? You documented the design, showed analysis, identified equations you used to model performance, and finally, all the graphs you generated showing the simulation results. This is exactly what you should be doing to document your design. I have seen engineers who have taken the time to maintain incredible design notebooks and their efforts paid off when it came time to design reviews. They simply presented all the modeling and simulation results contained in their notebooks. They were organized and came across as technically in control.

▶ *Lesson Learned.* A well-documented and organized design notebook is worth its weight in gold.

I have also seen engineers who thought they could keep everything in their head and maintaining a good notebook was not necessary. Consequently, when it came time for the design review, these engineers were unprepared and appeared technically unorganized. These engineers generally crashed and burned during design reviews.

During the design phase, the engineer may have to go through one or more formal design reviews. The intent of the design reviews is for the engineer to present his design for review to make sure he or she is meeting all the requirements and that all parts will fit together when the product is assembled. Design reviews are often attended by senior engineers and management in the company. Their purpose is to review the design and ensure that ideas will work. They will provide helpful hints and suggestions on how to improve your design. They are also making sure that lessons learned from other products are being applied and that mistakes from other products are not repeated. Often the customer will participate in the design review.

The end of the preliminary design stage is usually culminated in a preliminary design review (PDR). The PDRs are often held at the beginning of the program to make sure it gets off to a good start. At PDRs, the engineers present, to the customer, the plans for development and a preliminary design for the product. The purpose of the PDR is for all designers to review their design and show how the design is meeting the requirements of the specification. It is also to check that all designs are on track as well as a check that all subsystems and all interfaces will work. PDRs can be very short and informal lasting 1 or 2 hours or they can be very formal and lasting several days with customers in attendance.

Design reviews are an excellent opportunity to shine. A well-prepared PDR may contribute and shorten the time to your next promotion. It will make your supervisor look good and the team look good as well. It's worth the extra effort to make sure everything has been accounted for and the presentation is well organized. This is the time to show them your best. On the other hand, a poor PDR can significantly hurt your chances for a promotion. It's exactly the wrong time to look bad. So keep this in mind when you are preparing for a design review.

▶ *Career Tip.* A well-prepared and presented PDR presentation can help accelerate your career development. It will make you, your team, and your supervisor look good!

At most design reviews, there is usually a lot of heated discussion on what the correct approach should be. Your design will receive a lot of helpful criticism whether you like it or not. Remember not to take the criticism personally and remain calm! It's difficult when it's your hard work being criticized. But remember, others may have more experience and usually have good suggestions. The sign of a good senior designer is not taking the criticism personally and remaining calm. The best thing to do is find out from others how they would improve it and then use their suggestions. This will accomplish two things. First, you get them to participate in the design (this is especially good if a customer is participating). Second, you will show that you are cooperative and can deal with change. Both are excellent qualities for engineers to have.

▶ *Career Tip.* Gratitude expressed for suggestions or recommendations during a design review shows you have the right attitude.

If you really want to impress people, thank them for their suggestions! Everyone likes to hear the words "thank you." It makes them feel as if they have contributed something useful, a feeling even the most senior-level executives enjoy. Finally, when you are a senior engineer reviewing the design of a junior engineer, remember how it felt when you were in that position. Choose your words and criticisms carefully. You should be honestly trying to help the junior engineer and not just trying to make your importance known to others.

Oftentimes during the preliminary design stage, the engineer will be required to fabricate critical parts of the product to show that the concept they are proposing is feasible. One great design tool is use of prototype models of the final product. There are several different types of models that can be built. One prototype is called a nonfunctional mockup model that has the proper size and form factor of the product, but may contain little or no functionality. These models are great for determining size allocations and arrangement of major components and often used by the mechanical packaging engineers. Another prototype model is referred to as a breadboard prototype. Breadboard prototypes usually demonstrate functionality but may not be close in size, weight, and form factor. Finally, there is a brassboard prototype, which is as close as possible to the final product size, form factor, and functionality.

▶ *Career Tip.* Building and demonstrating a breadboard at the PDR is a career accelerator.

The next stage in the process is the critical design stage. During this time, the product design is completed and all the details are finalized. The approach has been optimized and all aspects of the design are modeled and documented. The drawing package or documentation package is completed. The documentation package contains all the formal or company-generated drawings and processes that will be needed to build and test the product. To document the design, part drawings and lists are generated. Plans are made for internal fabrication of parts or external procurement of parts. Everyone on the design team is finalizing their design to get it ready for build.

The critical design stage usually ends in what is referred to as the critical design review (CDR). At the CDR, the final design for the product is presented and approval is received from the customer and/or management to start the actual building and testing.

Once again the engineer must present his or her final design for review and approval to proceed into build. The CDR is the most critical review of the program; it marks the point where the program leaves the design stages and proceeds into product build stages. Or as we often say, it is the point where we leave the paper phase and enter the hardware build phase. Generally at this stage, the company is committing large amounts of funding to buy material and start fabricating parts. Therefore, the customer and management have a vested interest to make sure all aspects of the product performance and build plans have been covered.

Typically, management is looking to make sure the following has been completed:

1. All requirements are being met.
2. The design documentation is complete.
3. The design can be built and is testable.

4. All parts can be procured in a reasonable timeframe to meet the schedule.

5. Overly expensive parts are not required.

6. The customer is happy with the design.

Nearly every company I have interfaced with over the past 30 years has guidelines on how to hold and conduct CDRs. The companies have checklists and memos on lessons learned about conducting CDRs.

▶ *Career Tip.* To save time and significantly enhance your presentation, check into your company guidelines and policies regarding conducting a CDR. Review other programs' CDR packages.

Do you have a company website you could check? How about your mentors and other engineers who have participated in CDRs before; could you ask them for advice? Could you ask people on other programs for a copy of their CDR packages? Have you ever attended another programs' CDR? These are all time-saving and career advancement steps that you can take to help you when it comes time to present at CDRs.

Since the CDR is probably one of the most important reviews of a program and often the one review most attended by upper management, having an outstanding CDR presentation is critical for career advancement. Do not wait until the last minute to start on your charts. Complete your presentation early and then have your peers review the presentation. It is much better to discover your errors and left out items in private instead of in front of the customer and your management. Be prepared for the tough questions, and most of all, have action plans identified for any issues that might exist. As a senior reviewer for many CDRs, my danger alarms went off whenever the engineer was unprepared or claimed everything was great. The engineers I respected and knew had things under control were the ones who identified potential problems with the design and future risk reduction plans to mitigate the risks.

Critical design reviews are like combining senior labs, finals, and thesis dissertations all into one. Your work is reviewed in detail by everyone and your career depends on you getting approval or passing!

Granted that you pass the CDR, the next step is to complete the drawing or documentation package and start building and ordering parts. At this stage, the design portion of the program is over and the emphasis changes to build and assembly of the product. Some companies complete this documentation phase by having an official review of the drawing package to make sure all aspects of the design have been documented and we know what is going to be built. Oftentimes, the computer-aided design (CAD) department handles most of this and prepares the drawing package. Special notes are added to the drawings calling our standard processes and procedures to be followed during fabrication of the parts. The drawing package is released by submitting the drawings to a standard review process and then getting everyone to sign off on the drawing package. This signing off is referred to as "release of the drawing package."

To release drawings, there are several engineering standards utilized in industry. These standards define what should be on the drawing and the levels of control. The control levels define who should review and sign it before it is released as well as who has the authority to change it. At the lowest level of control, it is signed by the engineer and CAD person only. At the highest level of control, the CAD person, the engineer, the supervisor, the program manager, quality control, manufacturing, and even the customer, review and sign drawings. Are you aware of your company's drawing release levels?

▶ *Career Tip.* Knowing and understanding your company's drawing release process and levels for release is an essential career skill.

Most engineers look upon a drawing release as a genuine nuisance and avoid it, if they can. My recommendation is not to avoid it, but use it to your benefit. Once all your drawings are complete, schedule some time with your supervisor to go through them and sign off on them together. This is an excellent opportunity to show him or her all that you have accomplished. A neat, well-organized drawing package is very impressive. It also gives you a chance to show all the problems you've solved. In addition, you identify how the suggestions made during design reviews were incorporated and how you have things under control. These are all things that can highlight your contributions as an employee and shorten the time to your next promotion.

▶ *Career Tip.* Showing your supervisor a neat and well-organized drawing package is well worth the effort and a career accelerating move.

The release of the drawing package is generally accomplished by what is referred to as a configuration control board (CCB) review. The CCB is comprised of the program manager, lead engineer, quality, CAD designers, and others. The engineer must present their drawing package for review and sign off. This is another career advancement opportunity if the engineer is well prepared with an outstanding drawing package. Having a thorough drawing package is clear indication that the engineer understands the processes and can perform. If you are required to present your drawing package at a CCB for release, are you thoroughly prepared?

▶ *Career Tip.* Have senior people in the company review your drawing package prior to presenting to the CCB.

Another tip is to check with other engineers on the program who have already presented to the CCB to learn what questions and changes the CCB is making to drawings. Having all the typical corrections and changes already in your drawing package prior to showing the CCB is another means of looking like a top performing engineer.

As shown in Figure 20-1, during the next stage in the process, the parts are placed on order, and once they arrive the build process is ready to start. Oftentimes, the company will hold just prior to the start of build of the first product a build readiness review (BRR). The purpose of the BRR is to make sure all the resources are in place to start the build. The parts will all arrive on time, the parts will be inspected and brought to assembly areas. The assembly instructions have been completed and the tools and people to build the products are in place.

Once the product build is complete, the next stage is testing the product. Prior to start of actual product testing, a test readiness review (TRR) is often held. The purpose of the test readiness review is to again make sure all the resources for testing are in place and the test plans and procedures have been completed. The test department and test team understand the testing to be completed, the type of test equipment needed, and the data to be collected.

Upon completion of the test stage, the program and management has a difficult decision to make. The decision of whether or not the product is ready to go into full-scale production. To aid in this decision, the process calls for conducting a production readiness review (PRR). The objective of the PRR is to review all test results to make sure the product is functioning as planned. It is also a means of reviewing and making sure the factory is ready to begin production. Are the resources in place to build and test the product, have all the assembly and test procedures been defined? Are the product packaging and delivery systems in place? Usually upon successful completion of the PRR, production is started.

The engineering product design and development process described here is generic and greatly simplified compared to what actually goes on in most corporations. The purpose of sharing this was to alert the engineer that all companies have some type of process for designing and developing products. Some companies have no formal processes written down, but just the same, you are expected to ask and learn the process. Other companies have very formal processes that are well documented and their manuals contain thousands of pages describing in detail every little aspect of the product development process for the company. No matter what the case, informal or formal processes, your challenge is to become familiar with every step in the company's engineering design and development processes.

One of the best resources outside of your company for getting information on the product development processes I have found is through engineering societies. One organization is the International Council on System Engineering (INCOSE) whose website is http://www.incose.org/.

Another great resource is the Program Management Institute and their website is http://www.pmi.org/.

PRESENTATIONS AT PROGRAM REVIEWS: CAREER ACCELERATORS OR BREAKERS

I would like to point out to you that the engineering product development process as described in Figure 20-1 has a minimum of six to eight reviews identified. The engineer is required to present his design and plans at each of these reviews. Therefore, I highly recommend that the engineer take several classes in making technical presentations. The reason behind this is what I refer to as the good charts–good engineer, bad charts–bad engineer syndrome.

This means that the audience forms an opinion of the engineer's capabilities on the basis of their presentation skills and charts. If the engineer has great charts and has excellent presentation skills, the audience generally considers the engineer's work to be excellent and the design excellent.

On the other hand, if the engineer has poor charts that are confusing and hard to follow and also does a poor job presenting the material, the audience naturally assumes the design is inferior and the engineer does poor quality design work.

▶ *Career Tip.* Poor charts are equated to poor engineering designs; great charts are equated to great designs.

I do not understand this phenomenon, but I have seen this affect my entire career as an engineer. A brilliant engineer with superior product designs gets dismissed all because they could not make a simple presentation to management or the customer. I have also seen engineers with poor design skills and inferior designs make great presentations with excellent charts and everyone in the audience clearly took the design to be impressive (Figure 20-2).

▶ *Career Tip.* Invest time in learning the use of proper software to create successful technical presentations. Take courses in making presentations.

GUIDELINES FOR BETTER PRODUCT DEVELOPMENT

The design process varies from company to company and from product to product. I can only share with you some generalized guidelines that will help you design a better product. The following design guidelines are very useful and should apply to most design situations. Simply reviewing this list as you develop your design should help eliminate mistakes and identify problem areas.

FIGURE 20-2   Let people see you at your best.

Guidelines for better product development:

1. Before you start the design, write down all the requirements placed on the design or product. Summarize them in a consolidated table or matrix so that it is easy to review.

    This list should include as a minimum:

Performance requirements	All inputs or outputs
Clearly defined interfaces	Size constraints
Weight constraints	Safety constraints
Operating constraints	Special conditions
Reliability (MTBF)	Safety
Power constraints	Test requirements
Customer use	Rework and repair requirements
Human engineering	Maintainability

2. Get agreement on the product requirements from your system engineer, project engineer, program manager, and/or your supervisor before you start designing. This is a must! If you design your product to the wrong requirements, you will stand little chance of having it approved and you will waste valuable company time and money.

3. Generate a list or table of different approaches that you might use for the design. The table should contain acceptable and poor points for each design. Remember there is always more than one way to do the design. By generating different approaches you can make trade-offs to determine which design is best. Also, by generating different approaches you are not locked into only one.

4. Now show the various design approaches to people in your organization and find out what they think. You will get some good tips along the

way. Also, by showing your approaches to people, you will get a feeling for what is good and bad. You should show these approaches to your supervisor. If he or she likes one and disapproves others, you have a good indication as to which approach to take. It is best to find out which approach stands the best chance of approval rather than spend a lot of time and energy on an approach that will never be approved.

5. Analyze every detail of your design. Model everything you can about the design. The modeling should quickly identify the problem areas for you. By modeling the expected performance of your product, you can show how the product meets every requirement on your list. Engineers will often start reinventing the wheel by generating all their own models. Before you start modeling, find out which models exist in the company. A good place to start checking is with the senior-level engineers in the company. Chances are they already have a model you can adapt to your needs with minor modifications.

6. Build a mockup of the design if possible. A mockup usually does not function but has the correct form or shape. It will help you to visualize how the product comes together. It has been said that one picture is worth a thousand words. I have found that one mockup is worth a thousand pictures! Mockups can identify problems in advance and help you correct them early in the design process.

7. Identify contingency plans in case something goes wrong. If you base your design on a specific part and the part is suddenly no longer available, what are you going to do? Check to make sure all the raw material and parts are available to build it the way you have designed it. A good practice is to find two different suppliers for each part. This way, if one goes out of business or no longer produces the part, you have a backup.

8. Keep track of product costs as you design it. A well-designed product is no good if you cannot sell because it is too expensive.

9. Before you commit to the design, develop a build and test plan. This will help you to quickly identify whether you have all the resources to build and test the product. If you don't have the resources, you had better let the company know in advance so that they can make arrangements to get them.

10. Put plenty of design margin into your design. If your product must survive a five-foot drop, design it to survive a seven-foot drop. If your product must operate within five seconds of turn on, design it to operate within three seconds of turn. Design as much margin into your product as it will allow you. This can save a great deal of expensive and unnecessary redesign later on.

11. Write down and document every aspect of your design. Keep good design notes. You are the only one who has a complete understanding of the design but remember your coworkers must build and test it.

They need good notes and documentation to do this. It is better to make a mistake on the side of too much documentation than too little. Take the time to make sure your documentation is accurate. Check all numbers twice.

12. Build prototype models before you start to build the final design. A prototype will allow you to see any problems in advance and give you time to change things. It will also alert you to possible potential problems.

Remember, these guidelines are very general and you have to adapt them to your particular company or product.

KNOWING WHAT TO DO WHEN IT COMES TIME TO BUILD AND TEST YOUR PRODUCT

Before starting to build your product, it is good to generate a checklist of everything that you might need during the build. Some companies actually hold a build readiness review prior to the start of building the product. The following is a good checklist to go through prior to commencing the build.

1. All parts have been received and are in the stockroom.
2. A build procedure has been written showing the build process flow step by step.
3. All inspections and tests to occur during the build are defined and agreed upon prior to start of build. Datasheets for recording results of inspections and tests are available.
4. All hazardous steps have been identified and people informed of any dangers.
5. Technicians are available and trained for every step of the process. Training or practice on scrap parts for the more difficult assembly steps is a good way to reduce rework.
6. All necessary documentation (drawings) is available for the technician.
7. Persons to contact in case of problems have been identified.

These are some of the concerns that should be addressed prior to start of build of your product. You will have to generate your own list and tailor it to your company's or product's need.

The test and evaluation phase of the program follows the build process. The best way to prepare for this step is to generate a complete checklist of everything that should be done to test the product prior to shipping. The following is a generalized checklist that should help you in preparing your own.

1. Generate a test requirements document that identifies parameters to be measured, requirements of test equipment, and pass/fail criteria.
2. Generate a test plan that calls for verifying every requirement listed in the design specification you generated. Get inputs from test personnel on the tests that you plan to run.
3. The test plan should show the order of the tests to be performed, requirements to be verified, plans in the event of failures, and data-collected plans.
4. Generate a test procedure that defines all the tests to be performed in the exact sequence.
5. For each test to be performed have the following been identified?

   Test facility available

   All test equipment in place and calibrated

   Test objectives for each test identified

   Datasheets for recording results completed

   All test, quality, and inspection personnel notified

   Test procedure written, reviewed, and approved

6. Identify the test director or someone responsible for each test.
7. Establish contingency plans should failure occur during the test.

▶ *Career Tip.* A good engineering practice is to witness and monitor all aspects of testing.

The engineer should be readily available to answer questions as they come up and provide direction to the test team as required. In addition, he or she should be comparing the test results against the modeling results obtained from the design modeling phase. A summary of the documentation identified in steps 1 through 3 is shown in Figure 20-3. Generating this documentation is highly recommended, as it will be needed to successfully control the testing phase.

Generally, there are four types of performance verification methods that are utilized during the test phase. These performance verification methods are:

1. *Analysis.* This is usually a mathematical modeling of the product to show compliance with requirements. (For example, safety analysis of product handles.)
2. *Inspection.* This is usually a physical and visual inspection to verify performance of the product. (For example, inspection of labels on the product to ensure correctness.)
3. *Certification.* This is verification of performance by receipt of certification from manufacturing. (For example, certification on how pure certain chemicals were that were used in the build.)

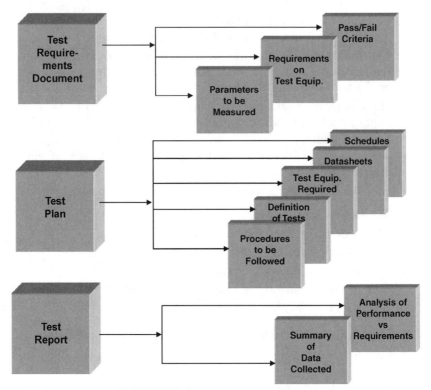

FIGURE 20-3   Test documentation.

4. *Test.* This is verification of product performance by operation and/or measurement of an item, usually requiring instrumentation to record and evaluate measured data. (For example, measuring the weight, size, or power consumption of a product.)

Documenting and Controlling Tests Are Essential

Often, it does not seem important to know exactly how the performance of a product is verified. However, if one considers the costs associated with the verification method, one quickly realizes how important the test method can be. Do not forget the cost factor associated with each test method. Usually, tests are most expensive and certifications are the least. Make sure you know the costs of performing the different tests in your company and minimize the cost of testing wherever possible.

In addition to defining what types of tests the product will be subjected to, there are different groups or combinations of tests that a product may be subjected to prior to shipping it to the customer. These groups of tests often include:

1. *Design Verification or Development Tests.* These are groups of tests that the engineer performs on the product to verify all aspects of the design. Most often, this is the largest group of tests. Some of the tests may be performed only once; since all the products are designed exactly the same, all others will pass it. Other tests in this group may be performed several times before the product is shipped and/or repeated for every product. (For example, measure the power consumed by each television set.)

2. *Qualifications Tests.* These are usually sets of tests that challenge the environmental performance of the product. (For example, vibration testing, drop testing, humidity testing, temperature testing, etc.)

3. *Acceptance Tests.* These are groups of performance tests that the customer usually specifies to be run on every product prior to shipping.

4. *Burn-in Tests.* These are groups of tests that the customer usually specifies to ensure that no instant or short-term failures will occur. These tests may include combining several tests into one. (For example, operate for 50 hours while cycling between hot and cold temperatures.)

A very good engineering practice is to develop a matrix showing exactly how and where every requirement of the specification will be met. An example of this is shown in Figure 20-4. The column on the left identifies the product specification requirement. The next four columns under "Test Method" identify the different test methods to be used to verify the specification requirement. The next four columns under "Test Group" identify the tests to be run. And the final column identifies the test plan paragraph that describes the test to be performed.

This type of table makes an excellent communication tool. It provides a condensed summary of all testing that is going to happen. It is great for the test team to have and can help clearly identify to the customer what is going to happen. Figure 20-4 is a generalized table that you should adapt and tailor to your product or company.

Finally, after the tests are completed, a test report and lessons learned summary should be written. The test report should contain a summary of the data collected as well as an analysis of performance versus requirements. Lessons learned should identify solutions to problems that were discovered during the test phase of the program.

▶ *Career Tip.* Taking the time to review test results with your supervisor can be very beneficial to your career.

Take the time to explain to him or her how the product performed against predicted results and the lessons learned. Make sure you point out to him or

Product Specification Paragraph	Test Method				Test Group				Test Plan Paragraph Describing Test
	ANALYSIS	INSPECTION	CERTIFICATION	TEST	DESIGN	QUAL TEST	ATP	BURN IN	
3.1 Power On/Off				X	X	X	X	x	4.1 Power On/Off
3.2 Paint Finish		X					X		4.2 Paint Finish
3.3 Material			X				X		4.3 Material
3.4 Hazard/Safety	X				X				4.4 Hazard/Safety

FIGURE 20-4   Requirements verification matrix.

her problems you overcame and the improvements you identified for future projects. Don't overwhelm him or her with data. Make sure it is neat and organized. Remember, they are forming ideas about your performance and your abilities. Take time to polish the report; now is the time to show your supervisor your best.

Career Benefits To Understanding How Your Company Does Business

Just as important as learning the engineering design process is learning how the product flows through the different departments of the company in its journey to your customers.

A typical department flow process is shown in Figure 20-5. This department flow process has been generalized and will vary from company to company as well as from product to product within any one company. The example was chosen since it is fairly typical of most companies. This example represents a starting point for determining the department flow process in your company. Study the example, then use it as a guide for developing your company's department flow diagram.

The engineering process starts with a product idea or a request for a proposal from the customer. If the product idea or business opportunity is worth pursuing, the company forms a proposal team. The team will write

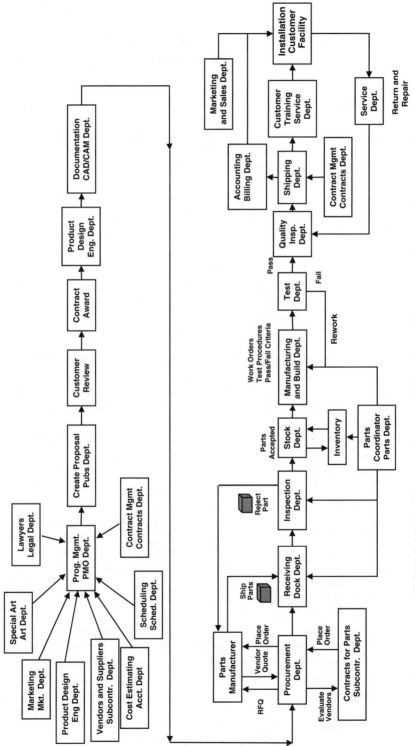

FIGURE 20-5 Diagram of a typical department flow process for a product.

239

a proposal to build a particular product for a certain customer, or to improve an existing product line, or to develop a new product. The proposal team includes people who are engineers, program managers, technical writers, accountants, lawyers, artists, contract specialists, and marketers, just to name a few. The marketing people provide information obtained from the customer about the product requirements and customer's desires. This information includes the customer's specifications and requirements, the time frame for completion, and the customer's budget. The engineers determine the technical approach to the product. The financial team assembles the cost associated with building the product.

The proposal is divided into various sections and the team members are assigned sections to write. Usually, the proposal effort is led by the program manager and run out from the proposal or publications department.

▶ *Career Tip.* If you become a proposal team member, your writing skills become critical. Take courses in technical and proposal writing. As a starting point, get copies of previous proposals for ideas on writing style, format, and technical data that need to be presented.

Once all the customer information is known, the team formulates a proposal to build a product to meet the customer's need for the least amount of money and deliver it within the time constraints. The engineers describe the product design and operation. The accountants compute all the costs associated with building the product.

Getting everyone to agree what should go into the proposal is no easy task.

Once the proposal is completed, it is submitted to the customer. The customer can then evaluate the proposal for its technical merit, cost, and schedule. This review process can take anywhere from several weeks for simple contracts to months and even years for large developmental programs.

If the proposal is accepted and your company is awarded a contract, the next step is the product design phase. For the product design phase, management organizes a team of engineers to develop the product. Often the teams are organized around a program management office (PMO). It is the responsibility of the PMO to execute the contract and make sure work is carried out as planned. These PMOs are led by a director who has profit and loss responsibility for the program. The director will have program managers to assist in running the program. The program managers are responsible for organizing the work. They determine the tasks to be accomplished, the order in which they are to be done, and the schedule for getting the work done by the design engineers.

After the program has been thoroughly planned out and the team members identified, the design work is ready to start. This step is known as product design. During this step, every detail of the product is designed and performance modeled. Design trade studies are performed showing different ways to build the product. The pros and cons of the various methods are identified. During the design phase, the engineer is utilizing everything learned in school and then some.

Once the design is complete, the hand sketches and rough drawings are brought to the drafting or CAD/CAM department to draw up the parts of the product. The CAD department documents the design by creating a drawing package. The drawing package contains a complete set of drawings that document every part in the product, how it is built, and how it is to be assembled. Drawing packages may contain anywhere from 10 to 10,000 prints, depending upon the complexity of the product.

If you have an opportunity to choose with whom you work in the CAD department, try to get the most senior CAD person. The reason is that he or she has seen many designs in their time and knows what has worked and what has not. The senior CAD designer has a multitude of little tips to make the design more producible, perform better, and have a better chance of succeeding. If the CAD person finds something that improves the design, make sure to highlight their contributions by enlightening their supervisor of their input. Believe me, I've seen the CAD department find a lot more of my mistakes once they knew they were going to get credit for it.

▶ *Career Tip.* Select the most qualified CAD/CAM designers to work on your project. Solicit their inputs and highlight their contributions to management.

After drawing release, the prints are used by the procurement department or "purchasing" as it is sometimes called. The procurement department uses the drawings to obtain bids from other companies to build the parts. The procurement person contacts companies that are interested in building the part and requests a quote from them. This is referred to as a request for quote (RFQ). The manufacturer sends the bid or cost quote back to the procurement department for review and approval. Usually, the procurement department and the engineer decide whom the bid will be awarded to. When a decision is made, a purchase order is placed with the chosen company.

Getting the purchase order signed off is an exercise that will try any engineer's patience. A typical purchase order will usually require the signature of four to five different people: your supervisor, your supervisor's boss, your program manager, your procurement agent, and the parts manager, just to name a few. On larger programs, it may also require the production engineer, the quality engineer, and the program accountant.

▶ *Career Tip.* Make sure you understand all the steps and forms required to get a purchase order approved for your parts. Getting your part orders through the system first can significantly shorten your time to build.

For those parts fabricated within the company, a drawing is brought to the department that does the fabrication and the department normally starts work immediately. No purchase orders are necessary. This is much simpler, so most engineers will try to fabricate everything in house.

To keep track of all the parts being fabricated, most program managers utilize a "parts coordinator" to keep track of everything. The job of the parts coordinator is to place the orders, track delivery dates, push the parts through incoming receiving and inspection, and get them into the stockroom as quickly as possible. If you want to have your parts built, first it pays to make friends with the parts coordinator. Remember, all other engineers on the program are also trying to get that same parts coordinator to order their parts. I have been on programs where part coordinators are handling 5,000–10,000 parts. So getting the one part you forgot to order and desperately need may not be on the top of their priority list.

Parts coordinators can make or break you. If you get on their good side, they can make sure your parts always get top priority. If you get your parts first, you stand a better chance to finish first. If you get on their bad side, you can expect serious delays.

The point I'm making here is that other people in the company can affect your performance. You need to understand the function of each department and how it can affect you. You need to stop and think: if I win the argument and make an enemy, am I really winning?

There is a saying, choose your friends well. It's also good to choose your enemies well so that they can do the least amount of damage to your career. Chances are you can make enemies with some departments and your career will not suffer. But make an enemy in a department that is key to getting your job done and you may be greatly limiting your career.

When parts ordered from outside the company arrive, they are usually logged in and sent to inspection/receiving. In the incoming inspection and receiving department, the parts are usually inspected to make sure that all parts were built to the required specifications. Most companies do this for two reasons. The first is to detect bad parts before they are used to build the product. This can significantly reduce rework later on. Second, it gives the company a much stronger case for return and free replacement of bad parts by the manufacturer. This is especially true if lawsuits can be the result of poor inspections.

In order for the quality inspector to examine the incoming parts, they must know what to look for. To get this answer, the inspector usually obtains a copy of the part drawing from the company print room. This is called "inspecting the parts to drawings." These are the same drawings that you had the CAD/CAM department create and release. As most quality inspectors say, "No print to inspect, then no parts get through." It pays to have all your prints completed and released prior to inspection.

There is a very good reason for the incoming parts inspection even thought it might slow things up. If the parts are inspected and found to be nonconforming to the requirements on the drawing, they are rejected. They are usually returned to the manufacturer for corrective rework. Sometimes this rework can cost a company significant overruns and result in large lawsuits that are settled in court. The legal judgments are often based on the drawings and whether or not they were fabricated to the drawing or not. Can you see why the documentation is now so important?

From receiving inspection the parts usually go to the stockroom where they are logged in and stored. Make sure your parts get to the right place once they leave the inspection department. Parts have been known to get lost between the inspection department and stockroom.

Most companies have an engineering build or manufacturing department that actually builds the product. These departments are often referred to as "model shops" for small quantity builds. This is where a breadboard or brassboard model of your product is first fabricated. Using the prints and assembly procedures that you have developed, these departments quickly assemble the product. There are two rules of thumb during assembly that can help you get through the build process easily and quickly. First, be available for the technician so that they can ask questions on how you want it assembled. Remember, if you are not around, the technician will assemble it the way they think you want it. This can lead to a lot of mistakes. Second, be ready to make changes if necessary during the assembly process. Chances are your design was not perfect and more than likely changes and modifications will be necessary.

▶ *Career Tip.* Make sure you are readily available during product fabrication. Check progress on a regular basis.

It's best if you are on the spot and can quickly remedy the situation; if you are not available and the technician has to go to your supervisor or other coworkers to find out how to solve the "glitch," it doesn't look good and can hurt your position. Solve problems quickly and as soon as you can. Spend as much time as possible in the assembly area during this time.

Just as in the CAD/CAM department situation, request the assistance of the most senior technician, if possible. He has seen a multitude of products and knows hundreds of shortcuts or helpful hints; make him your friend and not your enemy. Remember some technicians like to point out all the mistakes of the design without mentioning its good points. It may be tough to sit there and listen to criticism on your design, try not to take it personally, remain calm. Ask for ideas on what can be done to improve things. Be sure to give credit where credit is due. Again, it does help if you point out to the technician's supervisor all the good ideas that were contributed and how they are helping the design.

▶ *Career Tip.* Request the most qualified technicians support the build and test of your product. They bring a wealth of knowledge to your design.

At the completion of your product build, write up any lessons learned from the build process. Be sure to update your build process flow with any changes you have made to it. Updating and improving the build process flow is extremely helpful the next time you plan on building the product. Make sure you share the lessons learned with your supervisor and other people in the organization. This may save them from going through the same problems in the future. Also, as others share their lessons learned, you will learn faster and more efficient ways to build problem-free products.

After completing the build of the product, it is now ready to go to the test department. The test department typically has the responsibility for conducting all the tests you have specified. The test department will assign technicians to work with you to perform the majority of testing. Typically, the testing on the very first unit is done mostly by the engineer with the technician observing. By the third unit, the technician is generally doing the majority of the testing with little or no assistance from the engineer. Generating test plans and procedures from scratch is very difficult.

▶ *Career Tip.* Obtain copies of other test procedures developed and use as a model. Make sure you have datasheets showing the data to be collected during the testing.

Prior to shipping any products, the quality department joins the test team to witness and ensure quality products are going to be shipped. The quality department is usually required to witness all tests and sign off on all paperwork that the product is ready to be shipped. During my career, I have found that most companies require between 8 and 10 different types of paperwork to be filled out prior to releasing the product for shipment. Be prepared for this mountain of paperwork.

After successfully completing the testing of the product and filling out all the paperwork, it is time to ship it so that the company receives payment. Once you are ready to ship, several other support organizations get involved. As Figure 20-4 indicates, marketing, accounting/billing, shipping, and the service or installation departments get involved. Make sure they complete their job in a timely fashion so that you can still meet the schedule. The engineer should follow through and double-check everything.

▶ *Career Tip.* Team photos of the people who worked on the project generate an enormous amount of good will. Well worth the effort!

Here are some tips that may help company morale and benefit your career at the same time. First, before you ship the product, get photographs of the product for future customers. Second, if you can get the team together, have a team photograph taken. Give copies of the photograph to each team member.

See if you can have the company newspaper publish a story about the team and the product. Get a picture of the customer receiving the product if you can. And, finally, submit people who have helped you; they will be more willing to work with you the next time. Ask your supervisor for details on how to submit awards.

▶ *Career Tip.* Most companies have money for team awards. Submitting your team for an award is a very good career move and generates goodwill.

The product department flow discussed was a very general one. You can use it as a starting point for mapping out the product department flow in your company. By generating a flow diagram of the departments that become involved during a product development, you will have a better understanding of how your company does business, as well as which products and functions are key to its survival. By knowing the department flow process, you can contact other departments well in advance and schedule their help for your product. This allows you to get things done more efficiently, for less money, and on time, all good reasons for raises and promotions.

I have shadowed only three boxes in the department flow diagram. These were the only boxes/departments I had received training during my tenure at the university. Having to deal with all the other departments, people, and processes were all new to me. Learning what all these departments do and interfacing with all the people in these departments required I learn a complete set of new skills. Are you prepared to do the same?

You should have realized by now that 95% of product development is done outside your department and you must depend upon others to help you. Do you have the good interpersonal skills needed to make this happen? If not, take some classes.

▶ *Career Tip.* Take classes on dealing with difficult people. This is a skill required of all senior or lead engineers.

Pushing a product through large companies or organizations can seem impossible at times. Several senior engineers have shared this advice with me on occasions when I felt like giving up. Their words were

▶ *Career Tip.* "Don't let them wear you down; when you know you are right, keep on going."

Product design, build, and testing is a complex process. No one person can do it all, but with the help of teammates you can do it. Develop guidelines or checklists for each step to help you ensure that everything will be done. By generating guidelines or checklists jointly with your teammates, they get involved with the work and help you more. Let support groups make contributions to the product to get their commitment; you'll need it. Who knows, their contribution may be just what is needed to make the product really successful.

▶ *Career Tip.* Guidelines and checklists are well worth the time and effort. They can save hours of work!

Don't ignore documentation. This is a very important part of the design process. It helps you keep track of the design as changes are made. It provides a means to communicate to others all that is involved in your design, and it provides a record of what was built after the product leaves the company.

▶ *Career Tip.* Overdoing the documentation never hurts; underdoing the documentation always causes problems.

One of the benefits to knowing the product flow through the company departments allows you to quickly identify the departments that are critical to the company's success. Are you working for a critical department that is absolutely necessary for the company's survival? Or are you working for a department that is only there to solve a short-term problem and will disappear shortly after the problem is solved? People in critical departments get raises and promotions. People in short-term departments usually get reassigned or laid off.

▶ *Career Tip.* The more critical the department is to the company product development, the more leverage you have for your career.

Another benefit of knowing the product flow process in your company is that you can contact other departments well in advance and schedule their help for your product. This allows you to get things done more efficiently, for less money, and on time—results that motivate companies to give raises and promotions.

By not learning the engineering process, you are doomed to trial and error methods that usually cause delays and cost overruns, results that do not lead to raises and promotions. Remember, no one likes to have an unexpected job dropped on them without notice or time to respond. By going to these departments well in advance, you give them the opportunity to schedule and complete your job in a timely manner. This makes both the company and you look good.

Knowing the engineering process will aid you in determining which projects are best to work on and which projects should be avoided. If you get reassigned to a project that has nothing to do with the mainline processes of the company, you can be assured that you will also probably not be in the main line for career advancement.

By studying the engineering process and learning all the steps involved, you quickly become aware of the possible shortcuts and ways around all the red tape. In doing so, you should be able to accomplish assignments in a much shorter time.

▶ *Career Tip.* Getting assignments completed sooner and at less cost than expected usually assists in career advancement.

A side benefit to learning the engineering process is that you will also educate yourself about the company's products. Knowing both the company's products and the engineering processes provides a valuable insight into the company, a valuable insight that you can use to your advantage. For example, once you learn the products of your company, you can determine which product is best to work on for the betterment of your career.

To illustrate this point, let's assume you are working on the highest profit product in the company and you just figured out a way to produce it 10% cheaper. Chances are very good that someone is going to notice your improvement and reward you for it. Or you may be working on the most critical calibration and assembly process for the most profitable product of your company. Having knowledge of this is always great leverage when it comes time to ask for a raise. Supervisors like to reward key personnel who are absolutely essential for getting products out the door.

The point here is that you will not find out about these positions unless you study the engineering process and products of your company. You must analyze the process to determine which are the critical steps and which products are the most profitable.

▶ *Career Tip.* The optimum situation for your career is being responsible for the key process steps of the most profitable product of your company.

Failure to study and learn the engineering process and key products of your company will only sidetrack your chances for career advancement. A good example of this is working on a product that is about to become obsolete and phased out. Under these circumstances, you may soon be facing a layoff even if you do an outstanding job.

Another example of how ignoring the engineering process can hurt you is when changes are introduced into the process. Often companies will change the engineering processes to eliminate unnecessary steps, or they may send out the work to other companies that can do it cheaper. If you are working on a process step slated for being phased out, this will only hurt your chances for career advancement. Only by thoroughly understanding the process will you be able to determine how changes will affect you. With this knowledge you should be able to sidestep trouble and keep your career on track.

Knowledge of the engineering process is essential for successful career advancement. It provides you with so much valuable insight into which jobs are critical and which jobs are not, which process steps are important and which steps are not.

The engineering processes are different from company to company. Not only is it different from company to company, it is continually changing within any one company. Therefore, you must be continually updating yourself on the changes. Never assume that once you have determined the engineering process for your company that you are finished. Learning and understanding the engineering process in your company is a never-ending career activity. Now that you understand the extreme importance of learning the engineering process, let's look at all that is involved in the engineering process.

Generally, it takes a team of people of various backgrounds to get a product out the door. The size of the teams varies anywhere from 15 to 150 people depending upon the product. In addition, it takes several different departments working together to design, build, and test the products. It is extremely important that you identify the different departments and people who become involved in producing your product and the order in which it is produced. You must determine all the steps a product must go through in your company to get out the door. Next, you need to diagram the product flow. While they normally exist in most companies, you may have to do some research to find one. If product flow diagrams are not available, then you need to generate one.

If you don't know how to generate a product flow diagram, simply ask for help. Ask your supervisor, as he or she generally has a good feel for what it takes to produce the product. Listen closely to what your supervisor tells you, for they will be sharing with you how things should operate. I am certain your supervisor will also tell you about all the pitfalls they encountered and how they overcame them. Try and do more listening than talking.

After you have mapped out your company's engineering process, you should quickly come to the realization that 95% of the work done on your product is actually performed outside your department. The engineer must call on the support of other departments to help him design, build, and test the product. What this means to you is that you must have good interpersonal skills as the utilization of work done by others is extremely important. This brings us to a very important question.

How well do you interface with other people? Can you easily convince other people to support you? If you find yourself having problems with this, don't panic. You can get help developing interpersonal skills. Guidance or career advice is available through many colleges and universities. Oftentimes, evening classes are offered to better accommodate people working and going back to school. Contact your local college or university to obtain a course listing. Remember, career advancement depends on how well you interface with people. How good are you at requesting and receiving other people's help?

When dealing with other departments always try to create a win–win situation between your department and the department you are requesting help from. A win–win situation is one in which you get what you want and the department doing the supporting work gets what it wants. In other words, you both look good in accomplishing the work. Here's an example: you bring your job to the support department early enough so that they have time to respond and accomplish the work on time. You look good because you planned ahead and the work was completed on time and within the budget allocated. They look good because they were able to complete the work on time and with high quality.

▶ *Career Tip.* Win–win situations enhance your chances of a promotion or career advancement.

Stay away from a win–lose situation. This is where you win and the supporting department loses. A typical win–lose situation is where you are late bringing your job to the support group. You may even have to go over the head of the support group to get top priority for your product so that it is completed on time. The support group rearranges its priorities and gets your impossible job done, but this throws all other jobs in the department behind schedule. You win because you get your job done, but everything else looks bad and the support group loses. These win–lose situations will always come back to haunt you on the next project. That support group may put your job last and possibly keep it there. Unfortunately, I have seen this happen.

▶ *Career Tip.* Win–lose situations could possibly result in demotions or no career advancement

Sometimes projects require you to create a win–lose situation between departments. There are several ways around this situation. First, get the department supervisor to agree to rearrange the priorities in the department and try not to go over their head. Stay within the department to resolve the conflict. Raising it to higher levels usually creates more problems than it solves.

A great way to show your appreciation would be to nominate the work provided by the support team for a company award. This helps to smooth over ruffled feathers and will make it easier the next time you have to deal with the department. Still another suggestion is to apprise the upper management how the support department is handling your impossible situation. Nothing makes a support department move faster than knowing upper management is watching their efforts. It's their time to look good and they want upper management to see them at their best. Finally, don't take credit for their work. Nothing makes people more irritated than someone taking credit for their work. Make sure you give credit where credit is due.

▶ *Career Tip.* If a support group did all the work, give them the credit.

SUMMARY

The engineering processes changes continuously. Departments come and go. Policies within departments change, people change, and new ways of doing things are continuously being implemented. Always watch how the engineering process changes. It is an important part of your job.

Learning all the products and processes of your company is not an overnight task. It may take months and maybe even years before you fully understand all the products and processes. Don't become discouraged, you will find it was worth the effort.

▶ *Career Tip.* A shortcut to learning the process sooner is to study your company's "Policies and Procedures Manual."

All larger companies have policies and procedure manuals and guidelines. These manuals define the policies and procedures that are to be followed by the employees regarding normal business operations. It is a simplified guide to how the company/corporation does business. While it may lag at times, it is well worth the effort.

To summarize, the benefits of understanding your company's engineering process are:

1. By identifying the steps that products go through during design, production, and testing, you gain valuable insights into the company's operations.
2. Understanding enables you to determine what products, functions, and departments are critical to the company's success.
3. Knowledge of other departments' functions leads to better productivity.
4. Awareness that business requires a cooperative effort.
5. Realization that 95% of the work is done outside your department.
6. Realization that interpersonal skills are important and the key to getting people to support you is through win–win situations.

Have you identified any career actions you want to take as a result of reading this chapter? If so, please make sure to capture these ideas before you forget by recording them in the notes section at the back of the book.

ASSIGNMENTS AND DISCUSSION TOPICS

1   Map out the engineering process in your company. How does your department fit in? Can you identify a name or person for each department or function in the engineering process?

2   What are the key products of your division? Rank them according to the profits they generate.

3   What are the critical engineering process steps for the product you work on? Who controls these steps?

4   Write up a specification for your product. List all the requirements.

5   List the models you have available in your department that you can access help that you model your product's performance.

6   Generate a requirements summary and test matrix for your product.

CHAPTER 21

TEAM SKILLS
ESSENTIAL FOR SUCCESSFUL ENGINEERS

The complexities of today's products require teams of engineers working together in a unified, efficient, and productive fashion toward common goals. The teams are often composed of engineers with different expertise such as electrical, mechanical, chemical, and software. Engineers working in this environment need team skills to perform. Team skills do not come naturally to most people; however, these skills can be learned and with practice can be added to the engineers' skill set. In this chapter, we first discuss why team skills are important for your career and then specific team skills engineers can use to become a great team player.

WHY HAVING TEAM SKILLS IS SO IMPORTANT TO YOUR CAREER

The value of team work has been a subject of research for quite some time and was first recognized in books by Bradford and Cohen, *Managing for Excellence*, and Tom Peters, *Thriving on Chaos* [1,2]. Roger Allen from The Center for Organizational Design, Inc., puts it eloquently when he states "In spite of our technological advances, our competitive advantage lies in our ability to work together effectively" [3]. Technology and knowledge is proliferating at faster rates than ever before because of the Internet, and the ability to acquire new technology literally overnight whether it is next door or from another country on the other side of the earth. This means all engineering teams are starting out with the same basic knowledge and the team that best works together is going to be the most successful. I believe perfect examples of this are the Lego League and First Robotics where students compete with each other starting from a basic kit common to all. The team that usually wins is the team that has discovered how to work together productively to foster the best ideas, solve

The Engineer's Career Guide. By John A. Hoschette
Copyright © 2010 John Wiley & Sons, Inc.

problems, and discover new and improved methods of accomplishing work [4,5].

I firmly believe that the United States' ability to continue to be the leader in technology development is directly related to how well we train our engineers and scientists to work together in teams. I applaud both of these organizations for their efforts and encourage all engineers and scientists to volunteer some of their time to help support these activities.

Glenn Parker, in his book *Team Players and Teamwork*, identifies the top benefits companies receive from high-performing teams [6]. As you read this list, keep in mind, these are the same reasons engineers are promoted and advance in a company.

1. Greater productivity (getting more done than others)
2. Effective use of resources (doing more with less)
3. Better problem solving (ability to solve difficult problems)
4. Better quality products and services (more sales, less rework, and repair)
5. Creativity and innovation (develop new products)
6. Higher quality decisions (ability to know what is best)

The conclusion you can draw from this list is: what is good for the team is also good for your career.

▶ *Career Tip.* What is good for the team is also good for your career.

Sooner or later, all assignments come to an end and you have to transition to new assignments and/or new teams. Generally, managers who are responsible for putting together a new team will want to select the best members possible to ensure success, especially if the new assignment is very difficult. The jargon in industry is selecting "A" team members. The two key requirements of becoming a part of the "A" team are great technical skills and great team skills. If you have great technical skills and are also recognized as a great team player by management, you have career power and an advantage over others. Senior management will seek you out for the more difficult assignments and ask you to join new teams. In other words, career advancing opportunities seek you out rather than you having to go hunt for them. With great team skills, you are a more desirable worker and consequently more likely to stay employed during downturns.

▶ *Career Tip.* Develop great team skills and be recognized as an "A" team person.

Another benefit of being a great team player is hiring ability. James Challenger, in his book *The Challenger Guide: Job-Hunting Success for Mid-Career Professionals*, identifies that one of the best skills a person can possess are great team skills and this is a major trait employers look for when hiring people [6]. Highlighting your teamwork skills during an interview can be the factor that makes you stand out from the rest of the candidates and gives the employer the reason to hire you.

SUGGESTIONS ON HOW TO BE A GREAT TEAM PLAYER

The actions you take or the skills you practice will vary upon the stage of formation the team is in. The stages a team goes through while performing work have been highly researched and studied [7–9]. These stages are

1. Forming
2. Storming
3. Norming
4. Performing

Forming Stage. This stage usually occurs at the beginning when the team is first formed and they are meeting for the first few times. Most people are new to the project and just getting acquainted with each other and the objectives of the team. During this stage, most members are quiet and in learning mode, many are concerned about expectations and how much time this is going take. Are the assignments going to be difficult and who is responsible for what? Here are some skills and/or actions you can take during this phase to help the team out.

1. Ask the leader and/or other team members to introduce everyone and be prepared with your 30-second response in what you do and your skills. After the meeting, stay around and socialize with the key people on the team you will be interfacing with.
2. Ask the leader to share their vision of the team, its objectives, and the project tasks.
3. Discuss and ask questions about the team's roles, responsibilities, and logistics. For example, the organization chart, meeting times, the duration of meetings, and expectation of members for reporting progress.
4. Volunteer to help the leader in any way you can: taking notes, arranging rooms, or running the projector. Volunteer for action items.
5. When asking another member of the team a question, start out by saying the person's first name. For example, I would like to ask Bob a question about . . . . This helps everyone on the team learn and memorize each other's names.

▶ *Career Tip.* During the forming stage, start out on the right foot by helping the leader as much as possible. Volunteer for action items.

Storming Stage. This stage usually occurs after several meetings have taken place and people have a somewhat better grasp of the team objectives. Note I said somewhat better understanding. During this stage, team members usually feel comfortable enough to speak up about what is bothering them, what the problems are, and why progress is not being made. This, as the title suggests, is a turbulent or stormy time for the team. In some cases, everything about the team is challenged: the objective, organization, member roles, and methods of how work is accomplished and reported on. Everything is fair game to question and members want to know why they are doing it this way. Although turbulent, this is a good stage for the team to pass through because they are more unified as objectives, processes, and roles are agreed upon, and the members can focus on the real work at hand.

If the team fails to pass through this stage without issues resolved and agreements reached, more than likely the team is going to be divided and may end up as a dysfunctional team not accomplishing their objectives in an efficient manner or even failing. Here are some skills and/or actions you can take during this phase to help the team out.

1. Ask for clarification about anything you do not understand.
2. Remind the team of the need to find solutions to problems and ask for others' suggestions.
3. Be a facilitator and list the problems and help identify solutions.
4. When opposing sides seem to be in a deadlock, work to identify a compromise position both sides can live with.
5. Ask to list what is "absolutely necessary" and what is "nice to have."
6. Ask if two opposing sides can "agree to disagree that they have honest differences of opinions"; and move on for the sake of the team.
7. Ask for a recess or timeout for people to cool down and think through things.
8. Finally, one very bold move you can take to break the paralyzing arguments is to ask the team if there are any members who would like to resign since they feel too uncomfortable about continuing on the team.

▶ *Career Tip.* The art of compromise is a vital team skill.

Norming Stage. This stage follows the storming stage. In the norming stage, people realize they have to "go along to get along." At this point on the team, development members have learned all the complaining and heated discussions are no longer productive and they need to get down to real work. The team has settled on how they are going to operate, interface with each other, and deal with problems and the objectives. They are more into agreeing and getting along. A team spirit emerges and a sense of cooperation prevails in order to get at the real work at hand. Here are some skills and/or actions you can take during this phase to help the team.

1. Ask what is the near-term, most important items for the team to focus on.
2. Review the big picture and overall progress toward the final goals.
3. Encourage discussion of results and any issues people have identified.
4. Ask if anyone needs additional help.
5. Challenge the team to think of better methods, higher efficiency means, or ask for their help in streamlining.
6. Ask what can go wrong and how to prevent it.

There are two threats to the team's progress at this stage. The first threat is that people become complacent simply to get the work done and turn in poor work, thus the need for emphasis on high-quality work. The second threat is new members joining the team and pushing the team back into storming stage while the new members come up to speed on why the team has settled on its present operating methods and objectives. The team leader needs to spend special time with any new member discussing the progress to date and reasons why the team is operating the way it is.

▶ *Career Tip.* Don't settle for mediocre team work; challenge yourself and others to do the best they can.

Performing Stage. This is the final stage of team formation and at this point, the team is what some people call a "well-oiled machine" producing high-quality work. The team has made significant progress toward its final objective. The members are interfacing with each other as needed and problems are being raised and resolved. The team and meetings have settled into a highly productive routine. Here are some skills and/or actions you can take during this phase to help the team.

1. Volunteer to lead training efforts for new people.
2. Coordinate any new outside training needed for the team.
3. Ask if you can run the meeting, or take over any parts to help balance the workload of others. Redistribute the workload.
4. Volunteer to create a milestone chart that shows progress of the team.
5. Highlight and/or compliment team members' accomplishments during the past week.

6. Organize a team celebration and invite management when the project is completed.
7. Submit the team for awards.

▶ *Career Tip.* Turning in high-quality work on your team assignments and getting done ahead of schedule is a career accelerator.

So far, we have identified the four stages teams go through and great team skills you can practice at each of the stages. However, there are other team skills that should be used during all stages of team development. In the next section, the great team skills that can be and should be used during all stages of team development are discussed.

COURTEOUS TEAM SKILLS TO USE ALL THE TIME

The following team skills are great actions you can take ANY time you are attending a team meeting. They are relevant in any of the four stages of team development and are basically commonsense: polite and courteous actions that foster respect and understanding toward your fellow teammates.

1. Do not talk over people or cut them off in mid-sentence. Have the good manners and courtesy to wait until they have finished before responding.
2. When someone is sharing an idea with you use active listening and paraphrasing skills to acknowledge back to the person that you have heard them.
3. Praise in public, but do not overdo it.
4. Thank people for their contributions.
5. Ask to see the data supporting their conclusion before you pass judgment.
6. Live up to your commitments and deadlines; get your work done on time. When you are going to be late with your work, notify people affected in advance of deadlines and negotiate a new due date.
7. Show up to meetings early, be prepared to discuss your results, and show technical data to support your work if possible.
8. Find things you have in common with others on the team.
9. Take time to socialize with other members of the team either before or after meetings.
10. Volunteer for action items.
11. Go the extra mile to make sure things are working out for people who depend on your work.

12. Identify early on who are the people you are dependent on for work (your suppliers) and who are the people you are doing work for (your customers). Spend time getting to know your suppliers' and customers' needs, problems, and desires.

13. Ask quiet people their opinions.

DESTRUCTIVE TEAM ACTIONS TO AVOID

Please check any of your diversity biases, pet gripes, ego, and bad attitude, at the conference room door before going to a team meeting, especially if the meeting is going to be a stormy one. Being identified by others as a problem team member who is abrasive to others, and does not produce high-quality work on time, is very career limiting. You can win a single battle at the meeting and in doing so alienate others on the team from ever supporting you again. Your net result is that you won the battle but lost the war. Here are actions to stay away from when you are on a team.

1. Talking over others at team meetings before they are finished and not allowing other people to be heard. Dominating the meeting with your ideas and actions.

2. Attacking people personally at meetings. Here are some examples: I hate this idea, you're totally wrong, that's stupid, you know nothing about this, this will never work, and I think the correct thing to do is . . . .

3. Escalating the tone and harshness of your words when in a heated debate. Shouting at people to win an argument.

4. Surprising other teammates with unexpected results at meetings and embarrassing them just so that you can say "I told you so."

5. Always thinking of reasons why things will not work instead of thinking of reasons why they will work. For example, we tried that already, no budget, it's never been done before, what makes you think you can do it, that's too hard to do, no one will let us do that, the boss will never approve that, it's not my problem, so I am not going to worry about it, ...

6. Sidetrack the meeting by bringing up nonrelevant subjects that are fun to talk about but do not help the team.

From the two lists, it is easy to see there are twice as many positive and constructive team actions and skills you can utilize than destructive. Human nature for some people is to select the destructive actions at meetings. My challenge to you is to sit at your next team meeting and observe the team members in action. Count the constructive and destructive actions that occur without letting anyone know. I think you will be greatly surprised at the number of destructive actions. If you are on a team dominated by destructive acting members, there is something very simple you can do to help turn this round. Let me share with you what one of our team members recently did to turn a negative situation.

Our company is a proud contributor to the "Relay for Life" charity that raises money for cancer research. During the 4-week time period to raise money for the cause, one of the team members on a very difficult assignment became fed up with all the negative and destructive remarks of the team. So to help the team realize they needed to change their attitude in order to start performing, he got a large piggy bank and fined everyone a quarter (25¢) every time they made a negative or destructive remark at a team meeting.

The quarter was considered a contribution for the "Relay for Life" cause and the person had to instantly deposit the quarter into the piggy bank at the meeting in front of everyone. This was a very clever and non–threatening way to point out to people how negative they really were. People joked about being caught and paid the quarter at first. They even threw in dollar bills to be allowed to get all the negative responses and frustration out at once. During the first week of fining people, he collected nearly $80. By the end of the fund-raising, he had collected a little over $200 for "Relay for Life." If you quickly do the math, at a quarter a remark, this means the team members made over 800 destructive remarks during the 4-week time. By the end, most of the team members, and this includes myself, thought twice about making a negative remark, especially since everyone else on the team seized the chance to point at you during the meeting and identify you had just made a negative remark. The negative remarks subsided and the team moved to positive constructive actions rather than turning the meeting into complaint sessions.

SUMMARY

The great team player senses what contributions need to be made and automatically takes action without being asked. The great team player places the group's objectives, methods, and processes over their own personal goals and is more concerned about how he or she can best support the group in achieving its objectives.

The great team player plays many roles as the team moves through the stages of development from forming, to storming, to norming, and finally performing. These roles include facilitator, listener, cheerleader, challenger, collaborator, peacemaker, and even leader at times. The great team player is always cautious and respectful of others.

Practicing the team skills identified in this chapter is not hard to do, but requires a conscious effort and determination to simply "honor and play by the TEAM'S rules." People may feel that if they are not dominating the meeting, giving advice, or the leader of a team, they will never be noticed. However, as teams come and go, it will become obvious the successful teams follow you as you move from team to team. Management will definitely notice.

FIGURE 22-1   Do your research first.

1. What is the problem and how big is it? An easy mistake is to assume the problem is apparent and your supervisor perceives the need to solve it as you do. First you must make sure they are aware of the problem. To do this, describe the problem in simple terms. Get their agreement that a problem exists. If you are intent on solving a problem that they consider trivial, no matter how good your idea, they will perceive your solution as trivial. Next, explain the circumstances that exist that make the problem worth solving. Quantify how big the problem is in terms of work hours lost, dollars lost, or poor performance. Supervisors take immediate notice when the problems are in terms of cost overruns, failures, schedule delays, or rework required. Once you have clearly identified the problem and assessed its impact it's on to the next question.

2. What if nothing is done now? Quantify the short-term and long-term impact of the problem. Most managers will hope the problem will go away if ignored. Show the manager the consequences of ignoring it. Identify any new problems that will arise if the present one is not solved. Once the magnitude of the problem starts to settle in, he or she will want to know why your idea merits consideration. In order to answer this you must do some further research. Your objective is to establish credibility in your answer, and to do this you should have knowledge of the past history of the problem. This leads to the next question normally asked.

3. How did we get here? To answer this question you must do research on what was done leading up to the problem in the first place. You may discover the problem you are trying to solve is small in comparison to other problems.

4. What was suggested before? Try to discover any past ideas that may have been suggested to solve the problem. Your objective is to show

that your idea does not take them down a path that has previously been tried and failed.

5. What has not been tried? Here you want to summarize several ideas that show potential for solving the problem, your idea included. Your aim is to show that your suggestion is the best. If you really do not have a good solution, don't fake it. Drop the idea. Overselling the benefits of your idea and not thoroughly thinking them out will destroy your credibility.

6. What puts your idea ahead of the rest? Highlight your idea by researching what is great or different about it. Summarize in simple terms the benefits of your idea. Quantify all the benefits you can. Supervisors like to hear the benefits in terms of cost reduction, lower unit product cost, improved performance, quicker turnaround, improved producibility, product reliability, and product safety. These are a few of the buzzwords used to help sell ideas.

7. What are the potential problems with the idea? After you have completed all the research on the benefits of your idea, you need to also consider the bad points. A well thought out plan will not only present the good points but also the bad points. All new ideas have problems associated with them. Everyone knows that nothing is free and everything costs something. Your supervisor is no different, and will look for potential problems or snags. The best way to deal with this is to meet it head on. Do no try to hide it. Make it part of your plan. Identify the problems with your idea. If possible, show that the consequences are trivial and the benefits outweigh the negative effects.

8. What will it take to implement your idea? If you have a great idea, the first question your supervisor will ask is "what will it take to implement it?" This is where you can really establish credibility by having done your research. Do research and make estimates to show that your idea will not break the company or the budget. Explain how funds can be redistributed to handle what you propose or how long the payback will take. Is what you proposed feasible in terms of schedule? Are the resources and manpower available? How will it affect the group's morale and customer relationships? Are any special facilities needed? Identify the expertise and skills needed as well as where they can be found.

People often come up with excellent ideas that are not within the charter of the group or company. An advantage to selling your idea is to show how the idea supports the company's strategic plan or present product lines.

With your research completed, it is now time to put together the presentation or plan.

WRITING UP THE PLAN

The exact organization or style of the plan you put together will depend upon your company's products and organization, your supervisor's style, and your own preferences or style [1]. There is no right or wrong style. The length is entirely up to you. If you are proposing a multimillion-dollar plan with a high impact on the company, you may want a detailed and thorough plan. These plans may consist of 30–40 pages. On the other hand, if you are just trying to make a simple change in your department your plan may be only a couple of pages long. To aid you in writing up the plan I have put together the following outline:

1. Executive Summary
2. Statement of the Problem
3. Background/Previous History
4. Potential Solutions (Pros and Cons)
5. The Best Solution
6. Benefits and Impact of Implementing
7. Implementation Plan
8. Conclusion and Summary

Executive Summary. This is a summary of all the high points of the plan. It clearly defines all the benefits of the plan right upfront. Your intent is to get your supervisor so interested in the idea or plan by the executive summary that he or she has to continue reading it to find out more.

Statement of the Problem. In the simplest terms and shortest possible means, describe the problem as best you can. Draw diagrams if needed, show charts or graphs that pictorially describe the problem. Use the material and information that you have collected during your research to define the problem and show how large it might be.

Background/Previous History. Next present a summary of the background leading to the problem and actions taken previously to solve it.

Potential Solutions. In this section list or identify all the potential solutions to the problem including the one you are recommending. Explain what is beneficial about each potential solution.

The Best Solution. This section should contain a concise, detailed description of your proposed solution. This is where you expand upon your idea. Discuss the details and highlight all the benefits.

Benefits and Impact of Implementing. This is where you summarize all the research you have done on the cost and impact of implementing your idea. Organize it in a neat, concise manner to address all the issues.

Implementation Plan. In this section, you clearly identify the plan that must be executed to implement your idea. Here is where you present the research you did and what it will take to get the proposed solution or new idea working. The best method for doing this is through a schedule showing the tasks to be completed, the sequence and time phasing of the tasks, and the manpower and costs associated with each task.

Summary. This section should be a one- or two-page summary, at most, that describes all the highlights of the previous sections. It should be a section of the write-up that your supervisor can go to and get an executive summary of the points of your plan. It should clearly state that you recommend implementing the proposed idea.

HOW TO BEST PRESENT THE PLAN OR IDEA

Once you have completed the plan your first thought will be to rush into your supervisor's office and immediately start convincing him or her how great it is. You are just about ready, but you must complete other steps before you are ready for this. Your next step is to do some political selling before you enter their office. You now enter a new phase of getting your ideas accepted called "The Politics of Selling Your Ideas" [2].

You want to test out the waters before you go into your supervisor's office. The best way to do this is by showing the plan to your mentors or to other key people in the organization who are respected by your supervisor. In doing so, you can help the salability of your plan.

▶ *Career Tip.* Have a well thought out plan for presenting your idea. This is the time to let management see you at your best!

Ask their opinion and get a forewarning of how your supervisor may react. They will probably identify the same objections your supervisor will come up with. With this knowledge you have time to develop plans to work around these objections and easily handle them when your supervisor raises them.

If you are a politically savvy engineer, you will attempt to work out any differences between your viewpoint and that of the key people in the organization. You need to realize that the outcome of your effort to sell your plan will be based less on its technical value and more on the fact that enough people want it to succeed. A new idea can be less than perfect technically but still survive if key people involved want it to work. Where there is a will, there is a way.

These key people are also in a position to show you how to improve your plan. Ask for their support and any improvements they might have. By asking for their opinions and support you are effectively getting them to back your plan. They are less likely to shoot down the plan if their ideas are also included. It also provides a check to see if there are any fatal flaws or any major things you have overlooked. It's better to find out that your plan has a major flaw before you go into your supervisor's office.

You also quickly find out who supports your idea and who does not. Having the support of key people in your organization adds a large amount of credibility to your plan. Being able to say that the senior engineer backs your plan goes a long way to selling it. Also, being able to respond to the objections of the senior engineer and show how you plan on overcoming them effectively neutralizes the opposition. When it comes time to defend your plan you must know who to call upon for support and what objections other people are using to shoot down your plan.

▶ *Career Tip.* Be prepared to handle objections and negative feedback.

When you run into people who object to your plan, handle them with caution. First ask them why they are objecting to the idea or plan. Only when you allow them to talk freely are you able to air the differences. Empathize with the people objecting and explore their apprehensions. Get specific information about what exactly they are objecting to. Next, ask them how they would overcome the problems. Sometimes people are objecting because they really want to tell you the solution. Try to get them to help you define a solution. Nothing makes allies quicker than identifying that the senior engineer has identified a serious problem with your idea, but he or she is working on a solution so your group can realize all the benefits. If you win over resistors, periodically recheck their position to make sure they haven't changed their minds again. This is important.

Don't try to force the situation if you fail to get the support you need. Do not argue and counterattack; this is a lose–lose situation. Back off and recognize that you have an honest difference of opinion on the plan. In other words, you agree to disagree about your plan. This allows the objector to state their objections and you to continue; neither of you have lost credibility. By trying to force agreement you may win a short-term battle, but the first problem you run into implementing your plan will result in the person running around telling everyone how they told you so!

▶ *Career Tip.* Informally and ahead of time, discuss the idea with others to see what possible objections there may be. Be prepared to handle these objections.

When you present your plan to your supervisor, he or she will probably want to know the reaction of other people. Having already shown it to key people, and gotten their support before you go into the supervisor's office, allows you to identify who backs it. It also allows you to identify who is against it and the objections they have. In either case it only adds more credibility to your plan.

Show the plan to your mentors before you go into your supervisor's office. They can give you a good unbiased opinion of the plan. Their years of experience in the company can provide you with background on what has gone on in the past, what has previously worked, and how you might best sell your plan.

If your coworkers are open to new ideas, you will find a warm reception for your ideas. However, many workers feel new ideas create more work, something they are not interested in. To combat this they will claim that things worked in the past and there is no need to change things, "we always did it this way; no reason to change."

Another barrier you may encounter is the "sacred political laws" of your organization, which you may be unknowingly violating. For example, if you are in a PC company and recommend changing to a Mac you may be up against a sacred political law. Chances are some upper-level manager has already made the decision and you will be bucking the system.

Remember, new ideas or plans frequently upset the power structures of the organization and call for change. Your plans could be secretly saying, "Hey, supervisor, you are doing something wrong and here is the right way to do it." Your supervisor will not be very interested in implementing a new plan that may only highlight how wrong things were being done. New plans or ideas presented in this light often quickly end up in the trash before they can do any harm to people's careers.

After you have tried your plan out on other people, you should have a good idea of its strengths and weaknesses. Polish the plan with what you have learned. Incorporate new ideas when you have picked them up and be sure to give credit where it is due. Revise your approach if it needs it. Although it may be hard to change things, remember that your end objective is to benefit the company and its customers.

Here is a quick list of some of the helpful things you might have found out by sharing the plan with other people.

1. What are the weaknesses and strengths of the plan?
2. Is the timing or political climate right for my idea?
3. Are there any other issues that I can piggyback on to help sell my idea?
4. Who will make the final decision, and by what means? Committee vote? Supervisor approval? Manager-level approach?
5. Can I call upon other outside sources to help sell my plan? Coworkers? Consultants? Company publications? Technical journals? Competitors?

After you have shared your plan with other people, received their support and finally, modified and improved the plan, it is then time to share your plan with your supervisor.

MAKING THE BEST PRESENTATION YOU CAN

Making the big presentation to your supervisor or management may be one of the hardest things you will ever do. You will be nervous the first few times and you must practice several times before you actually do it. If possible, practice giving your plan several times to other people first. Run through it from beginning to end just the way you intend to give it. Your mentor would be a good person to practice your pitch on. After such practice you should be ready to present your ideas. Some people will tell you that you do not have to practice, just go in there and present it. However, I feel this is the worst thing you could do. Athletes practice many hours before they actually compete. Musicians practice hours before they perform a concert. Why not your let your supervisor see you at your best? Practice!

Make sure the setting and time of your presentation is good for your supervisor. The best way to do this is to talk to him or her in advance and block out a portion of time where they can give you some undivided attention with no interruptions. Announce you have a new idea you would like to discuss with them and specify how much time you are going to need. Generally, keep the meeting to approximately 45 minutes or less.

▶ *Career Tip.* Refine and polish your presentation through practice.

When the time comes, make sure you will not be interrupted. You have put a great deal of work into getting this prepared. Your final step is to get the utmost attention while presenting. Ask for all calls to be left for voice mail during your meeting if possible. If the supervisor answers his own calls, then ask if he would mind going to a conference room where you can shut the door. Your objective is to get undivided attention so that the value of your plan can be perceived clearly.

The style and type of presentation is best left up to your judgment. It may be as simple as a few typed pages or a complete report with a view graph presentation to highlight the important parts. A really effective engineer can personally produce computer spreadsheets, graphs, and view graphs for all types of presentations. If you need more training in this area, get it immediately. Poor view graphs can ruin any presentation. There is one very important concept that you must remember any time you are making a presentation: the goodness and value of your idea is being judged partly on how well you present (Figure 22-2).

▶ *Career Tip.* Your ideas will be assessed based on your ability to communicate, your presentation skills, and the quality of your charts.

This is unfortunate since your presentation skills do not necessarily reflect your technical ability. Nevertheless, people do react this way at

Formal Large Group

Small Group

Informal One on One

Which Presentation Style Is Best?

FIGURE 22-2   Different methods of presenting the idea.

nearly every presentation. Therefore, you must put on the best show you can.

Regardless of the size of the report or presentation, make sure your supervisor has a copy of it when the meeting is over. Invite them to make notes and keep it for future references.

As you present the plan make sure you pause and ask for feedback. If you're not getting feedback, ask people what they think. If they like it, you've succeeded. If they do not like it, find out why. Discuss objections but do not try to override or counterattack. You can always ask what it would take to make it a better plan. If they start to suggest things, you've got them hooked. Try to get everyone to buy into the plan and start supporting it.

If management likes your plan then it's on to the next phase: "How do we go about implementing the plan?" Remember, coming up with the good idea is only part of the solution. Implementing the plan and realizing the benefits is the other part. Only after you have started to realize the benefits will it positively effect your career. You must get your supervisor to share with you the details of how the company will approve and implement the plan. Once management has decided to move, only then can you start the implementation of your plan in earnest.

A FINAL WORD OF CAUTION WHEN SELLING YOUR IDEAS

Here's one final word of caution about selling your ideas. Remember that some people will think only of reasons why it will not work. You must have a strong will to overcome these objections and keep pressing on if you're certain you really have a great idea. Often it can only take one "no" to kill a great idea even if ten people have voted yes. Don't let one "no" vote ruin your plan; make sure it does not have a fatal flaw, then press on.

Finally, most people will say "If I have to go through all this trouble just to suggest something, then forget it." What they do not realize is the tremendous learning experience you go through doing all the tasks mentioned in this chapter. Maybe your first idea will not hit the jackpot, but perhaps you will be better prepared and more capable of selling your next idea. By putting the plan together you are showing your supervisor how you are trying to improve. If it is well organized and neat you are clearly demonstrating your ability to think through a problem in a well-organized and methodical manner. These are very valuable skills in an employee.

As a result of your efforts, your supervisor may be more willing to have you help on the next big cost-saving activity. These are the types of activities that help career advancement. If you follow the suggestions in this chapter you stand a better chance at career development regardless of whether your idea or plan is accepted or rejected. With this type of outcome, what are you waiting for? You have nothing to lose by suggesting something and stand a very good chance of career advancement.

SUMMARY

Just having a great idea is not good enough. You must have the skills to sell it to your coworkers and supervisor. The first step to selling your ideas is to get organized and write it down. Document the idea or plan. Part of documenting the idea includes doing research to support your ideas, find out what has been done before, and what benefits you can expect from implementing the plan. Next, you must write up the plan in an organized fashion. Discuss it with your coworkers and get their feedback and support. After you have obtained some backing and incorporated any changes or improvements, it's time to make the big presentation.

Have you identified any career actions you want to take as a result of reading this chapter? If so, please make sure to capture these ideas before you forget by recording them in the notes section at the back of the book.

ASSIGNMENTS AND DISCUSSION TOPICS

1 Brainstorm for a few minutes and try to think of some ideas that you could implement to save your company some money.
2 Pick the best idea and do some research on it.
3 Generate a plan or write-up that describes your idea and its potential benefits.
4 Share the plan with one or more of your coworkers and get their input.
5 Modify your plan and incorporate any good suggestions.
6 Present the plan to your supervisor.
7 Implement the plan.

REFERENCES

1. Fuller, O., "How to Write Reports that Won't Be Ignored," *Machine Design Magazine*, Jan. 11, 1979, pp. 76–79.
2. Raudsepp, E., "Politics of Selling Ideas," *Machine Design Magazine*, Nov. 7, 1985, p. 97.

CHAPTER 23

GETTING PROMOTED WHETHER YOUR PROJECT IS SUCCESSFUL OR UNSUCCESSFUL

During your engineering career you will encounter many so-called "truths" (actually fallacies) about engineering. If you blindly accept these so-called truths you will be severely limiting your career. One of these so-called truths is the belief "you can't get promoted if your project is unsuccessful." The fact of the matter is that most promotions are the result of outstanding performance during an unsuccessful project. The reason is that during an unsuccessful project there is usually quite a bit of upper management visibility. This extra visibility provides a unique opportunity for the engineer to demonstrate his or her capabilities to solve problems and show readiness for promotion. All the ingredients are there for a quicker promotion with outstanding performance and this extra visibility. Unfortunately, the converse is also true—all the ingredients are right for a quicker demotion if your performance is poor.

Another common fallacy is that most engineering projects are successful. The fact of the matter is, engineering projects are not all successful. Therefore, you stand a very good chance of spending most of your career working on unsuccessful projects. Another fallacy is that a successful project automatically results in a promotion.

If you blindly accept these fallacies, you have set yourself up for a very unsuccessful career. To overcome these fallacies and maintain career development you must develop talents or skills that lead to career development—regardless of whether or not your project is a success. If the project you are working on turns out to be unsuccessful, the question then becomes "How do I best survive and end up with a promotion?" If the project is a success the

The Engineer's Career Guide. By John A. Hoschette
Copyright © 2010 John Wiley & Sons, Inc.

question becomes "How do I best capitalize on the results to increase my chances for getting a promotion?" The specific actions you take depend upon whether or not the project will be judged as successful or unsuccessful. In this chapter, we will identify specific actions you should take when the project is unsuccessful and another set of actions when the project is successful.

▶ *Career Tip.* Successful engineers learn how to make their promotion independent of the outcomes of their projects.

Unfortunately, it is not always easy to determine whether or not a project is successful or unsuccessful. What one supervisor considers successful, another might consider unsuccessful. The first step is to determine if the project you are working on is heading for success or setbacks. To best do this let us identify some common indicators that will help you judge.

HOW TO DETERMINE IF YOUR PROJECT IS A SUCCESS OR NOT

One indicator that is always good to check on is your supervisor's opinion of how the project is going. When you have a few minutes alone simply ask how the project you are working on is going. Listen carefully to the answer; it will tell you a lot. If your supervisor responds positively then this is a good indication that things are moving along quite well. Try and get them to identify the specific things they see as significant. This information is very valuable and you may be able to capitalize on it later.

If your supervisor responds negatively to the progress or results of the project, this is a good indication that the project is not turning out as planned. In this case, ask what is going wrong and what should be done. Try and get them to identify the specific actions they think you should take to turn things around; this is very valuable information.

Another indicator is the project technical results. Is the project or design meeting the requirements or is it failing tests? Compare the test results to the original requirements and performance model. Are the actual test results coming close to the results expected or are they significantly off? Check the results in other areas of the project. Your area may be successful but if another area is failing miserably it could make the entire project unsuccessful. For example, a new car body design might look great, but there may be severe problems with the motor and braking system that can cause it to fail. If you are only working on the body design, you may not be aware of the setbacks in other departments that are hurting the overall success of the project. Overall test results are good indicators of a project's success.

▶ *Career Tip.* The actions you take for your career depend upon whether the project is headed for success or a setback.

Check the schedule for the project. Is the project progressing as planned? Is it ahead of schedule or behind? Being behind schedule indicates problems. Another indicator is cost. Are the costs running as planned or is the project overrun? Overrun projects are bad news and often get the president of the company's attention since they eat up profits.

Check the follow-up plans for the project. Is the project going to transition into the next phase? Is funding available to continue the effort next year? If so, chances are the project is getting the results expected and it will be considered a success. If not, you may be working on a dead-end project. Management may intend to end the project since the results everyone hoped for are not happening. By knowing the follow-up plans you can get a good indication of whether the project is considered successful or unsuccessful.

How are management reviews of the project going? If they are smooth and you don't hear much, then it is a good indication that the project is successfully progressing. If the management reviews result in more meetings and sudden changes in the direction of the project, then it is a good indication that the project is headed for trouble.

Another good indicator of project success is the customer's reaction. Does the customer seem pleased with the results to date?

The health or success of a project will change from day to day and week to week. As the project proceeds, unexpected problems will arise from time to time. Hopefully, they will all be solved. Therefore, it is impossible to determine whether a project will be successful on any given day. In order to get a good indication of how successful the project will be, you must constantly monitor all aspects continually throughout the project. Only by doing this will you get a good sense of whether the project is going to be judged a success or not.

CAREER ACTIONS IF THE PROJECT IS UNSUCCESSFUL AND HAVING SETBACKS

If the project you are working on is headed in the wrong direction you must take actions to minimize the damage to your career. These actions will be in response to the two most common questions asked by management when a project is headed for setbacks. These questions are: Is the project unsuccessful because the wrong people are assigned to it or are the people working on it the right and best qualified? If the answer is no, then the next logical question becomes: Is the project unsuccessful because the technical problems are insurmountable? As management searches for the answer to these questions you must be prepared to take action to minimize the damage to your career and hopefully use the opportunity to advance your career.

▶ *Career Tip.* Unsuccessful projects require you to go into a high-energy state.

Your first response to an unsuccessful project should be to go into a high-energy and high-output state. This means making your efforts visible to management to show how hard you are working on the problem. These visible efforts include working extra hours, being organized, making excellent technical presentations on the problems, identifying solutions, and projecting an "always willing to try" attitude. Let's explore some specific actions you can do for each of these efforts.

▶ *Career Tip.* Make sure your boss knows what you are doing!

The more time you put in, the better you look and the more you benefit. However, if you put in extra time and your supervisor doesn't know, you are setting yourself up for disappointment. It is a good thing to first approach your supervisor and ask about working overtime to help out. At the end of the week it is a good idea to stop by his office and let him know what you accomplished and how much overtime you worked. If you plan on working over the weekend, it helps to point this out also.

When you are working late at the office and you have voice mail or electronic mail, it's a good idea to leave your supervisor a message just before you go home. Both voice mail and electronic mail have a time stamp indicating the time you sent the message. It will impress the supervisor when she reads your message or progress report the next day and it indicates when you sent your message. Also, a message sent on the weekend can bring others up to speed on Monday morning and highlight how much effort you are putting in. The next thing you must do is get organized. This may be hard to do when things are falling apart all around you, but you must! Remember, management is constantly asking "is the project unsuccessful because of the people assigned to it?" If you are presenting an image of being organized, management will probably walk away saying "thank goodness we have them on the project—if anyone can turn this around it will be them!" These types of statements ultimately add to results in career growth and promotions.

▶ *Career Tip.* Successful engineers learn how to make their efforts and the efforts of other team members visible to management.

Now the question becomes one of how can you show that you are organized? It is simple, the first thing to do is generate a plan. What tasks are you going to be doing to solve the problem? What is the schedule for accomplishing the tasks and what results do you expect? The best way to communicate this to management is through a written plan documenting your intentions. By writing up a plan and giving it to management you are accomplishing two things. First, you are showing them you are organized and second, you are providing them with a "get well plan" they can share with the supervisors. If the plan is good, upper-level management will get the impression that your supervisor has assigned the right person to solve the problem, making them and you look good.

First, you must complete a thorough technical analysis. Second, identify the specific technical problems that must be overcome. Third, identify potential solutions to the problems. Determine the good and bad points of each solution. Fourth, rank the solutions and finally present a recommended approach. The worst thing you can do is present all the problems to upper-level management with no solutions. An engineer is paid to understand the problems, get help from coworkers, and identify solutions. Do your job if you expect to be promoted. Even a project that is considered to be unsuccessful will be looked upon as a success if you can explain exactly what went wrong and how to fix it.

There are two conditions for success as far as your career is concerned.

1. The project accomplished its goals and was successful.
2. The project encountered problems and setbacks, but we know exactly what went wrong and how to fix it next time.

In either case, the project results should be good for your career. There is only one condition for being unsuccessful: The project was unsuccessful and no one knows why.

When a project is heading for trouble there are often a series of meetings with management to ensure that everything possible is being done to make the project successful. If you are invited to one of these meetings never go unprepared. Make sure you have a well-organized and thorough report. We have already discussed the three major things you must bring to all the meetings: a plan, potential solutions, and good technical analysis. A well-organized and neatly prepared handout summarizing all three areas will be appreciated.

Management meetings will be tough and very stressful. It is important you be well prepared and have thought through your ideas. Your plan should identify key tasks and dates. It should show a logical sequence of events that you plan on following. Take the time to explain the importance of each task and how the task will contribute to the solution. Make sure the plan is realistic and can be accomplished within the time you have budgeted and with the resources available. If you need more time or resources, then identify the need.

▶ *Career Tip.* Let management catch you at your best, not your worst when things are failing.

The technical part of your presentation should describe the problems. It must be exact and any analysis presented must be correct. Use graphs and math

FIGURE 23-1   Having a plan and positive attitude are key.

modeling supported by any test results you have obtained. Compare modeling results to test results as much as you can. A thorough review of test results is always looked upon very favorably.

Remember your objective is to demonstrate that you have the technical knowledge necessary to successfully solve the problems and you are the right person for the job. Bring photos of anything that will help you illustrate the problem. The optimum career move is to bring the managers down to the lab and let them see, handle, or run whatever it is you are working on. People are more sympathetic when they see firsthand how difficult the problem is and will naturally become involved.

The attitude that you project at review meetings and during discussions with your coworkers also affects career development. The project may be unsuccessful in the end, but with a good attitude you can minimize the damage that can be done to your career (Figure 23-1). What you would like to hear your supervisor say at completion of the project is "The project was unsuccessful but you had such a good attitude that I want you on my team again!"

I strongly recommend not taking things personally and do not fight nor blame during meetings. This may be hard to do as people start to criticize your plans. You need a calm and level-headed approach when everyone else around you may be losing their composure. Rather than arguing with them, spend the time and energy in drawing out their ideas on how to make improvements. This will get their ownership in the plan and they will be less likely to blame you when their ideas don't work either.

▶ *Career Tip.* Be part of the solution and not part of the problem.

This attitude is one of always looking for solutions and volunteering help. Supervisors usually welcome someone willing to take on more work or try out something new after hours. Be willing to try out new ideas even though they are not yours. Often someone else will have a potential idea and need you to try it. Be willing to give it a try even though you may not agree with it.

Sometimes just following orders can benefit your career more than you realize.

Stay away from any negative statements. Being Ms. or Mr. Doom and Gloom does not help the team and may even contribute more to the setbacks and problems of the project. Some of the doom and gloom statements that indicate to management that they have the wrong people on the team are:

> That will never work because. . ..
> There is nothing we can do. . ..
> It's impossible. . ..
> No sense trying, the project will not be successful because. . ..
> Why won't it work? (As compared to: What will work?)
> It's not my idea. No way am I. . ..
> It's not my fault, it's their fault for. . ..
> It's not my department that. . ..

Some of the good attitude statements that reinforce management's conviction that they have the best person on the job are:

> I'd be willing to try that because. . ..
> The good points of the solution are. . . ?
> I think I might be able to make it work if we. . ..
> I'd be happy to put in extra hours to see if it works.
> How can I help the team out?
> What are the good points of the. . . ?
> It doesn't matter who caused the problem, we must. . ..
> Just tell me what I can do, I want to help with. . ..

The reason that attitude is so important is that most managers have worked on projects that have been unsuccessful. They realize that tough times require a positive attitude. It is the person who keeps on going who will eventually succeed and attitude has a lot to do with it. Besides, all projects come to an end and there will be the next project to work on. Supervisors will make assignments for the next project based on the performance demonstrated on the last project. Who do you think they will choose—a person who projects doom and gloom or one who has a positive attitude and is willing to work through problems? Who do you think they are going to promote regardless of how the project turns out?

▶ *Career Tip.* Having a positive attitude when setbacks occur isolates your career from unsuccessful projects.

At the end of an unsuccessful project one good action is to document the lessons learned. This may take the form of writing a simple memo that documents all the good and bad lessons learned during the project. Volunteering to write this memo is good for several reasons. First, management looks on this activity very favorably since it helps share with other groups those things that worked so that they can capitalize on them and avoid those things that did not work. Second, after you have written several of these memos you will have acquired an excellent library of things to do and not do on a project. This knowledge is power for future projects.

Finally, do not fix blame on any one individual; it was a team effort and everyone was unsuccessful together. The best thing you can do is learn from your mistakes and move on. Most people can handle success, but the *really* successful people in life are those who can learn to handle setbacks and problems only to overcome obstacles and turn an unsuccessful project into a success.

ACTIONS TO BENEFIT YOUR CAREER IF THE PROJECT IS A SUCCESS

The discussion thus far has focused on what to do if the project you are working on is experiencing setbacks and looking unsuccessful. Now let's look at what to do if the project you are working on is headed for success.

Most people believe that project success guarantees a promotion. This could not be further from the truth. Project success does not guarantee a promotion, but it does help. The reason that success does not guarantee a promotion is due to the fact that management often devote their time and energy to projects that are in trouble. Another common response by management is "Why should I promote you for a successful project? That's what you get paid to do. You are just doing the job I hired you to do in the first place. If it weren't a success you wouldn't be doing your job. We don't give promotions for just doing your job. You must, therefore, take actions that maximize the benefits from working on a successful project.

▶ *Career Tip.* Successful projects require you to go into a high-energy state.

As with unsuccessful projects, your response to a successful project should be to go into a high-energy and high-output state. This means simply making your efforts visible to management to show the excellent results you are obtaining. Again, these visible efforts include working extra hours, getting organized, making excellent technical presentations on the great results, identifying benefits, and projecting an attitude of success.

Do all these actions sound familiar? They should because they are the same actions you would be taking for a project that is unsuccessful, but with a different twist on them. Let's explore some specific actions you can take for a successful project.

The more time you put in, the better you look and the more benefit gained. If you put in extra time and your supervisor does not know about it, or they do not know how well things are going, you are setting yourself up for disappointment. At the end of the week, it is a good idea to stop by their office and let them know how much overtime you worked and the exciting results you obtained. Again, the same efforts an unsuccessful project required.

The next thing you must do is get organized. The question is, how can you show that you are organized? Again, it's simple. The first thing to do is generate a report showing all the good results, highlighting the things that went well. Compare modeling results to test results, show how the results met or exceeded plans. By writing up the results and giving them to your supervisor you are accomplishing two things. First, you are showing that you are organized and second, you are providing a "Good News Report" that can be shared with other supervisors. Upper-level management will get the impression that your supervisor has assigned the right person to the job, making your supervisor and you look good.

Oftentimes when a project is successful, management feels no need to review progress. In this case, you call the meeting. Make sure you have a well-organized and thorough report. The report should summarize the plan followed, any technical analysis, test results, and a benefits summary. A well-organized and neatly prepared handout summarizing all three areas will be appreciated.

The technical part of your presentation should describe the problems you solved. It must be exact and any analysis presented must be correct. Use graphs and math modeling supported by any test results you have obtained. Compare modeling results to test results as much as you can. A thorough review of test results is always looked upon very favorably. Remember, your objective is to demonstrate that you have the technical knowledge that contributed to the success of the project and you are the right person for the job.

Bring photos of anything that will help you show the success. The optimum career move is to bring the managers down to the lab and let them see, handle, or run whatever it is you are working on. People are more likely to appreciate the trouble you went through and the magnitude of the success when they see them first-hand.

▶ *Career Tip.* When reporting the project's success to management, report it in terms of "our" and "the team's" success, rather than "I."

The attitude you project at review meetings and during discussions with your coworkers is also important to career development. Make sure you give credit for success and report things in terms of "we" and not "I." The reason this attitude is so important is that most managers have worked on projects that were successful. They realize it required a team effort and no one individual did it all. Supervisors will make arrangements for the next project based on the performance demonstrated on the last project. Who do you think they will choose—a person who takes all the credit or one who is a good team player and willing to share the credit? Who do you think they are going to promote?

▶ *Career Tip.* Spread the credit for success around as much as possible.

If you are a team leader for the project there are several additional things you can do to help the team capitalize on the results. First, you can nominate the team for a company award. Or you might try to get an article published in the company newspaper. Make sure you get the team members' names in the article. Another good thing to do is take a team photo and pass out copies to the team members.

Photographs of the hardware or the test results are always a good thing to give your supervisor to show others or put them on an office wall. If your supervisor hesitates to do anything, you might point out the benefits received when it becomes known that they were responsible for assembling the team.

▶ *Career Tip.* Go all out to get the team rewards; it will pay dividends for years to come.

If it is possible, you may want to publicize the good results outside the company. Writing a paper and submitting it for publication or presentation at a symposium is an excellent idea. This gets your name known throughout the industry rather than just throughout the company.

At the end of a successful project, it is again a good idea to document the lessons learned. This may take the form of writing a simple memo that documents all the good and bad lessons learned during the project. Volunteering to write this memo is good for several reasons. First, management looks on this activity very favorably since it helps them share your work with other groups so that they can capitalize on it and avoid those mistakes and setbacks. Second, after you have written several of these memos you will now have acquired an excellent library of things to do and not do on a project. This knowledge is power for future projects.

SUMMARY

The so-called "truths" that you encounter in your engineering career should not be accepted at face value because they limit your career growth. This is especially true of the fallacies that successful projects always result in promotions and projects that are unsuccessful will limit or damage your career. There are actions you can take to ensure that your career continues to grow regardless of the outcome of the project. Hopefully, you have realized that the actions are similar for the successful project and the unsuccessful project.

For an unsuccessful project, you need to put in extra effort, get organized, develop a recovery plan, brief management, and identify the reasons for setbacks. For a successful plan, you also need to put in extra effort, organize the good results, brief management, and identify the reasons for the success. In either case you will be operating in a high-energy state and with positive attitude, giving credit where credit is due and not fixing blame. Excellent technical presentations showing theory, modeling, and test results, are musts following both unsuccessful and successful projects.

Have you identified any career actions you want to take as a result of reading this chapter? If so, please make sure to capture these ideas before you forget by recording them in the notes section at the back of the book.

ASSIGNMENTS AND DISCUSSION TOPICS

1　If you are working on a major project, determine if management considers the results to date to be successful or unsuccessful.

2　When can you be sure the project is successful or unsuccessful?

3　If it is successful, what should you be doing?

4　Name three ways to share the success.

5　If it is unsuccessful, what should you be doing?

6　Can you name any other fallacies that might be limiting your career that you are not even aware of? For example:

No raises or promotions are given in bad economic times!
They won't promote me because I'm. . ..
They never have before!

(Hint: Fallacies are usually great sounding reasons that put blame on some abstract or uncontrollable circumstance.)

CHAPTER 24

HOW TO ASK FOR A RAISE
IT IS EASIER THAN YOU THINK, EVEN IN BAD TIMES

Do you feel underpaid and deserve a higher salary? Are you thinking about asking for a raise and not sure how to go about it? Does the thought of approaching your boss for a raise conjure up fear in you? In this chapter, I provide guidelines to make the task less intimidating. By following these guidelines you will be maximizing your chances for success and hopefully get the raise you want.

Here are the basic steps to follow when asking for your raise [1].

1. Do your research and build your case
2. Prepare and plan your presentation
3. Set up the meeting
4. Ask for the raise
5. Handle rejection

DO YOUR RESEARCH AND BUILD YOUR CASE

There is a lot of research you need to conduct prior to asking for the raise. You need to build your case and clearly identify why you deserve one.

Bad Reasons. First of all, do you have the right reasons? If you are asking for a raise because you need the money and can't pay your bills, you are asking for the wrong reason [2,3]. If paying your bills is the justification for your raise, your boss is more than likely to tell you that you have a personal financial problem—not a reason for them to give you a raise. Giving you a raise is not the answer. Look at the situation from your employer's prospective and base your case on why they would consider you for a raise and what you can do for them.

The Engineer's Career Guide. By John A. Hoschette
Copyright © 2010 John Wiley & Sons, Inc.

Another insufficient reason is that others in your group are being paid more and you feel your performance is as good as theirs. This is not a good reason since you do not know of all the experience and extra skills they have for which the company is willing to pay more. Managers know if they give a raise for this reason and word gets out, your entire group will be in the boss' office asking for a raise. This is a management nightmare and for this reason, managers do not give raises just because an employee feels they are doing just as good a job as others.

Right Reasons. The truly only right reasons for deserving a raise are that your job performance is outstanding and you are underpaid. You will need both of these reasons to build a solid case.

▶ *Career Tip.* Raises are justified on outstanding performance and being underpaid.

Underpaid. To build the case that you are underpaid, you will need to conduct some research. Here is research you should consider to establish the case that you are underpaid.

1. If you feel you are underpaid because others in the industry are getting considerably more, then research the salaries of others in the industry and show how your salary is below these. There are several websites that can provide salary data based on your job description like www.engineeringsalary.com and www.payscale.com. You can use the data provided by these and other sources to show how much your salary should be under the norm.

2. If you feel you are underpaid on the basis of your company's internal salary ranges, then you can build your case on the basis of this. By checking with your Human Resources department you can determine if you are being paid for your level and the rating you received. For instance, if you are rated an E3 and have been consistently receiving a performance rating of excellent, but are only being paid at the E3 level for average performers, then you definitely have a case to ask for a raise.

3. Cost-of-living adjustments is another good reason for being underpaid. If you have not been getting yearly cost-of-living raises your salary has actually decreased. Managers understand this since they are also subject to this. In addition, there are yearly salary range adjustments companies make based on market conditions alone. A company will increase the pay range to attract new hires and may not adjust the employee's salary. After several years of this happening, the employee is underpaid relative to new hires. Managers recognize this and consider it a reason for deserving a raise.

Outstanding Performance. To build the case for outstanding performance you will need to seriously review your accomplishments and overall job performance [3,4]. If you have been receiving average ratings, then this is going to be an extremely tough sell. If, however, you have been receiving outstanding ratings and awards, your case is much stronger.

Make a list of your recent significant accomplishments for the company and how they contributed to the bottom line. Document costs savings, productivity improvement, important projects achieved, above-the-call customer service, and ways in which you have contributed more than your job required [4]. Some other reasons for giving a raise might include extra revenue you generated, tight deadlines you've met or beat, new initiative you took above your normal job, and extra hours you put in.

Make a list of any additional responsibilities you have added to your job. An increase in responsibility, more employees supervised, or special projects are often grounds for an increase. Documented outstanding performance where you clearly went above the norm will definitely support your case for a pay increase.

Research Company Policy. Find out your company's policy on salary increases. Read your employee handbook, look at company policies, and check with your Human Resources department. Doing this should reveal the process whereby salary increases are granted. If a policy or a process exists, your best bet when asking for a pay raise is to follow the process exactly.

Other things to consider are raise cycles. Are all employees reviewed at the same time each year and are raises given only at that time? Does your boss have a budget to give you a raise? For the highest chance of successfully getting the raise you want, you have to know the company's policies regarding compensation—if your boss alone has authority to grant the raises or do other departments like Human Resources get involved? Once you have conducted and documented your research you are ready to move on to preparing and planning your presentation.

PREPARING AND PLANNING YOUR PRESENTATION

Here are some good actions to take when you start preparing and planning your presentation.

Networking. Network with other employees or engineers in the industry who might have recently asked for a raise. Professional associations also do salary surveys and provide networking opportunities with people in similar jobs. Ask how they prepared and if they have any recommendations for you.

Have a Reasonable Figure. After you have done your research into salaries, you should have a good idea of how much you are underpaid and what a fair and equitable raise would be. When it comes time to discuss how much of a raise you want, talk of how much you are underpaid in terms of dollars per month. Using dollars to describe the amount underpaid makes it look better for you. When you discuss raises, put it in terms of percentage of increase; this number appears much smaller and easier for people to accept. This gives the appearance of being significantly underpaid and only a small percentage increase is requested.

Don't Use the Ultimatum Plan. Some people feel the right thing to do is give their boss an ultimatum. Give me the raise or I quit is their approach. Many managers faced with this situation will simply call your bluff. Considering the employment situation and job competition today, this approach is not recommended [5].

Practice. Practice and rehearse your pitch at least five times before you meet with the boss [6]. Practicing will help you appear confident and firm about your request [6].

Timing. Timing is everything, including how your company stock is doing, how the project you are working on is going, time of year, day of the week and hour. The ideal case is to ask just after the company announces record sales and profits just as you successfully completed a very difficult assignment. Studies have shown that at on Fridays, workers are in the best mood, and therefore more agreeable rather than at the beginning of the week, and especially not on a Monday [7]. Try to pick a time during month that is not your boss' busiest time. A time when your boss will not be distracted by deadlines and will have some extra time to work your request.

Familiarize yourself with your employer's pay practices. If increases only occur once a year, you are unlikely to receive a raise at any other time. If your company offers more frequent increases, you'll have more luck asking for a raise.

Asking During Tough Times. Your company might be losing money because of a downturn but if you can prove that you're vital to getting the company through the recession, then a raise is assured [8]. Also, if the company may have downsized and as a consequence you've taken on additional responsibility and people that warrants a higher salary or promotion.

Anticipate Objections. Your boss is going to give you objections and you should be prepared to have answers that overcome these objections. Here are some objections you might encounter and good answers.

"I can't give you a raise, I don't have the budget and I need upper management approval.": Your response to this needs to stay focused on two reasons: underpaid and excellent performance. Again repeat your best reasons and then ask what is needed to get the budget and upper management approval. Point out how by your excellent performance you overcame barriers like these and you are sure your boss can overcome these too.

"It's not raise time, therefore I cannot do anything.": For this objection you can focus on what you can do. First, the response indicates the boss is not arguing with your reasons and must believe they are good. State this obvious fact to get agreement that this is the case. Next, set the stage for the next raise period by asking if you can expect the raise during the next salary adjustment period. If the answer is still no, then probe further and ask what is the reason.

"I don't know if I can.". This is most common since the boss may not really know if it is possible to get you a raise. Your response is to ask what should we do to find out if you can. Who else should we be talking to?

Now that you are prepared, it's time to contact your supervisor and set up a time to talk.

SET UP THE MEETING

Set up a meeting a couple of days in advance with your immediate supervisor to discuss your compensation. It is a good idea to explain that the purpose of the meeting is your desire to discuss your pay in light of your recent accomplishments and pay relative to your overall performance on the job. This will show your boss how serious you are about asking for a raise [9]. Don't discuss your raise by e-mail, in the hallway, between meetings, or by telephone. Meet with your boss in person.

Give your supervisor time to prepare for the meeting. Your boss will want time to do their own research into company policies and consult with Human Resources. It is best to pick a neutral meeting place like a conference room with a door instead of the boss' or your office.

ASKING FOR THE RAISE

Start the meeting on a pleasant note and make some small talk to begin. After a few minutes transition to the reason why you are there. Start with reviewing the performance reasons why you deserve a raise. State the reasons slowly and directly and make eye contact. It is okay to talk from a note sheet. Let the boss respond to your reasons and try to have a two-way conversation.

Then follow your reasons with how you consider your salary low and that you deserve a raise. Show the evidence you have about how low you think your salary is and the size of the adjustment you consider reasonable. Keep it strictly professional with no shouting or yelling and discuss everything in a confident and firm voice.

Make it perfectly clear as you summarize that you want a X% raise and would like to know if the boss is going to work to help you get it. Then stop talking and let the boss respond. Listen closely to the reasons being given. Is the boss saying no, maybe, or yes? If yes, then you want to tell him you are glad to hear it, appreciate how your boss agrees with you, and how you are looking forward to the raise. If the boss is saying anything other than yes, then you are on to the rejection scenario.

HANDLING REJECTION

Getting a "no" from the boss does not have to be the final word. If your boss is not able or willing to grant your request for a raise then have a Plan B. You can ask your boss for other perks in lieu of a raise such as additional vacation days, more job flexibility, or a change of office, if this is acceptable to you. Another way to deal with rejection is to ask what you can do in the next six months to make this conversation successful the next time. Ask the boss to be as specific as possible.

Do not respond with anger or by threatening. You will need to continue to interface with your boss on daily assignments and putting up walls around you is never going to help your cause. Another natural reaction is to cut back on your work in retaliation for the rejection. Some people think if they are not going to pay me more then I am going to do less. Keep your performance up, continue to show them you deserve it, and get ready for the cycle of raises.

SUMMARY

There are just two basic reasons why a company will give you a raise. The first is that your performance is outstanding, and the second is you are underpaid. The wrong reason for asking for a raise is because you need money to pay your bills. The basic steps to follow when asking for your raise are:

1. Do your research and build your case
2. Prepare and plan your presentation
3. Set up the meeting
4. Ask for the raise
5. Handle rejection

If you follow these guidelines and present a compelling case, you are more than likely going to be successful. Set up a special one-on-one meeting with your boss and come prepared to present your case and handle objections. Look your boss directly in the eyes and speak with confidence about your desire for a raise. Do not have an emotional and bitter exchange with your boss since it is only going to seriously hurt your career. Be prepared for "no" and have a Plan B just in case.

Have you identified any career actions you want to take as a result of reading this chapter? If so, please make sure to capture these ideas before you forget by recording them in the notes section at the back of the book.

ASSIGNMENTS AND DISCUSSION TOPICS

1   What are the only two good reasons to grant a raise?
2   What are bad reasons?
3   Why is it important to control the timing?

REFERENCES

1. Storm, Alison, "How to Ask for a Raise," Web site www.bargainst.com (07/21/ 2008). http://www.bargainist.com/deals/2008/07/how-to-ask-for-a-raise/.

2. "How to Get a Pay Raise," Web site www.financialplan.com. http://financialplan. about.com/cs/personalfinance/a/HowToGetRaise%20.htm.

3. Heathfield, Susan M., "How to Ask for a Pay Raise, Steps in Asking for a Pay Raise," Website About.com: Human Resources. http://humanresources.about.com/od/ salaryandbenefits/a/ask_raise.htm.

4. Niznik, John Steven, "Asking for a Pay Raise," http://jobsearchtech.about.com/ od/salary7/a/pay_raise.htm.

5. "How to Ask for (and Get) a Raise Like a Man," http://artofmanliness.com/2008/ 02/17/how-to-ask-for-and-get-a-raise-like-a-man/.

6. Johnson, Tory, "How to Ask for a Raise — Women Have to Be Confident When Making Their Case for More Money," April 28, 2006 http://abcnews.go.com/ GMA/TakeControlOfYourLife/story?id=1898808&page=1.

7. Santiago, Andrea, "Asking for a Raise — Quick Tips, Increase Your Salary with a Few Easy Steps," www.About.com.

8. Weiss, Tara, "How to Ask for a Raise When Times Are Hard," 04.29.08, http://www.forbes.com/2008/04/29/raise-downturn-interview-lead-careers-cx_tw_0429bizbasics.html.

9. McKay, Dawn Rosenberg, "How to Ask for a Raise — Tips to Help You Ask for a Raise," www.About.com.

CHAPTER 25

HOW TO GET YOUR NEXT PROMOTION

Getting your next promotion is not as simple as just asking for a raise; it is accomplished with hard work and a plan. The reason it is more difficult is due to so many more factors that have to do with your skills, business, and the economy. In addition, the competition for promotions is more intense today than ever before. In order to get your next promotion you will have to control all factors. It is not an easy task to move up the corporate ladder and each successive promotion becomes harder and harder. In this chapter, proven successful guidelines are provided to make this clearer for you and significantly decrease the time to your next promotion.

UNDERSTANDING THE REASONS MANAGEMENT PROMOTES PEOPLE

The reasons why people get promoted are entirely different from the reasons why raises are given. When you get a raise it is because you are doing an excellent job and there is growth in your job level. When you get promoted it is because you have clearly demonstrated that you can perform successfully at a level above yours. Perhaps, for example, you took on the extra responsibility of a level above yours and successfully handled the challenges. Management wants a sure thing and the only way they feel comfortable about promoting you to the next level is when they see you can already handle the job before you are promoted. They do not promote people hoping the person will be able to do the job once they get there. With management it is "Show me first you can do the job then we'll talk about promotions."

Many people have just the opposite point of view and some real promotion killers are attitudes of "I wasn't hired to do that," or "when you start paying me more, I'll start working more" [1]. These attitudes clearly signal to

---

management that you are not a promotable person. These people simply don't get it.

THE PROCESS TO ACHIEVE YOUR PROMOTION

The best way to approach this complex problem is first understanding the process to follow and then the actions and variables you have to control all through the process. The recommended process to follow for getting a promotion is shown in Figure 25-1. It is recommended because it works.

Step 1. Plant the Idea and Do Research. Most people do not realize under the best cases that it is going to take a minimum one year to demonstrate the type of performance you need to be promoted. In most cases promotions normally take much longer. Therefore, you need to plant the idea of a promotion in your boss' head early in the process. The way to do this is by meeting with your boss and having a career discussion where you identify that you are looking for a promotion in the near future.

The purpose of the career discussion is to identify what you need to improve upon to qualify for a promotion. Research the next level as well as what you have already demonstrated as promotable accomplishments? Ask your boss to list what they consider as the necessary skills for the next level and have them comment on how close you are to possessing these skills.

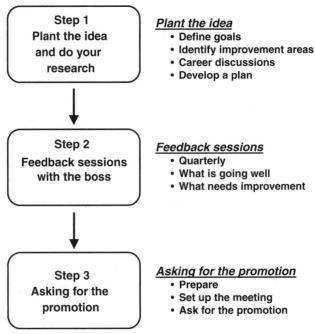

FIGURE 25-1   The process to achieve your promotion.

During the career discussion, record the actions you need to take and then outline a plan that leads to demonstrating these skills with the ultimate goal of being promoted [2].

Developing a plan creates two very positive things. First, it puts the boss at ease that you are not asking for a promotion immediately but are willing to work for it. And second, it plants the idea that you are going to work hard for a promotion, and it is only a matter of time before you demonstrate you deserve it. Once you and your boss agree this is a good plan, write up the plan and keep a copy in a place you can look at every day. Give your boss a copy.

Step 2. Feedback Sessions with the Boss. About once a quarter meet with your boss to get feedback on how you are doing against the plan. Your objective in every quarter is to complete a few more actions and identify what is left to work on. Clearly identify what is going well and what is not. Get approval from your boss to close out actions on the list that you successfully completed and acknowledge which actions are remaining and which needs improvement.

After a couple of quarterly meetings, you should have your list down to one or two remaining items. At this point, the boss should be realizing you are getting very close to completing all the criteria and will probably be asking for a promotion shortly. You might emphasize this point by highlighting during your feedback session that you are on track to complete all the criteria in the next quarter. Once you get the list down to this point you are ready to move on to step 3.

Step 3. Asking for the Promotion. This step in the process is identical to when you ask for a raise. You prepare for the meeting by collecting your supporting data. In this case, it is the list with all actions completed. You set the meeting up and go prepared to ask for the promotion. More on this later in the chapter.

Now that we have reviewed the process, we need to change the focus and determine which items and factors you will need to control and monitor all during this process. These factors are related to your skills, the business, and the economy.

WHAT IS REALLY IMPORTANT FOR GETTING A PROMOTION

As I discussed earlier in the chapter, many more factors come into play for a promotion and these factors are related to your skills, business, and the economy. Controlling all these factors in your favor is required for getting the promotion. Let's look at these factors and what you can do to control them in your favor.

Your Performance. The first and most important factor in determining your readiness for a promotion is your performance [3]. This is your performance as your boss perceives it and measures it [4]. To get your promotion, you will need at least one, if not two, performance reviews where your boss rated you at the top of the scale—either outstanding or excellent. You need consistently high ratings since this is the basic factor the company and Human Resources consider as evidence you are ready for promotion.

How have your ratings been? If they are just average or slightly above average, then during your feedback session you need to discuss how you can turn these average ratings into excellent ratings. During your feedback sessions detail out the tasks and results needed in order for you to receive a top rating. Once you have identified these actions work at them every day. Focus your time and energy on the tasks, determined by your boss, that are most going to help your ratings.

Now the real problem—in general you will need at least two consecutive ratings at excellent. If your company only does ratings once a year this means you are possibly two years away from your promotion. Get going now if you want a promotion soon.

The next is performing above and beyond your present level and successfully demonstrating that you have the skills to handle the next level up. This will only occur if that you take on more responsibility and do significantly more than your job calls for.

A Hint Here. During your feedback sessions talk about taking on more responsibility and tasks to perform at a level above yours. Stay away from the "promotion" word until you are at the end of your plan. If you are constantly putting everything in terms of a promotion your boss may get the wrong impression that "you're only interested in the promotion and not the work."

Skills and Education. As you move in the company each level brings more responsibilities and challenges. To successfully move to the next level and handle these additional responsibilities you may need further education or new skill development. For example, if you are moving to the level of a team leader you may need training in conducting meetings and handling difficult people. These may be new skills that you were never trained for in engineering. Or the next level up may require an advanced degree. The best thing you can do is, early on when you put your plan together, check the requirements for the next level up and make sure your plan includes learning these new skills or obtaining any education required.

Boss and Upper Management. In most cases, your boss, Human Resources, and upper level management, need to support your promotion. Your plan should address how you are going to convince all these people that you are ready for advancement. Showing up one day with all your actions completed will not result in a promotion. You will need to socialize and network your plan with everyone involved in the decision. This is a good discussion topic at your feedback sessions with your boss. Together, you will be able to work this process.

Economy and Budget. The next factors that come into play are the economy and the budget. You may have successfully completed everything on your plan but if the economy is bad and your company is losing money and laying people off, then the chances of getting a promotion are very slim. As you work toward your next advancement, monitor the economy and the company's situation to see if they favor promotions. If they don't, you may have to wait until things turn around.

There must also be a budget for you to be promoted. The reason is that when you are promoted most companies give a raise that costs the company more. Is there enough budget for your additional salary once promoted or do you have to wait for another fiscal year when there will be sufficient funds available.

Position Open. The next factor to consider is whether there are positions open that you can be promoted into [5,6]. If there is no position open, then it becomes an obstacle. If you and your boss work together, then the boss should be able to identify the need for the position you want to get promoted into. Upper management has checks and balance to make sure first-line supervisors are not just promoting without true cause. It is very helpful if your boss opens a new job requisition for the level above you just prior to completing your plan. However, opening a new job requisition opens the door to competition that gets us to the final factor.

Your Ability to Convince People You Are the Best Person. If you have successfully completed everything on your plan as defined by your boss during the feedback sessions, then you should be the best candidate. Of course, you will still need to convince others in the decision chain that you are the best person. At this point you are interviewing like everyone else for the new position, so having great interviewing skills and the ability to market yourself now becomes key. Think of it as interviewing for a job and prepare for it. Update your resume, develop a 30-second commercial, and update your portfolio. It helps to have personally networked, well in advance, with all the people involved in the decision.

Following these guidelines will significantly enhance your chances of getting that promotion; by taking control and shaping these factors in your favor you are maximizing your chances for success.

Networking is another key activity that should be occurring all during the execution of your plan. Networking with other employees, your mentors, or engineers in the industry who might have recently been promoted or involved in a promotion process can reveal many helpful hints. Inquire how they prepared and solicit any recommendations.

ASKING FOR THE PROMOTION

This is the final stage in the process. During all the feedback sessions watch and listen to your boss' reactions to the discussions. Their reactions are going to clearly indicate how things are going. If your boss is positive and acknowledging your progress during the feedback session you are on the right track. However, anything less than this is indication that something is not up to par. If this is the case then you have to use the career tools and skills

I discussed in Chapters 6 and 14 on career discussions and barriers to discover the reasons.

Preparing and Planning Your Presentation. In the final months of your plan is when everything has to come together perfectly. Schedule your last feedback session about three months in advance of the next cycle of promotions. It is the best timing to ask for the promotion. This gives the boss time to forecast your promotion in the next cycle of performance reviews.

Here are some good actions to take when you start preparing and planning your presentation.

Practice. Practice and rehearse your pitch at least five times before you meet with the boss. Practicing will help you appear confident and firm about your request.

Set Up a Meeting. Set up a meeting a couple of days in advance with your immediate supervisor to discuss progress on your plan. It is good to explain that the purpose of the meeting is to discuss how you are at the end of the plan and you feel confident that you have demonstrated everything to qualify for advancement. Plan together the next steps. Don't discuss your advancement progress in the hallway between meetings or by telephone. See the boss in person for this.

Give your supervisor time to prepare. Your boss will also want time to do their own research into company policies and consult with Human Resources. It is best to pick a neutral meeting place like a conference room with a door instead of the boss' office or your office.

Asking for the Promotion. Start the meeting on a pleasant note and make some small talk to begin with, then after a few minutes transition to the promotion discussion. Start with reviewing the plan to show how you have met all the criteria and deserve a promotion. State the reasons slowly and directly at the boss while making eye contact. It is alright to have a note sheet to talk from. Let the boss respond to your reasons and try to have a two-way conversation.

Keep it strictly professional with even voice tone and discuss everything confidently. Make it perfectly clear, as you summarize, that you feel you have completed everything successfully in the plan and you qualify for a promotion.

Then let the boss respond. Listen closely to the reasons being given. Is the boss saying no, maybe, or yes? If yes, then you want to tell him that you are glad to hear it, and appreciate how your boss agrees with you, and how you are looking forward to the promotion. If the boss is saying anything other than yes, then you need to listen carefully and be prepared to discuss further. Move the conversation toward determining what specific actions are needed to get the promotion approved. If the boss is saying anything other than yes, then you are on to the rejection scenario.

Don't Use the Ultimatums Plan. Some people feel the right thing to do is give their boss an ultimatum. Give me the promotion—I deserve it or I am going to quit. Many managers faced with this situation will simply call your bluff. Considering the employment situation and job competition today, this approach is not recommended.

Anticipate Objections. If your boss gives you objections you better be prepared to have answers that overcome these objections. Here are some objections you might encounter along with good answers.

"I can't give you a promotion, I don't have the budget and I need upper management approval." Getting a "no" from the boss does not have to be the final word [7,8]. If your boss is not able or willing to move forward and recommend you for a promotion, then have a plan B. Plan B is to ask what you can do in the next six months to make this conversation successful the next time. Ask the boss to be as specific as possible.

"It's not promotion time, therefore I cannot do anything." For this objection you can focus on what you can do. First, the response indicates the boss is not arguing with your reasons and must believe they are good. State this obvious fact and get agreement this is the case. Next, set the stage for the next promotion time by asking when this occurs and what do we need to do to get ready for it. If the answer is still no, then probe further and ask what other things are stopping it.

Handling Rejection. Do not respond with anger or by a threatening reaction. You will need to continue to interface with your boss on daily assignments and putting up walls around you is never going to help your cause. Another natural reaction is to cut back on your work in retaliation for the rejection. Some people think if they are not going to get promoted then I am going to do less. Keep your performance up, continue to show them you deserve it, and get ready for the next feedback session.

SUMMARY

Getting your next promotion is not as simple as asking for a raise, but with hard work and a plan you can make it happen. The reason it is more difficult is due to so many more factors entering into the mix. Some of these factors have to do with your skills and others are related to the business and economy. In addition, the competition for promotions is more intense today than ever before.

The reasons why people get promoted are because they have clearly demonstrated performance at a level above theirs. Management wants a sure thing and feels comfortable about promoting you to the next level when they see you can already handle the job before you are promoted.

There is a three-step process to getting your next promotion. These steps are (1) plant the idea and develop a plan, (2) conduct feedback sessions, and (3) asking for the promotion.

The key factors you have to control during the process are

Your Performance
Skills and Education

Boss and Upper Management
Economy and Budget
Position Open
Your Ability to Convince People You Are the Best Person

Preparing and planning your presentation to ask for the promotion has to occur at the end, after you have already successfully demonstrated all the actions required to be recommended for a promotion. When asking for the promotion be firm, speak confidently, and state your case in a positive tone. If the answer is yes, then move to the next step of helping your boss get the approval. If the answer is no, go to plan B and ask what else is needed in the next six months to make this happen. Don't use ultimatums. You can get the promotion you want but it takes planning and controlling all the factors to the best of your ability.

Have you identified any career actions you want to take as a result of reading this chapter? If so, please make sure to capture these ideas before you forget by recording them in the notes section at the back of the book.

ASSIGNMENTS AND DISCUSSION TOPICS

1 Identify the three steps in the process.
2 What are the factors involved in getting a promotion?
3 Why are feedback sessions good? How often should they occur?
4 Is it important to document your plan?

REFERENCES

1. Lenihan, Rob, "Promote Yourself at Work," http://money.cnn.com/2000/02/15/career/q_promotion/, February 15, 2000.

2. Jenkins, Linda, "Raises and Promotions," Web site Salary.com, http://www.salary.com/personal/layoutscripts/psnl_articles.asp?tab=psn&cat=cat011&ser=ser031&part=par173.

3. Chambers, Harry E., *Getting Promoted: Real Strategies for Advancing Your Career*, Perseus Books, 1999.

4. "Asking for a Promotion: Never Done It Before? Here's What You Need to Know," Web site http://www.hundredsofheads.com/37-241-1.Article/Asking-for-a-Promotion.

5. "Time to Advance Your Career? What You Need to Know About Asking for Promotion," Web site http://www.articlesbase.com/careers-articles/time-to-advance-your-career-what-you-need-to-know-about-asking-for-promotion-263641.html.

6. Agarwal, Abhishek, "Asking for a Promotion—Tips to Get It Easily," http://ezinearticles.com/?Asking-For-a-Promotion—Tips-To-Get-It-Easily&id=1659934.

7. Weiss, Tara, "How to Ask for a Promotion in a Recession," http://www.forbes.com/2009/01/07/job-career-promotion-leadership-careers-cx_tw_0107basics.html.

8. Thompson, Steve, "How to Ask for a Promotion," November 22, 2006, http://www.associatedcontent.com/article/87279/how_to_ask_for_a_promotion.html?cat=3.

CHAPTER 26

PEOPLE NETWORKING
A VITAL AND INDISPENSABLE CAREER SKILL

People Networking 101 is not taught in engineering schools, yet it is one of the most basic, vital, and necessary career skills an engineer must acquire. Knowing how to network is on the same basic career skill level as learning how to differentiate or integrate mathematical functions. An engineering student who cannot perform these basic mathematical functions has little or no chance of graduating with an engineering degree. An engineer working in industry must also have basic people networking skills to advance.

People networking is something you learn with practice and time. You may not do it perfectly at first but with guidance and practice you can learn how to effectively network to advance your career.

WHAT IS PEOPLE NETWORKING AND WHY IS IT SO CRUCIAL?

Many engineers do not realize they are probably already networking in their job. Have you gone to lunch with others and discussed problems you are encountering at work? This is networking. Have you found great time savers when using Microsoft Word, Excel, Power Point, or MathLab, and shared this with others? Have you also asked for help and gotten helped with any of these? Are you a member of an engineering society committee? Or have you recommended another engineer as an expert in their field and who people can get help from? These are all networking activities.

Networking is simply the ability to exchange beneficial information with other engineers in your department, company, city, state, or even other states and nations. It is not only limited to you exchanging information with only engineers, but all types of people at work and in your personal life. You need other engineers and people in your life to be successful. Others can, and will help you, if you know how to properly network. Networking can save you time, money, and energy in dealing with career problems or technical problems at work.

The reasons for networking according to Diane Darling in her excellent book *The Networking Survival Guide* [1] are

- Sharing of beneficial knowledge and contacts
- Getting and receiving help
- Getting more done with less effort
- Building relationships before you need them
- Helping others

Diane Darling identifies two basic styles of networking; they are strategic and serendipitous [2]. Strategic networking is a conscious and deliberate effort upon your part to meet people and develop a mutually beneficial relationship for you as well as their career. Good examples of strategic networking for the engineer are trade shows, marketing promotions, and client visits. Here you plan in advance to meet new people and network to build relationships. Hopefully these new relationships will benefit your career and your company.

Serendipitous networking is unexpected or not planned. It occurs without preplanning and thought. Often this occurs in more of a social setting rather than a work environment. Good examples of this are at engineering conference social nights and after-hours social events. People meet in a relaxed environment with casual, informal conversation. The participants soon realize they have things in common and start to exchange ideas and thoughts. Usually by the time the conversation is over both have realized they have gained a new friend or acquaintance who could provide some help in the future.

Note that true networking is not a one-sided relationship where only you receive and everyone you network with gives help to you. True networking requires you to share and help others as much or more than you receive. One-sided networking relationships where you only interact with people to get information from them without giving or sharing information or help in return are referred to as schmoozing. Don't become a schmoozer. Also, networking is not selling. Giving a potential client a sales pitch on your company's products, services, or technology is not networking.

▶ *Career Tip.* Networking is NOT schmoozing or selling.

Networking is one very important career skill that every engineer needs to develop. Here are some hard facts supporting this:

1. Networking to get technical help may be the only solution when you are faced with problems above your skill capability. Networking with

the technical experts or subject matter experts in your company can significantly reduce the time to getting problems solved and products to market quicker.

2. Networking with other engineers who have similar interests is the premise of why engineers join a society. Networking for mutual benefits is the reason why many engineering societies continue to gain members in the age of the Internet. People prefer face-to-face discussions. Engineering societies are an excellent means to find out how others solved technical or career problems. It is an excellent means to find out what career openings are available at other companies.

3. Networking with cross-functional team members opens up resources and helps when faced with difficult problems that are outside your area of expertise, or deal with the business aspect of the job.

4. Many key jobs are filled in corporations without ever being posted or advertised. These un-posted job openings are filled through networking with key people in the company. By networking people can find out in advance when people are leaving and it allows an aggressive engineer to secure the job before it is posted.

5. Karen Susman, author of "102 Top Dog Networking Secrets" identifies that greater than 80% of jobs are filled by referrals, people networking over cold calls, and just submitting a resume [3].

WHO SHOULD YOU NETWORK WITH?

You do not know the next time you will need someone's help nor the type of help you will need. So the simple answer to this question is network with everyone you can. You should be networking every opportunity you have at work and out of work. To help you identify potential people to network look at the network diagram in Figure 26-1. Shown in the center is you and surrounding you are potential places or people to seek out for networking.

Universities are an excellent place for networking. At the university you have the opportunity to network with alumni, classmates, professors, and even students. Attending alumni events and keeping in touch with old acquaintances is great for networking. If you are taking a class, your classmates are good candidates for networking with. As always, networking with professors will keep you in touch with the latest advances in technology. Even volunteering and helping students is a good place to network. These students are the future employees and leaders of our companies and networking with them will give you insights into what they are thinking and what motivates the next generation of engineers. It is also an excellent means of recruiting the top talent of the class for your firm.

Networking Opportunities Are All Around You

FIGURE 26-1   Potentially great places and contacts to network with.

Job placement centers, job fairs, and recruiters are other great networking opportunities, especially if you are looking for employment. Getting to know these people and taking the time to develop relationships keeps you in touch with the latest career opportunities in the industry. If you are solidly networked in this area, your contacts will be contacting you first as new opportunities arise.

Networking with your suppliers is another good career move. Your suppliers will have knowledge about new and improved ways of building products and new processes. Networking with this group will provide you with valuable insight into the next generation products and processes that are becoming available. Early insight into this technical data can keep your company's products on the leading edge technology. Suppliers are also a great source for ideas on how to improve your production processes and reduce costs and at the same time improve the quality of your products.

Another excellent reason for networking with suppliers is the fact that suppliers are in contact with other people who are using their products. These other people who already have bought the products are successfully using them, are great sources of help when you run into problems. More than likely they have already run into problems you are experiencing and have solved them. One prime example of this is new software programs and tools. Software user groups are formed for this specific purpose: to solve problems and get better performance out of the software.

▶ *Career Tip.* When procuring products from a supplier, also ask for names of people who are already using the products. This way, if you encounter problems, you will have other users of the product who you can refer to for help.

Networking at engineering conferences and trade shows is putting you in contact with the movers and shakers of industry. You will be in contact with the people, technology, and new products. You will be exposed to new methods of producing products as well as the latest engineering design capabilities and tools. Networking with people who publish engineering magazines and trade journals is another great career advancing opportunity. Publishers have an unbelievable contact network that includes hundreds of people. Writing an article and getting it published is a lot easier when you personally know the publisher and the review board. Their helpful suggestions can significantly improve your articles and help establish you as an industry expert.

Do you have a favorite hobby or recreational interest out of work? If you are a member of a club you have an excellent opportunity to network with other people who have similar interests to you. These clubs might include golf leagues, bowling leagues, book clubs, and exercise clubs. When people meet and network at a social event it is usually more relaxed and people are willing to share their knowledge and experiences. Who knows? Your next job lead may come from a recreational club contact.

Networking with people who have the same technical background as you, will provide you with people you can refer to for help when you run into technical problems within your engineering discipline. Most likely they will have previously faced the problem you are working or may know someone who has, giving you a short-cut to solving a problem and hopefully a short-cut to success.

Networking with other engineers who have different backgrounds is another excellent career move. For instance, if you are a mechanical engineer, then networking with electrical or software engineers can yield contacts and potential sources of help when you face technical problems outside your technical area of expertise.

▶ *Career Tip.* Networking with other engineers of similar and different technical backgrounds is an excellent career move.

Another great area for networking is with your customers. More than likely your customers are going to know others who could use your products. These users might be in the same company or different companies. If you have a good customer network, new customers, new sales, and new work will be finding you instead of you searching them out.

Another excellent place to network is at engineering society meetings. When you join an engineering society and attend meetings, you are in the company of people who all have very similar interests to you. You are in the company of other engineers who come to these meetings to meet people who have similar backgrounds, have a desire to share knowledge, or receive help, and hopefully expand their technical knowledge base. By using these meetings to network with other members, you are developing a large network of people who offer the greatest potential for you to achieve a full and rewarding career.

▶ **Career Tip.** Joining an engineering society and networking with other members is an excellent career move.

Finally, the last area to consider networking with is voluntary organizations. You need to keep balance in your life and volunteering to help others and networking with this group is one way to do it. Sometimes our jobs can overtake our lives to the point that we have no life outside of work. One way to counter this and keep balance in your life is to join a volunteer organization and network with others in the organization. There are many volunteer organizations you can become a member of such as United Way, Habitat for Humanity, Green Peace, Food Shelf, and church groups, blood drives, student tutors, and the list goes on and on. Joining these groups and networking with these people, will provide the necessary balance to your life.

Now that you understand the importance of people networking to your career and where to network with people, let's move on to some simple tips on how to network.

SIX EASY STEPS TO FOLLOW FOR SUCCESSFUL NETWORKING

Networking is basically having a conversation with people. It is about opening up to others and sharing something about you and taking the time to learn about others. It is a two-way interaction for the purpose of simply getting to know each other and learn about each other others lives.

Step 1. Introduce Yourself and Shake Hands or Use an Ice Breaker. Start out with the very basics when you meet someone for the first time and introduce yourself. By introducing yourself, it means sharing your name and some small fact about yourself.

Make sure you look directly at the person and, if appropriate, shake their hand. Note that this is an American custom for greeting and may not be suitable for other cultures. The Japanese culture bows instead of shaking hands and do not look into the eyes of another person until they get to know each better. Other cultures have different methods of greeting people for the first time. In Terri Morrison's book, *Kiss, Bow, or Shake Hands,* she describes how to greet people in 60 different countries [4]. She has a very creative writing style. It is an easy to read book, and well worth your time.

If you can reveal a small fact about yourself, it gives a little added personal touch and opens up the potential to start a conversation. Some small little facts you may reveal about yourself include such things as:

1. What company you work at.
2. Where you are from, town, city, or state.
3. What department you work in.
4. What type of engineer you are—electrical, mechanical, software.
5. The reason you are at the meeting.

If you are too uncomfortable about starting out with a formal introduction, and many people are, there is an alternative approach to starting up a conversation. This technique is to use ice breakers. Ice breakers allow you to start a conversation without giving away personal information. Many men and women feel too uncomfortable about striking up a conversation with a complete stranger, or giving their personal information, like their name and where they work, before they get to know them.

Ice breakers are simple statements that offer the other person to comment and generally are very neutral and noncontroversial in nature. Here are some great ice breakers.

1. The weather is great (or bad)?
2. Did you see any of the (sporting event) last night?
3. If you are at a presentation or class, you can compliment the speaker, and ask their opinion.
4. Ask for directions to the next meeting room or presentation, or ask about the program planned for the day.

Ice breakers are great because people do not feel like you are purposely out to get information from them, and it gives them an opportunity to interact with you without revealing too much. Ice breakers can also be used to bridge the awkward silence that comes once you have introduced yourself.

You do not always need to start a conversation with a stranger by introducing yourself. You can start out with an ice breaker and after you are feeling more comfortable and would like to get to know this person better, introduce yourself and get their name.

Step 2. Body Language and Active Listening. Body language says many important things about you. For instance, the best body language is to maintain eye contact, smile, nod your head in agreement, and be a good listener. This body language tells the person you are talking with that you are genuinely interested in what they have to say.

When you are looking around, not making eye contact, appear to be distracted with other things on your mind, you are using the wrong body language and sabotaging your networking efforts.

▶ *Career Tip.* Good body language and active listening are networking skills an engineer acquires and uses often.

Being a good listener is a great networking skill to develop. Good listening means you do not interrupt the person or cut them off in mid-sentence. You let the person talking finish expressing their ideas. If you agree with the person, you can actively listen by repeating what they just said and how you agree. This provides positive feedback to the person you are networking with that you heard them. Susie Cortwright's website identifies 10 tips for effective listing [5]. Richard Weaver and John Farrell identify active listening as a critical skill for managers in their book *Managers as Facilitators: A Practical Guide to Getting Work Done in a Changing Workplace* [6].

When networking there will be occasions where you find yourself disagreeing with the opinions expressed by the person you are trying to network with. This will require you to invoke your diplomatic skills. You might repeat back what they said and then offer your own opinion on the subject matter. However, when networking, it is not advisable to start arguing the first time you meet a person.

Step 3. Making Small Talk and Sharing Conversation. People networking is not about interrogating people for your benefit. It is about having a conversation with a person and getting to know them. Finding out what you have in common and sharing information. Most conversation starts out slow and with small talk. Small talk is simply talking about something trite that all people have in common. Here are some good and safe small talk conversations subjects:

1. The weather
2. Person's commute to work or the meeting
3. Newspaper headlines for the day
4. Any great vacation plans
5. The immediate surrounding and conditions of the room, too hot, too cold, just right, easy to see, hard to see, which way to the rest room, what is planned for lunch.
6. If you are at a meeting and it is break time, you can ask for clarification on a certain point that you are not sure about. Ask a simple question, what they meant when they said . . . .

During the course of the small talk you are trying to establish a relationship with this person so that they feel comfortable about sharing information with you. You are assessing them on their willingness to share information with you. Debra Fine explains very well in her book, *The Art of Small Talk*, "How to start a conversation, keep it going, build networking skills, and leave a positive impression," and points out the value of small talk to networking and how mastering this skill can be a great door opener [7].

When trying to initiate small talk or network in general, there are definitely certain subjects to steer clear of and not bring up at all. These include such things as race, politics, religious beliefs, age, weight, sex, and any inappropriate jokes. By inappropriate jokes, I mean any joke that you would not tell in front of your spouse, children, parents, or gathering of civil leaders.

Step 4. Sharing Conversation and Exchanging Information. After you have had an opportunity to make some small talk with a person and they are open to further discussions with you, then you can probe further into their background and experiences. The next level in the conversation is to exchange basic information about yourself and find out basic information about them. Hopefully each of you is a great reference or source of knowledge for the other. The idea is to discover, through conversation, what the valuable knowledge each person has and share it with each other.

Let's look at a very subtle and interesting fact. Assume the person you are networking with knows at least 25 other people. And these other 25 people in turn know at least another 25 each (25 × 25). This means the person standing in front of you represents a knowledge base of 625 people. That is quite a vast network of people.

Moving beyond small talk starts with you both sharing basic information about yourselves. These may include technical background, company you work for, and your job. To move the conversation in this direction you can either share your background first or ask about their background. A word of caution here is not to share too much personal information. Most people do not want to hear about your hardships nor any gory details of your medical health. Keep the exchange of personal information more toward your positive career accomplishments, or your technical background and experiences.

The information exchange should be two-way. The information exchange should be facilitated through conversation with both of you doing the talking and listening. If you find yourself doing all the talking, stop and ask some questions.

There are basic reasons for networking with people. The first is to find someone to help you with a particular problem. The second is to make an acquaintance who you could call upon in the future if you need them and vice versa. And the third is to share your background and knowledge with them in case they should need your help in the future. Ideally, you network for all three reasons since not every person you meet will be able to help you.

If your purpose is to find help, then you want to gradually steer the conversation toward asking for help. After you have both exchanged some basic background information, steering the conversation is as simply as stating, "I have this problem I am trying to solve at work and involves . . . . Do you have any help you can offer or know of anyone who could help me?"

If you are not networking to solve a problem you should be networking to expand your acquaintance network and put valuable contacts into your contact database for future use when you really need them.

Finally, an excellent way to increase your network is by helping others get what they want. If you have the answer or solution to their problem then by all means you should share it with them. If you don't have the answer, but know of someone who would, then share this contact information. Once you help someone out, these people generally never forget the help you gave. They enter you into their databank as a valuable resource and refer people to you. Second, they are more than willing to help you in any way they can with your problem including doing networking on your behalf to help solve your problem.

Zig Ziglar in his book, *Top Performance, How to Develop Excellence in Yourself and Others,* points out that "You can have everything in life you want if you will help enough other people get what they want." [8].

▶ *Career Tip.* Helping others get what they need is a primary goal of people networking.

Stephen Covey in his book, *The 8th Habit: From Effectiveness to Greatness,* identifies that crucial challenge of our world today is this: to find our voice and inspire others to find theirs [9]. Helping others get what they want is key to success.

Step 5. Exchanging Contact Information and Ending the Network Session. All networking sessions come to end. Social etiquette dictates it is your responsibility to make sure the session ends on good terms and in a polite manner.

Prior to ending the session, if you feel the person you have met is a "good fit" for you and you would like to enter them into your contact base, then you need to make sure to obtain their contact information before you end the conversation. The best thing to do is exchange business cards. Your business card should have your name and some sort of contact information? The best thing to do is simply ask them for their contact information in the form of a polite question. "May I ask you for your contact information. Here is mine. I would like to be able to contact you in the future in the event I may need help. Would you mind?" Or after providing some help to the person, you could give them your contact information and tell them to contact you if they have need of any more help.

If for some reason you do not feel this is a good networking contact, or you are apprehensive about sharing your personal contact information, you should politely end the conversation and move on without requesting or sharing your contact information. In the event they ask you for your contact information, and you do not want to share your contact information, simply say, "I do not have any cards," and ask if they would just give theirs to you.

▶ *Career Tip.* Always end a networking session on a positive and polite note. Leave the door open for future contacts.

Ending the networking session is very easy. You should always thank them for talking with you and how nice it was to meet them. Here are some simple methods.

1. Thank you so much for talking with me. I enjoyed meeting you. I need to get going; there are several other people here I need to talk with. Please contact me if you need any help.
2. Thank you. I enjoyed meeting and talking with you. I have to leave and get back to my desk or go to another meeting.
3. Thank you. I must get going. Your helpful ideas are wonderful and I can't wait to try them. Thanks for sharing them with me. May I contact you if I have any further questions?
4. Thank you. It was very interesting talking with you. I especially liked it when you said .... then repeat something they said which you found interesting.

When the networking session is coming to an end shake their hand. Note again that this is an American custom and not appropriate for all cultures. The handshake at the end signals you are leaving and you are happy to have met them. You are pleased with what they have to say and hope in the future to continue networking with them.

Step 6. Follow Up with a Thank You. As Barbara Safani points out in the article titled "Seven Rules for Networking Success," "Always thank your contacts in person and follow up with a letter." If your handwriting is legible, the personalized touch is always appreciated [10]. I would like to add that a note or e-mail is also acceptable and widely used today.

GREAT TIPS FOR MAKING NETWORKING EASY

Practice, Practice, Practice. The first tip for making networking easy is to simply practice. Start your practicing outside of work with friends, neighbors, and people you meet as you go about your normal activities. Practice starting up conversations and entering into small talk. Make a game of it see how many new people you can meet in a single day. You do not have to become friends or tell them your life story. A simple 3 to 4 minute conversation allows you to practice some ice breakers and become comfortable with the process of meeting new people.

Let other people talk about themselves. Another great tip for making networking easy is to ask some simple questions and get people talking about themselves or opinions they hold. Stay away from political and religious questions but rather concentrate on the weather, hobbies, sports, or the morning traffic. Sometimes just a simple question like "How are you today?"

can go a long way to get things going. Who knows, other people may be wanting to network with you, but too afraid to start up a conversation.

Ask open-ended questions. Open-ended questions are questions that people cannot answer with a simple yes or no. These are questions like: What is your technical background? What do you enjoy most about your work? What was your most interesting assignment? What did you think of the presentation? In doing so you are providing them with the opportunity to talk and share.

▶ **Career Tip.** Most people like to talk about themselves and what they have accomplished. Simply ask questions.

Be prepared and do your homework before going to events. If you are attending an industry or work-related event do some research on the people who will be attending. If possible, do some researching of the attendees and their companies. Have a rehearsed one minute commercial ready to say something about yourself or about the subject that you are there to network for. The commercial should be based on your goal. If it's to get a new job, then speak about that. If it's to volunteer or whatever the case, be prepared.

▶ **Career Tip.** Create a one minute commercial about you or the reason you are networking.

Here are some simple questions to ask yourself to help you get prepared for the event.

> Why are attending the function?
> Are you attending to network for a new job?
> Are you attending to share news about something or the organization you belong to?
> Are you attending on someone else's behalf and what were they looking to accomplish?
> What is the best possible thing that could happen for you at the event?

Sometimes, even with all the preparation you may simply not feel like networking. If this is the case here are some ideas to help you overcome these feelings.

Find a friend or coworker to network with. Making yourself do things you are uncomfortable with or prefer not to do is always hard. Just like many other tasks in life, having a friend or coworker to do something with makes the task a lot easier. You can team together and network jointly, taking turns talking and sharing. A team member may think of items you might not think of. They are also there to compare notes or fill-in the blanks you may have missed.

Get yourself pumped up. Just like the athletes do before the big event, they spend time preparing and thinking about the event to come. They envision themselves completing the perfect pass, scoring the touchdown, or hitting a home run. They envision themselves being successful and accomplishing their objectives. Envision yourself successfully networking around the event, meeting people, carrying on conversation, learning about others and their experiences.

Overcoming Your Fear. In Zig Ziglar's book, *Top Performance,* he points out that fear is simply false evidence appearing real [8]. If you examine your fears about networking you will realize most are based on preconceived notions that more than likely will not occur. You are simply sabotaging your chances for success. Here are some common fears and counter actions you can take to control them.

I don't know what to say ---- Start by saying hello and how are you ----- Have a small commercial ready about yourself

I'll look stupid—you will only look stupid if you do not network. Everyone there is thinking how not to look stupid; you are not alone. Practice and being prepared are the key to overcoming this fear.

I don't know what to wear—contact the event organizer and ask about dress attire. Pick out something that makes you feel great when you wear it.

I am just not good at this—If you truly believe this then you will be terrible at networking. Networking is an easy skill you can learn with practice. If you believe you can learn this and improve, you will be highly successful. It is up to you what you want to believe.

In addition to these tips, Jeanne Martinet in her book, *The Art of Mingling,* has an excellent section on "tricks and tips for the tongue tied" to help people who are having trouble getting the networking skill developed [11]. I highly recommend you read her book.

NETWORKING AT YOUR WORK PLACE

Networking at your work is a career must. You are dependent on other people in your company to get your job done. Stop and think about all the people you interface with in a day's or week's time. Establishing a good networking relationship with these people is critical for your career advancement. Whether it is supplying material for products, checking and documenting designs, marketing products, finance, or getting technical advice, networking with other people in your company is all part of your job.

The best places to network at work are meetings, hallway conversations, and the company cafeteria. Networking at work can lead to better job assignments, promotions, and finding solutions to very difficult technical problems and even people problems. It should be a natural reflex when meeting new coworkers to ask them to share what department they are in, what projects they are working and correspondingly, for you to do the same. By sharing your work information you are essentially marketing your skills and opening the potential for others to ask you to join their project or team which, of course, is excellent job security.

Make sure you take advantage of every meeting you attend to make new acquaintances and network with other participants. These people attend meetings hopefully with the same desire as you—to help the team accomplish their goals. If you network during meetings, you will soon develop a people network that contains contacts from nearly every department in the company. If you network properly, you will also have these contacts for future projects that you can call upon when in need.

In an article from IEEE USA *Today's Engineer* on "Networking: Getting to Know You," author Amina Sonnie shares some important tips about networking on the job [12]. She identifies that during meetings you should share what you know about the company or project that might interest people in attendance. By sharing and participating you are opening the doors to further valuable exchanges. If possible, make sure you meet and know everyone's name and their background. If you don't know, then simply ask them.

▶ *Career Tip.* Network at company meetings to build your contacts for future work.

However, stopping the meeting so that you can learn about people is not an efficient way to run a meeting. So to overcome this you might want to arrive early at the meeting and introduce yourself to people as they arrive. Another tip is to wait until the end of the meeting and introduce yourself and start a conversation. If you do not know people in the meeting, you can start an attendance list and make sure all attendees write down their contact information (phone and email). At the end of the meeting obtain a copy of the list so you can easily follow up with people who you think could be a good networking candidate.

One easy way to learn the people in your department is to get hold of your department's organization chart. Then when you have a department staff meeting, network with everyone. Find out their backgrounds, schools they graduated from, and the projects they are working on.

Networking with nontechnical people in your company is also another good career move. Oftentimes, you are going to be faced with problems that are not technical in nature and will require you to get help or guidance from people who are not engineers. For example, these might include parts procurement, or legal matters when implementing a new contract. Other examples include financial help when determining costs and revenues.

HOW TO CONDUCT A NETWORKING SESSION FOR A NEW JOB

In an article from IEEE USA Today's Engineer titled "Build Your Network Purposely (Before You Need a Job)", author Debra Feldman points out the best time to start networking for a new job is before you need a job [13]. Networking for a new job should be a conscious focused effort to make contacts in companies or industries that have job openings for your skills. Attending trade shows, engineering conferences, or engineering society meetings are all excellent places to focus your networking talents with the intention of finding new job opportunities. In an article from IEEE USA Today's Engineer titled "Engineers: Work on your Networking Skills" author Nancy Salim points out that face-to-face networking is the path to approximately 80% of all new job opportunities [14].

Networking for a new job requires you to go prepared to an event with a well-rehearsed 1 or 2 minutes commercial about yourself, contact information, copies of your resume, and the ability to quickly record or recall people you meet and the opportunities so that you can follow up. Here is an example of how to conduct a networking session for a new job.

1. Start out by greeting the potential contact, saying hello, and introducing yourself. Let the contact introduce themselves.

2. Create some small talk and assess whether or not the person is really interested in talking. If they respond unfavorably to your small talk then it is time to move on. Tell them it was nice meeting them and politely move on.

3. If they respond favorably to your small talk then it is on to the next step. Inquire about what company they work for and what they do. Listen closely and use the active listening technique to make the contact know that you are paying attention and have heard them. When they are finished, share with them what company you work for and what you do.

4. Next inquire about whether their company is hiring or not.

   a. If the answer is no, then ask if they know of any other companies hiring with your type of skills. If they identify other companies hiring, you might ask if people from these companies are in attendance and could they provide introductions, if possible. If not, you might want to thank them for their time and express how nice it was to meet them and politely move on.

   b. If the answer is yes, their company is hiring, you can probe further by asking what type of people and skills they are looking for. If there is a fit between your background and skills with their job openings, then share your two minute commercial about your skills and ask if they think your background would be a good fit.

5. At this point, if they respond favorably to your two minute commercial, you have several options. Your objective at this point is to get an interview. Here are some actions you can take.

   a. Ask them how to best make a contact in their company to get an interview. If they provide a point of contact, ask if it is okay if you use them as a reference.

   b. Offer to send them a copy of your resume. Could they share their email address? In return, give them your contact information.

   c. Ask if they would like a hard copy of your resume and give it to them on the spot.

   d. Make sure you have their contact information (business card) in the event you need to get back to them or the point of contact does not work out.

   e. If you are really forward you can ask if they have any hints or recommendations on how to present yourself to the point of contact they just provided. Are there any special skills the hiring individual is looking for or anything else they can tell to improve my chances.

6. If the individual wants to talk more and starts going through your resume, you might suggest getting a separate table in a quiet area off the beaten path where you both can talk more. At this point, you are essentially going into the interview mode. You should be ready to impress.

7. After you have shared all the pertinent information and sense it is time to end the networking session, make sure you end on a positive note. Thank the individual for spending time with you and sharing information. Mention how helpful it was and how much you appreciate it. Finally express how nice it was to meet them. Smile, shake hands, and gracefully move on.

When you connect with someone and have a good networking session, it is good etiquette to follow up within a few days of the meeting with an email or snail mail note, thanking the individual for their time and sharing information. Offer your help and request they call upon you if they should ever need it. Make sure you store their information in your contact base for future reference.

HOW TO BE A FANTASTIC NETWORKER

As Jay Levison and Monroe Mann point out in their book, *Guerrilla Networking*, to become a super networker, you do things that will want people to meet and network with you [15]. You stop trying to meet people and start doing activities that others will see as valuable activities and make them to want meet you. What are these activities? Mann suggests such things as write a book, be a network hub, and sponsor networking sessions. Other activities include doing something radical, solve a problem, name drop, and help someone else become successful. In other words, go way beyond just meeting people, take matters into your own hands, and get people to want to meet you.

SUMMARY

People networking is a vital and necessary career skill an engineer must acquire. It is not taught in engineering schools and it is something you can learn on your own. It simply takes practice. You should be networking with nearly everyone you meet since you do not know when the next time you will need someone's help. Build your network on the job and at other functions you are attending. Potential networking contacts are all around you.

Networking is not only about getting information but also about sharing information to help out others. Networking can open doors to wonderful new opportunities both at work and outside of work.

Create a two minute commercial about yourself and be comfortable with using it when networking. Make sure you can create small talk when the conversation drags or hits an uncomfortable pause. Networking can be easy and fun once you have mastered the skills.

For further information on networking I recommend you conduct an Internet search on "people networking" and "meeting people."

Have you identified any career actions you want to take as a result of reading this chapter? If so, please make sure to capture these ideas before you forget by recording them in the notes section at the back of the book.

ASSIGNMENTS AND DISCUSSION TOPICS

1 Develop a two minutes commercial about yourself highlighting your skills and key achievements.
2 What are some benefits of people networking?
3 Name the six steps in successful networking.
4 What is active listening? Give an example of how you use it.
5 What makes a super networker?

REFERENCES

1. Daring, Diane, *The Networking Survival Guide*, McGraw-Hill, 2003, p. 16.
2. Daring, Diane, *Networking for Career Success*, McGraw-Hill, 2005, p. 3.
3. Susman, Karen, "102 Top Dog Networking Secrets," www.karensusman.com.
4. Morrison, Terri, Conway, Wayne, and Borden, George, *Kiss, Bow, or Shake Hands*, Adams Media, 1994.
5. Cortwright, Susie, "10 Tips to Effective & Active Listening Skills," http://thelife. com/students/people/listen/.

6. Weaver, Richard G., and Farrell, John D., *Managers as Facilitators: A Practical Guide to Getting Work Done in a Changing Workplace*, Berrett-Koehler Publishers, 1999.

7. Fine, Debra, *The Art of Small Talk*, Hyperion, 2005.

8. Ziglar, Zig, *Top Performance*, Baker Publishing Group, 2007.

9. Covey, Stephen R., *The 8th Habit: From Effectiveness to Greatness*, Simon and Schuster, 2004.

10. Safani, Barbara, "Seven Rules for Networking Success," http://www.quintcareers.com/networking_success.html.

11. Martinet, Jeanne, *The Art of Mingling, Second Edition*, St. Martin's Press, 2006.

12. Sonnie, Amina, "Networking: Getting to Know You," *IEEE USA Today's Engineer Magazine*, April 2003.

13. Feldman, Debra, "Build Your Network Purposefully (Before You Need a Job)," *IEEE USA Today's Engineer Magazine*, April 2006.

14. Salim, Nancy, "Engineers: Work on Your Networking Skills," *IEEE USA Today's Engineer Magazine*, February 2006.

15. Levison, Jay, and Mann, Monroe, *Guerrilla Networking*, Morgan James Publishing, 2008.

CHAPTER 27

ACCOMPLISHING MORE IN LESS TIME AND GETTING ORGANIZED FOR SUCCESS

Efficiency studies show that people without organizational skills waste more time [1–3]. The more unorganized you are the less efficient you will be in accomplishing objectives.

Over a longer period of time, organized people will accomplish significantly more, making them extremely valuable employees and candidates for career advancement to the more significant roles in the organization.

Most days you are paid for 8 hours of work. This amounts to approximately 480 minutes. This may sound like a lot, but if you consider the real productive time you have available, it is much less. Let's assume during a typical day that you take a 30 minute lunch, four 5 minute bathroom breaks, 10 minutes for your computer to boot up and shut down, 15 minutes for time cards and other daily work-related activities, 10 minutes deleting out old email and emptying your junk mail, 15 minutes socializing with your co-workers, another 10 minutes going to and from the printer and 15 minutes a day walking to and from labs or conference rooms. This leaves you with $480 - 130 = 355$ minutes of actual time to do work or approximately 6 hours a day of actual time for productive work. No wonder so many people are working late at the office every day. To put in a full 8 hours of work requires spending 10–11 hours a day at the office. Add in a 45 minute commute either way to work and this means that average workers are gone from the home about 12 hours. To put in a full 8 hours of productive work means you are literally spending 12 hours to do it. Why not improve your organizational and time management skills to accomplish tasks in less time?

Getting organized and becoming more efficient is simpler and easier than you think. It requires establishing a routine and development of some easy

The Engineer's Career Guide. By John A. Hoschette
Copyright © 2010 John Wiley & Sons, Inc.

organization skills. These basic skills are time management, priority ranking of tasks, and organizing your work and day into manageable portions. In this chapter, we address how to accomplish more in less time. An excellent place to visit is the Franklin Covey Stores for a complete line of Planners, Calendars, Daily Planners, and Personal Organizers. In addition, they also have excellent software to support your efforts to become better organized [4].

IDENTIFY AND ELIMINATE YOUR TIME-WASTING ACTIVITIES

The first step to getting better organized is eliminating those activities that waste your time. Let's look at some big time wasters.

1. Not planning out your time
2. Attending unnecessary or nonproductive meetings
3. Not creating a "To Do List" and not prioritizing work
4. Drop-by visitors
5. Telephone interruptions
6. Procrastination
7. Email
8. Constantly changing priorities
9. Not delegating work
10. Lack of daily routine and self-discipline
11. Taking on others' work and not being able to say no

These are not all the time wasters but some of the key ones. The best indication on whether or not an activity is a time waster is to ask yourself: "Is this the best use of my time or could I be doing something else that is far more productive or important?"

If you answer "yes, there is something more important," then stop what you are doing and go do it. Move on to the most productive use of your time and stay focused on this task. Now let's look at how you can avoid these time wasters.

Not Planning Out Your Time. The first step is to get a date book, if you do not have one, so you can plan your daily activities. Using your computer calendar is another good method. If you do not take control and plan your time others will do it for you. Record all your daily key events and meetings you are invited to. Then prioritize the meetings that are the most important to attend.

Another good technique is to block or schedule your own personal office time to work. This way you plan in time for just your work. By controlling your calendar, you will have more time to complete your work and not end up spending all your time in meetings that eat up your day—leaving you with no real time to complete your own work except for after-hours.

Attending Unnecessary Meetings or Nonproductive Meetings. Just because you are invited to a meeting does not mean you have to attend, nor does it mean you have to sit through the entire meeting. If you feel you have nothing to contribute to the meeting, either don't attend or leave early. If you have something to contribute ask the meeting leader if you can speak first so you can leave. If you are finding yourself stuck in a meeting that is going nowhere, excuse yourself and leave. You can always come back later at the end to see if there was anything of significance that you should be aware of or need to take action on.

Bring work to the meeting. It has become socially acceptable if you are not a main participant to bring your laptop and get work done during the meeting. However, if upper management is there, this is frowned upon.

When you have overlapping meetings, ask if you can give your input at the start of one so you can attend another meeting. In advance of the meetings, let the second meeting know you are going to be late and ask if the leader can schedule your items later in the meeting when you will be able to attend.

Not Creating a "To Do List" and Not Prioritizing Work. Most people do not create a "To Do List" nor prioritize their tasks. This results in wasted time and energy that often culminates in missing key deadlines. "To Do Lists" allow you to identify and track all the important tasks and help you focus on the most important ones first. See the "Creating To Do Lists" later in the chapter on how to effectively do this and save yourself time and hopefully significantly reduce the stress in your life.

Drop-by Visitors and Hallway Conversations. Do you know how procrastinators waste their time and yours? They stop by to see what is up with you and chitchat about what are the latest and greatest things going on in the company. Some employees believe this is part of their job, to walk around talking to others to find out the latest gossip around the office. You have to be very proactive with these types of people and control them or they will control you and wreak havoc on your time and work.

When these types of people show up at your desk unexpectedly and want your valuable time, the first way to control them is ask why they are there. If it is an emergency you can drop what you are doing and start on it immediately.

If they express that they are there simply wanting to talk, then you need to set a time limit and end the conversation. One method of doing this is to identify that you are extremely busy and have some critical deadlines in about an hour and you really don't have the time if you are going to make the deadline. Ask the person if it is possible to get back to them later in the day. Most people have been in this situation themselves and will leave you alone. Or, if you feel it is politically correct to talk for a few minutes then identify that you only have about 3 minutes of spare time and need to keep it short. After 3 minutes announce that you have to get back to work and would they mind if you end this conversation?

Another option is, after several minutes, you announce you have to get ready for your next meeting or you need to go to the bathroom. Get up out of your chair; this is a nonverbal message indicating the conversation is quickly coming to an end. However, some people may not take the hint, so you will actually have to walk away.

Hallway conversations are another great time waster. As you walk down the hallway, people who you have not seen in a while like to tell you all that has happened to them recently. They like to share with you the great vacation they recently took, the problems in their family, the amount of work they are getting done or not getting done, or simply catch up on old times. If you slow down your pace and stop to talk, you are doomed to a 10 minute conversation. Again you need to take control. Be polite and keep on walking. Share with them that you are in a hurry, and will stop by later when you have more free time.

Telephone, Cell Phone, and Blackberry Interruptions. Most people believe that if the phone is ringing you need to answer it. Considering most phone conversations last at least 5–10 minutes, and you may receive 8–10 calls a day at the very least, this means you could be spending up 2 hours on the phone per day. Ask yourself, is this the best use of my time right now to answer this call or return this blackberry message? Is the person calling critical to accomplishing my tasks today? If the answer is no, let it roll over and let the person leave a message. That is why they invented answering systems. Return the call when it is convenient to you. But, of course, if it is the boss calling make sure you answer it unless you have a good reason why you could not.

▶ *Career Tip.* Schedule a time during the day when you return calls and make calls.

Procrastination. Are you a procrastinator? Do you put off work until tomorrow simply because you do not feel like doing it today? Or do you often need more time to think everything through prior to making a decision? Or do you not know what is the correct action or path forward, so no action seems like the best course of action? If you are answering yes to these questions more than likely you are a procrastinator. We all do it from time to time, but if this is a habit or standard method of operating then you need to make changes. Burka and Yuen in their book, *Procrastination: Why You Do It, What to Do About It*, provide excellent tips in dealing with the problem of procrastination [5].

Some of the underlying causes of procrastination include

- First and primary, the cause is often depression. If you feel this is the cause, get help with the depression immediately. Seek out professionals and get mental and medical help if necessary to help you deal with your depression.

- Planning and staying focused is difficult. (People with Attention Deficit Disorder (ADD) may be part of this category [6].)
- Tasks seem overwhelming, so it appears to be futile to even try.
- Resentment or anger toward the person requesting your support and subsequent actions. You want to "get even" by ignoring or putting action off.
- Being organized and in a routine causes you to rebel against authority.
- Fear of doing something wrong and failing, disapproval from others, or embarrassment.
- Guilt-driven about not accomplishing what is expected of you and rather than correcting the lack of action, continue to put things off.

Once you recognize the root cause of your procrastination, you can take steps to stop it. One of the most important things you can do to combat procrastination is get organized. Make lists, take a class in organization, and break tasks into very simple smaller actions that are easier to accomplish. Start small and grow into larger tasks and actions. Don't overwhelm yourself and take on too much. If a task is not obtainable in a single day, then break it down into small tasks that are easily accomplished in a single day. Once you break the procrastination pattern into smaller tasks it becomes easier to take on more.

Email. In electrical engineering, designers are worried about the bandwidth of a communication system or how fast the messages and data are transmitted. The faster the messages are transmitted and understood the better. The list below identifies the communication techniques used between people for communicating and their estimated words per minute or bandwidth. A relative factor is estimated for each method as compared to email communication.

Method of Communicating	Words/Minute	Relative Factor
Text messaging	5 words/minute	0.30X
Typing an email (computer)	20–30 words/minute	1X
Average person speaking ability	100 words/minute	3X
Average person listening ability	300 words/minute	10X
Pictures	2000 words/minute	100X
Video	>5000 words/minute	300X

When you compare these methods of communicating information, you quickly realize that email and text messaging are the slowest means of transmitting information back and forth, and consequently, the most time wasters. For good time management, you want to control how much time you spend using these slow techniques of communication in your work. By comparing the relative factor for each of these communication techniques, you quickly realize that in most cases picking up the phone and talking directly to the person is going to save you significant time over text messaging and emailing. Of course, this is not possible all the time hence the value in email and text messaging. People can leave you a message when they cannot reach you.

Do you have a plan for controlling the amount of time you spend text messaging and emailing? You will need a plan to limit the use of these communication techniques if you are going to control your time and free up more time to do more important tasks. Here are some simple actions you can take to limit the time you spend on emailing.

1. Check your email at specific times during the day.
2. Do not respond immediately to every email that comes in. This is like answering the phone every time it rings.
3. Set a limit on the time you are going to spend responding to email. When you hit the time limit move on to over work.
4. Set up your junk mail filters and delete junk mail only once a day.
5. Ask the question, "Do I need to really respond to this email?" Or is simply reading it enough?
6. Ask people to remove you from their email list if you have nothing to contribute.

Email is a time-eating monster. Have you ever been in an email war? This is where someone sends out a simple email to several people stating their opinion. Something in the email is controversial and the recipients all start hitting "Reply All" so they get their opinions in. The number of emails sent grows geometrically and soon you start to receive emails so quickly that it is hard to keep up with reading them all. What starts out as simple little email explodes into a full-blown war where everyone wants to get their opinion heard. Pretty soon you are spending hours reading all the replies, most of which have no real value.

On a large distribution email message you can have 20 or 30 people all hitting "Reply All." Suddenly your computer can be flooded with up to several hundred emails and if you spend the time reading each one you will be there for hours. This is totally a waste of your time.

▶ *Career Tip.* Limit your time and number of text messages and emails to only those that are absolutely necessary.

▶ *Career Tip.* Stop emailing people after the third reply message. Pick up the phone and talk.

Email does have a purpose and good use, but it should NOT be your main method of communicating with people on your team or in the company. Email is good for instant long-distance communication outside the plant, across the nation, or across the world. It is also good when people cannot be reached but can pick up your message later in the day. Emails are also good when files are attached for further information exchange. Another good use of email is when a written record of the conversation is needed.

▶ *Career Tip.* Emailing should NOT always be your primary method of communicating with people.

Constantly Changing Priorities. This can be a real time waster if you let it get out of hand. You no sooner start on a task and the priorities are changed. Then you start on the new task and before you can complete this task, the priorities are changed again. You end up never completing a task and everything is half done. You have to break this cycle by letting people know you will shift priorities after you complete the present task. This may be hard if your boss is the one asking, but he or she should understand if you take the time to identify the danger or amount of work that will be wasted by switching tasks prior to completing the present one. Oftentimes, it will take more effort to restart the present task once you drop it, than the time it would take to complete the existing task in the first place.

Not Delegating Work. Not delegating work and trying to do everything yourself is another time waster. Most people with this affliction believe "if you want the job done right, you have to do it yourself." As you move up and become a team leader or manager, you must realize your job is to accomplish work by getting others to do the work. A simple question you can ask yourself is, "Do I really need to do this or could someone else on my team do the work?" If the answer is someone else can, then delegate the task and switch your attention to those tasks that only you can do.

▶ *Career Tip.* Put your time and efforts on those tasks that only you can accomplish; delegate to others the rest if you can.

Lack of Daily Routine and Self-Discipline. Many people start out developing an excellent time management plan only to have it fall apart at the first crisis. You develop the self-discipline to execute your plan and when things start to fall apart self-discipline and structured work habits quickly go by the wayside. Daily routines and staying focused are two very good habits to acquire for time management. Look at your work routine and the flow of activity you undertake during the day. Can you identify the time wasters and eliminate them? Can you substitute a good routine or habit for a bad one? Try to visualize what would constitute a perfect day for you and how you complete all the tasks.

▶ *Career Tip.* Start each day visualizing your perfect day and strive to make it happen.

In the beginning, to help you get in a routine, write the routine down that you want to follow and post it where it is visible. For instance, you might post on the wall above your computer:

1. Check calendar for day and determine meetings and critical activities
2. Review "To Do List" and update
3. Set priorities for day
4. Return critical calls or emails from yesterday that you were not able to get to
5. Start on highest priority on the "To Do List"

Taking on Other's Work. One great time waster is taking on other people's work just to help them out. Being a nice person and helping someone else is not entirely a bad thing and we all do this from time to time. But when this becomes a habit and others use you to get their work done then it no longer becomes beneficial to anyone. Are you taking work from others that you never planned and is it slowing your progress? If so, you need to stop taking on others' work and return the work back. This habit of helping with others' work should be a rare exception and not the norm.

In the event that you do take on others' work, make sure the boss approves. If you help someone else out and you missed your deadline, the boss may be upset.

GREAT ORGANIZING SKILLS AND TOOLS TO SAVE YOU TIME

Creating "To Do Lists," Setting Priorities, and Budgeting Time. Creating "To Do Lists" and setting priorities can significantly help you accomplish more work in less time and with less effort. The first step to creating a "To Do List" is simply brainstorming the tasks that need to be done and write them down. DO NOT worry about the order in which you write them down, we will prioritize the list later.

To help you determine what needs to be done, refer to your plans, deadlines, and objectives for the week. Do you have a project plan that shows the activities you are working on and the critical ones for the day or week? Have you coordinated with your boss or engineering lead and determined what is most critical? Have you determined the near-term and long-term

To Do List for _____	To Do List for ___Monday___
Return Calls to Jane & Bob	2 Return Calls to Jane & *Bob - 20 minutes*
Generate Agenda for Friday's Customer Meeting and email out notices	1 Generate Agenda for Friday's Customer Meeting and email out notices - *10 minutes*
Work in Lab on Product	3 Work in Lab on Product- *3 hours, this morning*
Check Emails	4 Check Emails- *20 minutes after lunch*
Meet with Boss and discuss actions for customer visit & lab work	5 Meet with Boss and discuss actions for customer visit & lab work- *30 minutes - 2:00 today his office*
Conduct Vibration Test	8 Conduct Vibration Test - *Two days from now; plan on Wednesday*
Write section of test plan	6 Write section of test plan- *After 3PM today - 1 hour*
Analysis of Circuit Design	7 Analysis of Circuit Design - *3 days, start Tuesday*

FIGURE 27-1   Example of generating a "To Do List" and prioritizing.

actions you need to take. What must you get done today to keep you on track for tomorrow? An unprioritized list of actions for the day is shown in Figure 27-1 on the left-hand side.

In Stephen Covey's book, *First Things First*, points out that prioritizing your "Things To Do List" is key to getting organized [7]. So once you have the list of tasks and everything is covered, the next step is to budget a time for each task and when it needs to be completed. Next comes the priority ranking of the tasks. What is the order to best accomplish the tasks? Simply number the tasks in the order that makes the best sense to accomplish work.

There are many methods and reasons for setting priorities. You might prioritize some tasks since they are fun to do or others may be time critical. Other tasks might be extremely important to the boss. Some tasks may take days to complete and progress needs to be scheduled in parts. This is where the skill in developing a great "To Do List" comes in. With practice you will acquire this skill.

▶ *Career Tip.* Make sure the tasks the boss gives you are always at the top of the list.

The completed list with priority ranking and order of tasks to be accomplished is shown on the right-hand side of Figure 27-1. For the example shown, it was decided that the most critical item for the day, and first task to be completed, was generating an agenda for Friday's meeting and emailing out the notices. This task only takes 10 minutes, is critical to get out on Monday, and can easily be accomplished. This is a simple but critical task that can be quickly accomplished. Get it done and out of the way. The next task involves coordinating with teammates via returning phone calls. Limiting this task to 20 minutes gets all the important coordination done quickly in the morning and frees the rest of the morning for concentrating on lab work.

▶ *Career Tip.* First Brainstorm Actions
Then Prioritize the List

After getting top priority tasks completed quickly, it frees up the rest of the morning to work in the lab. This planning takes care of the morning. After lunch, limiting time to check emails to 20 minutes allows a person to see if anything important came up and if they need to respond quickly. The remaining tasks for the afternoon are either multi-day tasks or activities that are planned to start later in the week.

"To Do Lists" can be done on your computer or written out on simple sheets of paper. Many office supply stores carry formatted pads of paper specifically for "To Do Lists."

Three Week Look-Ahead Calendar. Using a "To Do List" is only part of the equation to becoming organized. Another component is having a good calendar and specifically a 3 week look-ahead calendar. The purpose of this tool is to make sure you plan and take action for the tasks that are coming up beyond the present week. An example of a 3 week look-ahead schedule is shown in Figure 27-2.

The purpose of the 3-week look-ahead schedule is to map out the key tasks for the present work week and provide foresight into key tasks coming in the following weeks. This tool allows one to plan his or her time for key tasks of the present week and allows also to initiate early activities in advance that will support upcoming future tasks. In the example shown here, each week is started out with a staff meeting to start the priorities for the week. In the first week, the key tasks are to prepare for and present at the executive staff meeting. The week ends with preparing a presentation for the staff meeting on the following Monday of the second week. In the second week, preparations are started for the customer visit. The customer visit actually occurs in the third week. In addition, in the second week, a job review is planned with the boss. A very significant item one should not go to unprepared.

In the third week the major event is the customer visit on Wednesday the 23rd, and all preparations are complete and checked the day before. Also, the final task of the third week is to ship products on Friday the 30th. It appears to be quite a heavy workload in the third week and anything one can do to prepare ahead of time in week 1 or week 2 will greatly help reduce the workload in week 3. Therefore, the default tasks when time becomes available

Three Week Look-Ahead Schedule

Mon	Tue	Wed	Thu	Fri
7	8	9	10	11
10:00 AM Weekly Staff Meeting		Prepare for Executive Brief	Executive Brief on Progress	Prepare Presentation for Weekly Staff
14	15	16	17	18
10:00 AM Weekly Staff Meeting		Start Preparation for Customer Visit	10:00 AM Job Review with Boss	2:00 PM Team Meeting
21	22	23	24	25
10:00 AM Weekly Staff Meeting	Final Check on Preparations for Customer Visit	Customer Visit		Ship Products

FIGURE 27-2   Example of a 3-week look-ahead schedule.

in weeks 1 and 2 are to prepare for the customer visit and check how the products are coming for shipment.

The real value of the look-ahead calendar is to make sure you are aware of the present week's tasks. When there is slack time in the present week, you can use this time to prepare for the following week's tasks.

▶ *Career Tip.* Maintaining a 3-week look-ahead calendar greatly reduces stress and significantly reduces the number of unplanned surprises and panics.

Naturally, at the end of each week, the task is to update the look-ahead schedule since the present week drops off and a new third week will be added.

Notebooks. One of the best organizing tools for your career is a well-kept notebook. Even though computers are a great tool, they cannot match the ease and convenience of a good old-fashioned hand-written notebook.

A diagram for creating a great notebook is shown in Figure 27-3. Starting with the cover, make sure it is labeled with your name and how to contact you in the event you inadvertently leave it some place or lose it. Pasting on a pocket on the inside cover is a great place to store your business cards. You will always have them with you in case you should need them.

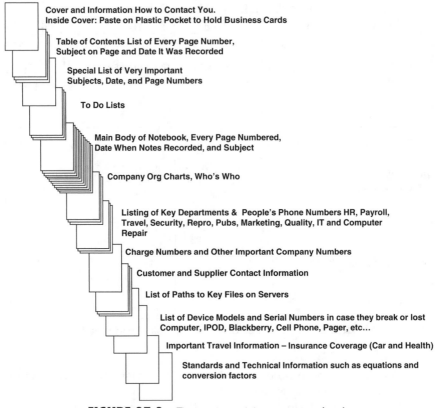

Cover and Information How to Contact You.
Inside Cover: Paste on Plastic Pocket to Hold Business Cards

Table of Contents List of Every Page Number,
Subject on Page and Date It Was Recorded

Special List of Very Important
Subjects, Date, and Page Numbers

To Do Lists

Main Body of Notebook, Every Page Numbered,
Date When Notes Recorded, and Subject

Company Org Charts, Who's Who

Listing of Key Departments & People's Phone Numbers HR, Payroll,
Travel, Security, Repro, Pubs, Marketing, Quality, IT and Computer
Repair

Charge Numbers and Other Important Company Numbers

Customer and Supplier Contact Information

List of Paths to Key Files on Servers

List of Device Models and Serial Numbers in case they break or lost
Computer, IPOD, Blackberry, Cell Phone, Pager, etc...

Important Travel Information – Insurance Coverage (Car and Health)

Standards and Technical Information such as equations and
conversion factors

FIGURE 27-3 Tips to organizing your notebook.

Next numerically paginate every page in the notebook. Then create a table of contents that identifies a one-line listing for every page with the date, and title of the subject matter entered. Update the table of contents periodically to reflect the notes you have taken.

Following this table of contents, create a page for listing just the special items. This is a great time saver. After your notebook starts to get full, searching through a 100 or 200 line table of contents to find that special data you recorded, can take quite a bit of time. On this page enter critical or key items you know that you will be referring back to in the future.

Make the next section of your notebook, the "To Do List" area. People often create "To Do Lists" on separate sheets and then end up losing them. Keep them in your notebook and you will know where they are and stand less chance of losing them. Also having previous lists to refer back to, allows you to quickly recall when and what you did for weekly status reports.

Following the "To Do List" area is the main body of your notebook. This is where you record your meeting notes, calculations, phone notes, and all other important items.

Keep the title of the page very specific to the subject matter on the page. If you have very nebulous and vague names, when you enter all the subject titles in the table of contents, you will have no clue as to what is really recorded on the page. For instance, instead of naming a page "Phone Call," you might want to entitle it "10:00 AM Phone Call to Supplier X about Cost Impact." The more descriptive and precise you can make the title the easier it will be to determine the contents of the page.

▶ *Career Tip.* Sign and date each page that will support any patent applications. You will have an official record of your inventions and chronological evidence when it comes time to file for a patent.

After the main body of the notebook, toward the back, you can paste in copies of your reference data. For instance, one good section to make is a compilation of organization charts. This makes a great section to refer to when attendees at the meeting refer to different departments and people in the department.

▶ *Career Tip.* Be an avid collector of organization charts. These charts reveal how your company is organized and who is in charge of what. Great time savers.

Another good section to create is a listing of key departments and people's phone numbers, Human Resources, payroll, travel, security, repro, pubs, marketing, quality, IT, telecom numbers, IP addresses, and computer repair.

In most companies, the project costs and your time spent on a task are tracked by charge numbers, so a good section to create is one listing all the charge numbers you use. Your customers and supplier contact information is another good section to have in your notebook for quick reference.

With the information overload going on these days and the drive to become paperless, every document is stored on servers. The only problem is remembering the path to these key documents that you only reference once in a great while is very time consuming. It is extremely difficult to find documents when the documents are stored on servers that you do not normally access. One way to overcome this problem is having sections in the notebook where you record the path to folders containing the files. This is a list of paths to key files or folders on the servers. Storing a short-cut on your computer desktop is another method, but this can quickly lead to a much cluttered desktop, although usually it is a good back-up.

The final two sections deal with recording those very important numbers for device model numbers and serial numbers in case they break or you lose your computer, IPOD, Blackberry, cell phone, pager, or credit card. Other important information to record here is your travel information and your insurance coverage for such things as car and health care coverage when you are traveling. This comes in handy, and I always take it with me when I travel.

Last, but no means least, is a section containing technical information on such things as industry standard, equations, and conversion factors/tables that you constantly reference. One of my favorite things I have included is the binary to hexadecimal conversion table.

DEVELOPING A DAILY ROUTINE

One of the best habits in getting organized, is developing a daily and weekly routine, and sticking to it. A daily routine to help improve your productivity is shown in Figure 27-4. You should start out each day with some quiet office time to get yourself organized. From 7:45 AM to 8:30 AM spend some quality time planning and establishing your priorities. During this time you should be scheduling the important meetings you want to happen and contacting key people you will need to interface with during the day. If you feel really ambitious, you can even make calls during morning commute to help get prepared for the day.

▶ *Career Tip.* Establish a daily and weekly routine and stick to it.

From 8:30 AM to 11:00 AM is your highest productivity time. This is where you want to tackle those complex, critical, or potentially dangerous activities. You are the most alert and usually have the most energy. Dig in and go for it. Eliminate all distractions during this time period and remain focused. Do not let email, phone calls, or others distract you, and waste your high productivity time. Shut your office or lab door and totally focus on what you need to get done. Some people actually block out their calendars during this time to stop other people from scheduling meetings and robbing them of this valuable time.

From 11:00 AM to 11:45 AM is wind-down for lunch time. During this time you should be bringing the morning tasks to conclusion, and returning to your office to check messages and emails. In general, you are slowing down for lunch.

Supply a lunch when scheduling a meeting between 11:00 AM and 12:00 PM. People are hungry at this time and, as a result, they usually have less patience, are more irritable, and less willing to compromise.

DEVELOPING A WEEKLY ROUTINE

Starting the week out prepared and organized is a great productivity booster. A sample of how to organize your week to make sure you start the week highly productive and finish it on a positive and strong note is shown in Figure 27-5. The days of the week are across the top of Figure 27-5 and below each day is an AM and PM box for that day.

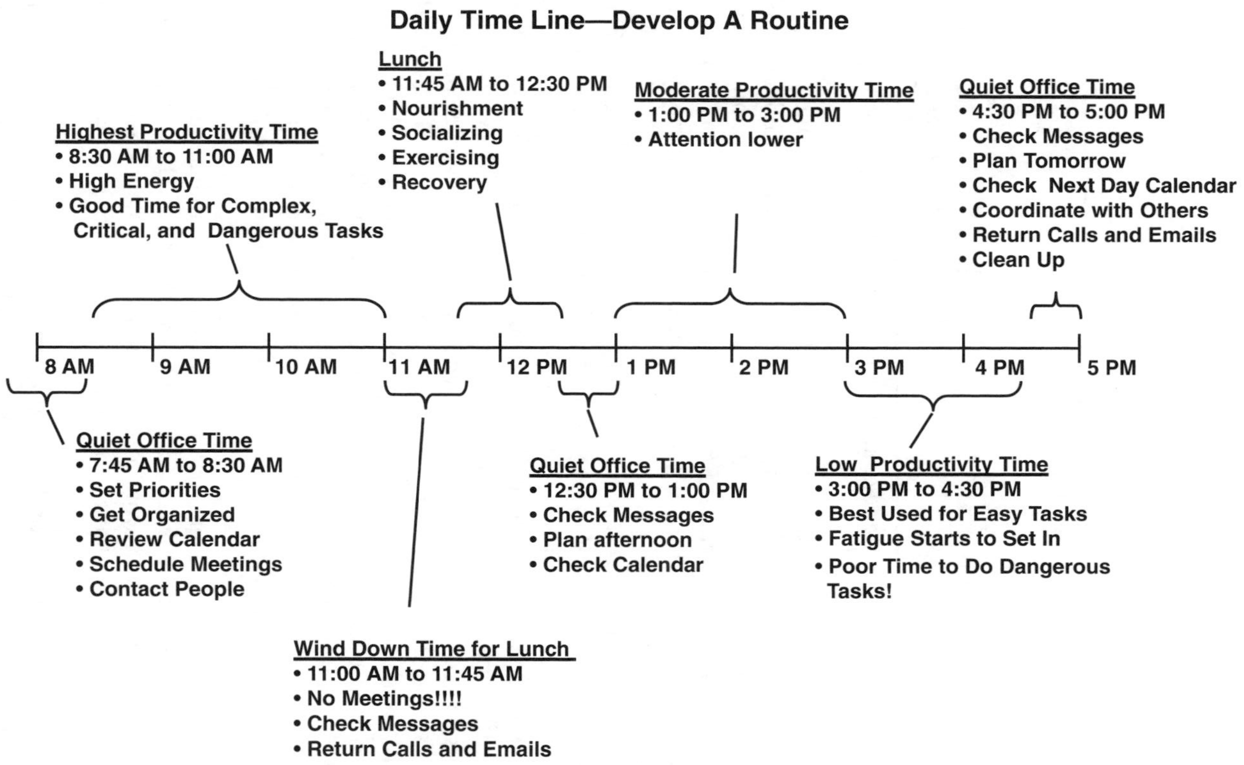

Daily Time Line—Develop A Routine
Lunch
• 11:45 AM to 12:30 PM
• Nourishment
• Socializing
• Exercising
• Recovery
Highest Productivity Time
• 8:30 AM to 11:00 AM
• High Energy
• Good Time for Complex, Critical, and Dangerous Tasks
Moderate Productivity Time
• 1:00 PM to 3:00 PM
• Attention lower
Quiet Office Time
• 4:30 PM to 5:00 PM
• Check Messages
• Plan Tomorrow
• Check Next Day Calendar
• Coordinate with Others
• Return Calls and Emails
• Clean Up
8 AM
9 AM
10 AM
11 AM
12 PM
1 PM
2 PM
3 PM
4 PM
5 PM
Quiet Office Time
• 7:45 AM to 8:30 AM
• Set Priorities
• Get Organized
• Review Calendar
• Schedule Meetings
• Contact People
Quiet Office Time
• 12:30 PM to 1:00 PM
• Check Messages
• Plan afternoon
• Check Calendar
Low Productivity Time
• 3:00 PM to 4:30 PM
• Best Used for Easy Tasks
• Fatigue Starts to Set In
• Poor Time to Do Dangerous Tasks!
Wind Down Time for Lunch
• 11:00 AM to 11:45 AM
• No Meetings!!!!
• Check Messages
• Return Calls and Emails

FIGURE 27-4 Sample daily routine to improve productivity.

Weekly Routines Are A Good Thing!

	Mon	Tue	Wed	Thu	Fri
AM	8:00–8:30 AM Plan Your Week Set Priorities for Week 9–10 AM Get Team Organized For Week		Check in with Management to Report Progress and Discuss Issues		
PM	Coordinate with Management Write Previous Week's Accomplishments Report			Perform Low-Priority and Low-Value Administrative Tasks	Perform Low-Priority and Low-Value Administrative Tasks Plan Next Week's Activities Clean Office and Lab Coordinate with Management

FIGURE 27-5   Sample weekly routine to follow to improve productivity.

The best way to start out the week is by getting prepared on Sunday night. By spending 45 minutes on Sunday night to review your accomplishments from the previous week, and identifying the key tasks for the coming week, you start out ahead of the game. Walking in the door on Monday morning fully prepared with a plan for the week, lets you start out significantly ahead of others.

If you cannot prepare on Sunday night, then as shown in Figure 27-5, block out 8:00 AM to 8:30 AM to get prepared and come up with a plan for the week. Shut the door to your office or spend some quiet time at your desk to organize your thoughts and plans for the coming week. Identify your goals and objectives for the week and a simple plan for each day of the week. Once you are organized and have a plan, then you can start to organize others.

If you are a team leader, Monday morning 9:00 AM to 10:00 AM is a good time to hold a team meeting and set the plan and objectives for the week. If everyone comes to the meeting as well prepared as you are, then the meeting should be short and everyone can quickly jump to their important tasks for the week. Following this routine, you and your team should be organized and have agreement on what is important and what tasks should be accomplished for the week.

Monday PM is an excellent time to coordinate with management or the boss on your plans for the week. Since you are organized, have a plan for the week and even meet with your team and obtain their status. You can then provide the boss with the latest and up-to-date status of your progress and plans for the week.

Another good activity for Monday afternoon is to write the weekly progress report summarizing the accomplishments for the previous week and your plans for the present week. There is nothing better for your career than a condensed well-written and concise progress report highlighting your significant accomplishments.

▶ *Career Tip.* A well-written weekly progress report highlighting signifi-cant accomplishments is a career accelerator.

Monday is usually a day of coordination and getting things set up for the week. Tuesday, on the average, is one of the most productive days of the week. By mid-week, Wednesday, a significant amount of work has occurred to warrant a quick update to the boss to let them know that things are still on track and you are making good progress. If you're not on track it is good to let them know about the problems and delays you encountered and the lack of progress. A mid-week update to the boss is a good career habit.

▶ *Career Tip.* Updating the boss mid-week is a good career habit.

As you approach the end of the week, you start to lose energy and these days are best for the lower priority tasks. Thursday afternoon is great for getting all those little things done that you have been putting off all week. Friday is wind down day. Friday afternoon is great for wrapping up the week's tasks and getting prepared for the next week. Closing out the week by reviewing the next week's calendar and generating a new "To Do List" is an excellent activity. Also, making a few notes on your accomplishments for the week will help in writing the progress report on Monday. Other good activities for late Friday afternoon are organizing your desk and the lab.

USING LUNCH AND AFTER-HOURS FOR RECOVERY TIME

Working through lunch time everyday is not a good thing. However using a planned lunch hour once in a while to recover and be prepared is good to incorporate in time management. Make sure you close the door to your office or go to some place quiet to work. Otherwise, people passing will stop to talk and eat up your valuable time. One quiet place most people do not realize is available is a conference room. Also staying late or arriving early in the morning is another recovery technique.

▶ *Career Tip.* If you constantly arrive early, work through lunch, and stay late, you are doing something wrong. Get organized and manage your time better.

If you remember back to your school years we all had times where it was necessary to put in extra hours. Work is no different, but if you are doing this constantly, something is wrong. You need to take control, get organized, and manage your time better.

SAVE TIME BY ORGANIZING YOUR COMPUTER AND SERVER FILES

Most people just start creating folders to hold files randomly and place them on the server or their computer without any thought on how to best organize them for quick and easy retrieval. The modern method of file by piling up the folders is now done electronically. Do not keep storing folders and files on your computer desktop until the entire desktop screen is covered with folder and file icons. Have a system and organize your files for easy retrieval.

▶ *Career Tip.* File by pile does not work.

The real challenge to storing files is setting up a structure that will allow you to quickly retrieve them. I have seen several different types of file structures. The first one is where you only have one or two folders and store all your documents in just these two folders. This will work until you have accumulated over 20 or 30 files in each folder, and soon you end up spending more time looking through all the files.

On the other side of the spectrum, one can create a folder that contains only two or three files. This method causes one to create folder after folder within folders. Using this method when you have to file 40 or more files results in paths to files being 6 to 8 folders deep. One wrong click along the way and you are possibly hopelessly lost and again wasting more time.

The best method I have observed is holding the number of files in a folder to about 10. This is a good number since, on most computers, 10–15 different files in a list will easily all fit on the screen at once. The next thing is limit the folders in any given folder to about 10 folders. This means that you can store approximately 100 files only 2 folders deep. Or you can get to any file with just two clicks of the mouse.

Working a major program with several hundred engineers, all setting up folders and storing files in different manners based on their personal preferences, is a recipe for disaster. Without someone assisting you to find where the files have been stored you can waste hours a day just looking for files.

Computers were made to shorten or reduce our workload and improve productivity. Think about your productivity and how you store files. Can you reduce the amount of time wasted looking for files by simply organizing your files on the computer better?

HOW TO CREATE YOUR OWN TECHNICAL LIBRARY

With the massive amounts of information we now have at our fingertips, it becomes practical to assemble a very comprehensive technical library. There are two very easy ways to do this. One is to set up your search engines for key technical topics and then store the results. The search engines will quickly give you access to all that is written on special subjects. With the results stored, you are just a few clicks away from massive amounts of technical data to help you with your work.

Another method of assembling a technical library is to review trade journals. Oftentimes, trade journals or trade magazines will run articles describing the state of the art on various technical subject areas. These articles are written to bring the reader up to speed quickly and often have references to the key websites in the article, if perhaps, the reader wants more information.

If you simply tear out the articles and place them in a 3-ring binder each month, after a couple of years you will have an unbelievable collection of technical articles. Now the real interesting aspect is that most trade journals are offered free if you qualify. Your own technical library is available to you free. By simply doing a search for trade magazines and the subject matter of interest you can quickly locate their website. Fill out the information they request and you will be on their mailing list within minutes.

I personally receive 30–40 free journals a month in various subject areas related to my work. My collections of technical articles contain over 200 + articles. When I or a coworker needs information, I can retrieve the latest information in minutes. With this technical library so quickly at hand, I have become quite popular with my coworkers at times, who are desperate to quickly find information.

SUMMARY

The ability to organize your work and manage your time is critical to career advancement. You will receive many dividends by changing bad habits into good habits when it comes to organization and time management. Simply audit and record where you spend your time for a week. Identify the time wasters and eliminate them. Utilize a "To Do List" to help you focus on the most critical tasks at hand. Prioritize your task list and then schedule the tasks and actions for the week.

Watch out for the slow and time wasting methods of communication: text messaging and email. Pick up the phone and call a person. Utilize the highest bandwidth means of communication whenever possible.

Next, develop a daily and weekly routine where you start each day and organized for success. Start out each week with a game plan week. Organize your weekly activities and protect your valuable high-productivity time.

Create a 3 week look-ahead calendar, so you are not surprised and caught off guard quickly approaching critical tasks. Get notes by organized and properly recorded in a notebook. Arrange your notes so it's easy to retrieve important information in less time. Organize your files on your computer and servers so information can easily be recalled.

▶ **Career Tip.** Being organized allows you to accomplish more work in less time and for less effort. This is career advancement.

Here are some great words to use an Internet search for if you are interested in learning more in this area:

Get Organized
Organized for Life
Time Management
Productivity Improvement
Time Savers

Have you identified any career actions you want to take as a result of reading this chapter? If so, please make sure to capture these ideas before you forget by recording them in the notes section at the back of the book.

ASSIGNMENTS AND DISCUSSION TOPICS

1 Review the time wasters and determine how many you are guilty of doing.
2 How and what are you going to change to be better organized?
3 Map out how you presently spend each day. Can you improve it?
4 Audit your computer logs and see if you are guilty of over-text messaging and emailing. (Compute what your total time per day doing these is if you assume 3 minutes per email and 5 minutes per text message?)
5 What "To Do List" can you create right now?
6 How big is your technical library?

REFERENCES

1. Gleeson, Kerry, *The Personal Efficiency Program, How To Get Organized To Do More Work In Less Time*, John Wiley & Sons, 2003.
2. Aslett, Don, *The Office Clutter Cure: Get Organized, Get Results*, Adams Media, 2005.
3. The National Association of Professional Organizers (NAPO), http://www.napo.net/.
4. Franklin Covey Store — Planners, Calendars, Daily Planners and Personal Organizers, http://shopping.franklinplanner.com/.
5. Burka, Jane B., and Yuen, Lenora M., *Procrastination: Why You Do It, What to Do About It*, Da Capo Press, 1988.
6. Institute of Attention Deficit Disorder, www.add.org.
7. Covey, Stephen R., *First Things First*, Simon & Schuster, 1994.
8. Covey, Stephen R., *The 7 Habits of Highly Effective People*, Free Press, 1990.

CHAPTER 28

YOUR INTERNET IMAGE
IT COULD BE SABOTAGING YOUR CAREER

The Internet with all the knowledge available on an instant notice is a wonderful thing and can really help accelerate one's career. Have you ever considered what information about you is available to others on the Internet? Does your Internet image paint a very positive image and help support your career efforts or does it sabotage them? Never thought about it? If not, you should check to see what is out there. You may not realize it, but your private information from the past may be an instant career killer if someone was checking. In this chapter, we will explore some of the do's and don'ts about putting your private and professional information on the Internet.

SEARCH YOURSELF PERSONALLY ON THE INTERNET

If you have not done so already, the first thing to do is Google yourself and check every hit. Find out what information is readily available about you. Does it paint a positive image about you or are there some really embarrassing personal items you may want to get rid of? Many people put very personal items on the Internet in such places as myspace.com, youtube.com, facebook.com, twitter.com, linkedin.com, and personal web pages that can be searched. Have you ever considered who has access to any of your private information and how this could be used?

BIG BROTHER IS WATCHING

In the 1970s George Orwell wrote a book titled *1984*. The book was about how computers and video cameras were used by the government to track every

---

The Engineer's Career Guide. By John A. Hoschette
Copyright © 2010 John Wiley & Sons, Inc.

move and action you made during the day including what you did in the privacy of your home. Well let me tell you, *1984* is now happening at work. Big Brother is watching you. Here are some examples.

Internet Searching of Applicants' Background. The Human Resources and/or Security departments of most companies do a background check of potential new hires. Some companies hire independent firms for background checking. A background check is no longer just a phone call to the references on your resume, but a full-blown Internet search using several different powerful search engines. These search engines, in a matter of minutes, pull up all the Internet information out there about you. This includes pictures, videos, tax records, criminal records, financial and legal information, and any information posted on the Internet that has your name associated with it. Remember, Human Resoureces has the specific task to check out every individual applying for a job.

Given this knowledge, what do you think they are going to find out about you? Do you have some rather embarrassing information still on the Internet from past years? Old college party pictures? Personal dating information?

▶ *Career Tip.* Do an Internet search of your name and delete or remove all the improper information immediately, especially if you are interviewing for a job. This information could be sabotaging your chance for getting hired.

Have you ever considered that there might be people out there with your exact same name and when Human Resoureces does the search, they find others with your name and rather embarrassing information? You better be aware of this just in case you are asked during the interview process about the images and information Human Resoureces found on the Internet that is from others with your same name.

Turn off the instant email preview feature on your computer for group meetings and presentations. Many people use the instant email preview feature so that they can see if the email they just received is important and they need to respond immediately. What makes this feature a real embarrassment is when an engineer has their computer hooked up to the conference room projector, sharing their desktop during a net meeting, or when you are at their desk discussing something on the monitor with several others. The instant preview pops up and naturally everyone in the meeting looks and reads it. This can be extremely embarrassing when the engineer receives a very personal email with information never intended for others to see. I have seen some instant preview emails pop up that were viewed by all the people in the meeting resulting in quite an embarrassing moment for the engineer.

The company is monitoring your email and instant messaging. Don't put it in an email. Many people believe the company will never look up their email or instant messaging. The company would not store all the emails/instant messages sent, so there is no way they could recall anything. Let me warn you that this is not the case; in fact, the company specifically stores all emails/instant messages oftentimes for legal protection. Lawyers have tuned into this and the best way to win a case is by obtaining copies of all the emails/instant messages sent. The first thing lawyers typically do is subpoena the email and instant messaging.

This comes into play often in ethics cases especially if harassment is involved. Every email/instant message sent to the person being harassed is recalled and examined. In addition, all the emails of people sending email to the harassed person are also recalled and examined to determine if they have a bearing on the case.

▶ *Career Tip.* Always assume every email/instant message can be recalled at a later date and examined by the company.

Don't even think about visiting inappropriate web sites. Some engineers think they can visit inappropriate web sites and simply delete the files, cookies, and history, and they are safe. Well this is not the case; the files are still on the computer hard drive but only listed as free space. Using special software companies can recover these files. In addition, many companies have a policy that when the computer is turned in for repair or upgrades, the hard drives are scanned for any inappropriate material that may have been stored.

Companies can recall every web site you visited and the time spent. Companies have at their discretion the capability to produce a log of every web site you visited on their computers and how long you spent at the web site. This data is often used as a reason for dismissal. The data supports that the engineer is failing to do their job and addicted to surfing the Internet.

GPS and Blackberries. Many people receive a Blackberry with Global Positioning System (GPS) capabilities from their company so that when they are out of the plant, they can be instantly reached via cell phone, email, or text messaging. Do you realize the Blackberries are recording and sending your location at any given time? So if the company really wants to check to see if you are at the conference or business meeting, and not out on the beach or golfing, they can. Also where are you going after hours when you are out of town? Would the company approve of these places or would it be grounds for dismissal?

Cell phone and video cameras. Many people have cell phones with a camera and video storage capabilities. Are you using these appropriately at work? A simple photo of yourself at work or at your desk could reveal in the background what you are working on and even company secrets. When posting these on the Internet or sending to your best friends you could be leaking important company information without even realizing it.

Video phone teleconferencing. The new capability that most people are trying to exploit is video teleconference via the camera in their laptop or PC. Remember that in order to collect and transmit the video, it all has to be stored first and reformatted. This means that those video teleconferences can all be recalled at a later date. Are you letting them see and record you at your best? This brings us to the next topic: using the Internet to your advantage.

USE THE INTERNET TO YOUR ADVANTAGE

Now you know how Big Brother can be watching you. How can you use this to your advantage? The first and obvious answer is: don't put anything inappropriate in an email, text message, or instant message.

Compose your messages assuming Big Brother is watching. You should assume that every email you write could be recalled at a later date and the contents examined for its appropriateness. Keep this in mind as you compose your electronic message and let it guide you in your selection of words. This is especially the case if you receive an email that is quite upsetting and wish to respond about the anger and hurt feelings you have. A good rule of thumb when you are angry is to apply the 24-hour rule. Simply do not respond when you are in a state of anger and wait 24 hours before sending a response email. Most times after waiting 24 hours the anger passes and you are in a much better state of mind to compose a response.

▶ *Career Tip.* Apply the 24-hours rule when you are angry and tempted to instantly respond to an email.

Let them find you at your best when searching the Internet—knowing people are going to search the Internet to obtain information on you, why not put the best and most positive information you have out there. For example, do you have any information you can put on the Internet showing how you volunteered at a homeless shelter, helped raise funds for food shelves or charity organizations? Awards you received from the company? You could put together a portfolio on the Internet of yourself showing all your key accomplishments instead of the great beer party you had last summer. The key is to let people see you at your best versus at your worst. I am sure there are many items you could put on the Internet to enhance your image.

SUMMARY

The key to remember is that Big Brother is watching you 24/7 at work and at home if you are using company equipment. You must assume every electronic message you compose and send at work can be recalled and examined for appropriateness. All the web sites you visit are recorded and files do not go away just because you hit delete. Companies do extensive Internet searches on people, so all the personal information you have on the Internet is available for them to examine. Conduct an Internet search using your name and find out what others see. Delete the inappropriate material and replace it with material that will enhance your image. Let people see you at your best and not at your worst when they do an Internet search on you.

Have you identified any career actions you want to take as a result of reading this chapter? If so, please make sure to capture these ideas before you forget by recording them in the notes section at the back of the book.

CHAPTER 29

THE IMPORTANCE OF MAINTAINING A COMPANY CALENDAR

One of the most interesting phenomena I have observed is the total lack of interest in the company calendar by most employees. Most engineers have a calendar on their desk or carry an appointment book to keep track of important events. However, if you ask them if they track significant company events or have a company calendar, the answer is almost inevitably negative.

People have calendars for school events, sporting events, and even for birthdays and anniversaries, but rarely have company calendars. Is it important to develop a company calendar and keep track of significant events? The answer is yes!

▶ *Career Tip.* Failing to plan is simply planning to fail! And not having a company calendar is planning to fail.

COMPANY CALENDAR, HOW DO YOU USE IT?

To show you how important developing and maintaining a company calendar is, let's look at the following example. An example of a company calendar is shown in Figure 29-1. The activities are listed in the left-most column of the calendar. The time when this activity is planned to occur is shown by the black bars.

Let's study the first significant event on the calendar: salary considerations. In this sample calendar, raises and promotions are given out once a year, in May. However, a significant amount of activity occurs before the May raises.

The Engineer's Career Guide. By John A. Hoschette
Copyright © 2010 John Wiley & Sons, Inc.

Activities	Company Calendar											
	Jan	Feb	Mar	Apr	May	Jun	Jul	Aug	Sep	Oct	Nov	Dec
Salary Considerations												
Salary Forecast by Mgmt.			■	■								
Totems			■									
Promotion Rev. Boards			■									
Job Reviews				■	■							
Handout Raises/Prom.					■							
Department Budgeting												
Training Budget	■					■						■
Capital Equipment	■					■						■
Workload Forecast						■					■	■
Company Financial			■									
Profit Reviews			■			■			■			■
Supervisor Vacation								▧				
United Way										■		
US Savings Bond			■	■								
Company Picnic						■						

FIGURE 29-1   A sample company calendar.

In the March through April time period, the supervisor puts together a preliminary salary forecast. This is the supervisor's first estimate of the raises and promotions intended to be given out in May. Occurring also in this time frame are the company rankings and promotion review board meetings. Toward the end of April, job reviews occur. Last is the actual awarding of raises and promotions. Now let's examine why keeping track of these events is so important.

Let's assume you have been working for several years without a raise or promotion and you finally decide that it's time you do something about it. However, it is September and promotions or raises are only given out in May. What good do you think will come of you storming into your supervisor's office in September and stating, in no uncertain terms, you deserve a raise now. The supervisor's immediate response will most likely be to stall. Your supervisor knows there is no possibility you will receive a raise at the present time but the worst thing to do is tell you that now. How much better it would be if you had saved your energy and waited for March as the calendar indicates?

▶ *Career Tip.* Identify when raises and promotions are determined in your company and plan to have your major achievements occur just before.

The next thing to consider is job reviews. Your supervisor is busy handling a multitude of problems. Suddenly it's April or May and job appraisals must be written for 10 employees. If one appraisal write-up takes 3 hours, it may take over a week just to fill out the forms. Most supervisors have several employees and can probably only remember about 20% of what they accomplished during the past year. This time period is not the time to make waves or agitate your supervisor. If you recently had a setback, chances are this will be foremost in their mind. The best thing to do is have your breakthroughs and successes timed so they occur in March.

Send your supervisor a memo in March summarizing all your accomplishments over the year and then review them in a meeting. Remind your supervisor of your tremendous effort during the past year and highlight all you accomplished. Inform your supervisor you filled the forms out in advance to help save time. The supervisor need only file it away until the time is ready to do your appraisal; this should help save time. Chances are he or she will do just that. Most likely they will welcome and endorse this activity since you are making their job easier. Make sure you keep a copy in case they should lose or misplace their copy.

▶ *Career Tip.* Fill out the evaluation forms for your supervisor and send them in advance. Save him or her time and make sure all your significant accomplishments are noted.

Let's discuss some other important activities that are occurring at this time. Company totem is one and promotion review board is another. Do you know who is on the totem board? Do you know who is on the promotion review board? If you do, it is a good time to beat your drum a little bit so they remember your accomplishments over the past year and not just the most recent. Those rated highest on the totem are usually the ones who just recently made a major accomplishment. Do you know the criteria the promotion review board will be using? Generating a ghost memo written by you for your supervisor that highlights your accomplishments against the criteria is very handy to have in advance.

▶ *Career Tip.* Timing is very important. Successful career people do not depend on luck; they make their own luck. One way to do this is through the use of a company calendar.

Another important activity on the company calendar is the distribution of internal research and development funding. Do you know when this occurs? For the example shown, initial planning is done in January and progress reviews are held quarterly throughout the year. Do you have a project you would like to get funded? What projects are funded for the year and who is working on them? Are you on a pet project of great importance or something to keep you busy? Knowing when the project reviews are planned is important information. Nothing helps like having a breakthrough on your project just in time for the senior management review.

When do United Way and the annual Saving Bonds Drive occur in your company? These are excellent projects to donate your time to. These activities are often organized and executed by upper-level managers, the same managers who will be sitting on totem boards, promotion review boards, and handing out raises. These are all excellent ways to get the extra visibility you need. While working for a noble cause, you have the opportunity to visit with your lead engineers and supervisors, and meet their supervisors on a one-to-one basis and get extra points for doing so. As you sit there planning out the activities with the upper-level management, it doesn't hurt to tell them about your achievements.

Other examples of activities like these are the company picnic, the company holiday party, and any charitable things your company may sponsor, like food drives, shelters, clothing drives, or marathons. If you keep a company calendar, you will soon realize that there is something happening nearly every month that could result in giving you the extra visibility that you need. So much in fact, that you may not be able to take part in it all. It's important to make new contacts as well as friends along the way. Developing genuine friendships is as important as developing one's career. You can do both.

▶ *Career Tip.* Volunteering to work on a company committee will allow you to interact directly with the decision-makers of the company.

Another important activity you need to be aware of is the workload forecast your supervisor must do on a periodic basis. These are simply forecasts showing the work planned for you and your group over the next year. When does your supervisor do these? It may be a good idea if you paid attention to these since they will tell you if you will have a job in 6 months. If your supervisor is forecasting enough work then, come job appraisal time, you stand a better chance of getting the raise. If, however, in 3 months the forecast is no work for you, it may be a good time to start looking for work elsewhere.

One annual activity is updating the business plan or the company 5 year plan. This plan identifies the key business thrusts that the company plans to execute over the next 1 to 5 years. Knowing that such a plan exists, when it gets updated, and obtaining a copy, can be very beneficial to your career.

Department or supervisor budget planning is another activity of which you should be aware. A supervisor must forecast or estimate the expenditures the department will have over the year. Periodic updates and adjustments are made throughout the year and stopping by your supervisor's office when a budget is being done is a good idea. Was the tuition for classes you planned on taking included? Has the budget allowed for your new computer or the test equipment that you need?

Another important activity that usually occurs on a weekly basis is staff meetings. Do you know when your supervisor must meet with superiors and report progress? If you just happened to stop by and brief your supervisor hours before walking into a staff meeting, they may be more inclined to highlight your accomplishments. It does not hurt to have your supervisor highlighting all your accomplishments at the weekly staff meeting. Another way to do this is to submit a weekly written report to your supervisor highlighting your accomplishments for the week. Remember to save these weekly accomplishment lists. At the end of the year they become a handy summary of all your accomplishments. These weekly reports become a diary from which you can write your yearly summary.

If you don't write down everything weekly, you will probably only remember about 50% of what you accomplished for the year. This situation gets worse when you fail to summarize your yearly accomplishments in a written document for your supervisor. Without something written down, your supervisor will only remember 30% of your verbal message when you next meet. This means that without disciplined and documented summaries of your accomplishments, your supervisor will at best only report 15% of what you accomplished. No one gets raises or promotions when the supervisor only remembers or documents 15% of your efforts.

▶ *Career Tip.* Do a weekly activity report regardless if they are required or not. They become an excellent record of your accomplishments during the year.

SUMMARY

The important point about developing and keeping a company calendar is that it allows you to keep track of important career activities; activities that can benefit your career or allow you to take actions necessary to maintain your career. Developing a company calendar is no easy task and it may take you several years before you are recording all significant activities. Once you do this you will soon realize all the opportunities that are available to you. It will quickly become apparent that the only person limiting your career growth is you!

Have you identified any career actions you want to take as a result of reading this chapter? If so, please make sure to capture these ideas before you forget by recording them in the notes section at the back of the book.

ASSIGNMENT AND DISCUSSION TOPICS

1   Start a company calendar.
2   From the calendar, determine the most significant task/item you can do in the next 2 months to enhance your career.
3   Volunteer for a meaningful or noble company activity. Get involved!

CHAPTER 30

JOIN AN ENGINEERING SOCIETY
ENHANCE YOUR CAREER

Joining an engineering society and being an active contributing member is one of the best career moves you can make. Engineering societies offer multiple benefits to their members and provide excellent training to develop your technical, managerial, and people skills. They are more than willing to help you with your career development and offer training programs specifically for engineers.

A contact list for engineering societies is given in Appendix C. Please take a moment and review this list if you are not presently in an engineering society and find out how to join. The web sites listed provide an excellent starting point for gathering information. Most engineering societies have local chapters in all major cities that meet on a regular basis and cost little or nothing to attend. Looking back over my career, one of the best actions I took was joining an engineering society and getting involved. In fact, because of my involvement over the years, I was able to network, develop new skills, and find career advancing jobs. Don't wait. Join an engineering society today.

▶ *Career Tip.* Join an engineering society today; it is an excellent career move.

A quote from Paul Kostek, former IEEE USA President, "Many people ask what membership in a professional society will do for them. I always tell them the following—it's an easy place to network, members will have common interests and backgrounds; by volunteering you can develop skills, try new things and take risks that you might not want to at work. A failed weekend

---

The Engineer's Career Guide. By John A. Hoschette
Copyright © 2010 John Wiley & Sons, Inc.

program is not as career damaging as a project failure at work. Too many people see the initial cost of membership and don't consider the long-term gain they can get from being active members."

The Institute of Electrical and Electronic Engineers (IEEE) is the world's largest technical community with more than 375,000 members in 160 countries. IEEE establishes that by joining their society you gain access to the latest technical information and research, global networking, career opportunities, and exclusive discounts on financial and educational products. Students of any branch of science and technology can join IEEE. You can do anything you want in IEEE: learn, research, develop, teach, organize events, meet like-minded people, lots of possibilities. Benefits to IEEE members by joining are:

- Stay up-to-date on the latest technological innovations with IEEE publications and products. IEEE Society publications and conferences are the world's most respected sources for accurate, up-to-date information in electrical engineering, electronics, and computer science.

- As an IEEE member, you automatically receive *IEEE Spectrum* magazine. *IEEE Spectrum* is the monthly must-read magazine for technologists of today, exploring the applications and implications of new technology.

- IEEE Xplore™. As an IEEE member, to aid you in your research you can access table of contents, abstracts, and indexes, from IEEE publications. Optional access to *IEEE Potentials* magazine to members and various society memberships.

- International IEEE membership card, unique PIN and ID which lets you access unlimited information like technical journals, Proceedings of the IEEE, web issues of publications, and so forth at IEEE web site (www.ieee.org).

- Exchange information and interact with leaders in your area of expertise. As an IEEE member, you can attend any of more than 300 major IEEE technical conferences and 6,000 IEEE meetings and symposia available to members at special reduced rates.

- Chance to create your resume at the highly acclaimed IEEE jobsite and free mail account at www.ieee.org.

- Numerous scholarship opportunities for research projects and innovations. Up to $2,500,000 are awarded to student members each year.

- Upgrade your technical and professional know-how. Geared for engineering professionals who want to stay ahead in their career, IEEE books, videos, and self-study courses are cost-efficient, time-effective solutions for maintaining your technical acuity.

- Exhibitions, paper presentation contests, technical symposia guest lectures, and much more.

Gradschools.com is one of the leading and most comprehensive online graduate school guides for information on the best graduate schools and graduate degree programs. They have the following to say about joining a society:

> Thanks to ever-growing digital technologies, information abounds. But sometimes sifting through the tons of information available can be mind-boggling, not to mention time consuming. Instead, professional associations provide journals, newsletters, and web sites with invaluable information on up-to-date issues and developments in your specific field of interest.

Prospective employers seek out individuals whose field knowledge is not solely dependent on college studies; therefore, association memberships are excellent supplements for your resume. Memberships convey to an employer that you are dedicated to your field of study, while your savoir-faire during an interview will have you shining above other applicants. Your membership could also open doors of opportunity as you are provided with greater exposure to the job market—organizations provide "members only" job listings on their web sites, journals, newsletters, or other publications.

On the Women's Engineering Society (WSE) web site, the following benefits to joining a society are identified:

- Gain useful management experience.
- Build practical skills by getting involved in our activities (editorial board, web management, fund raising, organizing events, and awards).
- Meet like-minded people and enjoy working to build a stronger profession, build your profile at events, and network into your next job.

The Society of Women Engineers (SWE) identifies the following long list of benefits to joining their society.

- A strong network of women engineers from academia, government, and industry, corporate representatives, government officials, and other individuals, supporting and advocating for the mission of SWE Scholarship.
- Awards and recognition programs honoring outstanding accomplishments of women in the engineering profession as well as those who have contributed significantly to the advancement of women in engineering and technology professions.
- Scholarship opportunities for all educational and professional levels, including undergraduate and graduate students, and women re-entering the workforce as engineers (at national and section levels).

- Reduced fees for admittance to national and regional education conferences that offer networking, professional development, seminars, workshops, and leadership training.
- *SWE Magazine*—the Society's award-winning magazine, covering issues of interest to women engineers including the achievements of women engineers, career development, career guidance, activities within the Society, and technical topics.
- Webinars and podcasts.
- Online career center where you can manage your resume and search hundreds of jobs posted monthly by SWE's sponsors.
- Develop leadership experience in a non-threatening environment through leadership roles at the sectional, regional, and national levels.
- Publish articles in the *SWE Magazine*.
- Outreach programs that encourage girls to pursue careers in engineering through SWE's alliances with Girls Inc., Girl Scouts, and US FIRST (Foundation for the Inspiration and Recognition of Science and Technology.)

Have you identified any career actions you want to take as a result of reading this chapter? If so, please make sure to capture these ideas before you forget by recording them in the notes section at the back of the book.

PART 5

SUCCESSFULLY DEALING WITH DIFFICULT PEOPLE AT WORK (PEOPLE SKILLS)

CHAPTER 31

DEALING WITH A DIFFICULT BOSS

One common complaint expressed to me by engineers throughout the United States is "My Boss Is Driving Me Nuts!" This was a complaint expressed by all types of engineers at all levels in the company. Well, if you feel this way, you can take comfort in knowing that you are not alone [1–4]. In this chapter, I try to address this common complaint and some simple things you can do to handle and control your boss so that they do not drive you nuts any more.

EFFECTIVE COMMUNICATION IS BEST

In all of the cases discussed in this chapter the first and best solution is to try and sit down with your boss and talk things out. Effective communication with the boss is best. Often the boss may not know that their habits are driving you nuts. The most professional thing you can do is simply have a talk about the things bothering you. In most cases, this can quickly solve the problem.

When you talk with your boss, you can set up the situation for success by getting your boss away from their office. Meet at some neutral place like the cafeteria or conference room. Some place quiet and peaceful where you can both talk openly and yet not be afraid of being overheard by other employees.

You should have a professional and sincere tone in your voice. You can start out by explaining why you asked to meet. You are bothered by your boss' habit of . . . . and you would like to let him or her know it is bothering to the point where it is affecting your work.

The hope is that open and honest communication can quickly resolve the issue. You should not threaten nor put down the boss. Simply state the problem and what you would like changed. Stay away from emotional and put-down statements. Remember nobody, including you or your boss,

The Engineer's Career Guide. By John A. Hoschette
Copyright © 2010 John Wiley & Sons, Inc.

likes to receive negative feedback. Experts recommend that you start with first discussing several things you like about the boss and then transition to the problem. Make sure you have specific examples of when this occurred and how it made you feel. Do not talk in vague generalities and hypothetical situations; good solid examples are the best.

Have a recommended action to correct the problem and ask if the boss will do this. Talk through issues and make sure both of you leave with clear expectations of each other's needs. Finish on a positive note thanking your boss for taking the time to listen and understand.

I truly feel this is the best way to deal with these annoying habits that are affecting the quality of your work. Now let's look at the classic examples.

MY BOSS IS A MICRO-MANAGER

"My boss is a micro-manager and he or she wants to micro-manage every task I do and it is driving me nuts." Does this sound like something you are encountering? If so, relax. There are simple actions you can take to quickly turn this around.

First you need to understand the reasons why the boss tries to micro-manage employees. It stems from two basic problems. The first is they feel insecure about your ability to handle a problem. The second reason is they are unsure about their own ability to do their job and therefore want to make sure everything gets done right. Once you understand the root cause is insecurity, then it is easy to take steps to counter this.

Most people re-act to a micro-manager by running in the other direction and avoiding contact with the micro-manager. This reaction only makes the situation worse since it fuels the insecurity of the boss. The boss feels even more left in the blind and unsure of your progress. A very insecure position to be in. The avoidance tactic only causes the boss to want to micro-manage more.

▶ *Career Tip.* Micro-manage a micro-manager to turn the behavior around.

Instead of avoiding the micro-manager, the best thing to do is micro-manage back. In other words, turn the situation around where you are constantly seeking out the boss's advice before you do anything. Every little thing you can think of, go to the boss and discuss the opinions and what makes "best" sense. This way you are allowing the micro-manager to see how you think and how you arrive at solutions or recommended actions. Discuss everything, the pros and cons, the order of doing tasks, which are most important, which are least important, what to do if things fail.

For most people this is too painful to do and they would rather avoid than take a direct frontal approach. You do not need to be abrasive nor argue. You calmly and without emotion, in a very professional manner, discuss options and let the micro-manager direct you to the best approach. You do this for every little task you have. Be polite, ask for their time to discuss, make sure you drag out all the conversations, and thoroughly discuss all the unimportant details.

After about a week of this you will start to see how the micro-manager thinks and what insecurities are bothering them. By doing this, the micro-manager starts to build trust in you and your decision process. The micro-manager's insecurities will start to subside. As you go through options and the micro-manager provides you direction, comment on how you were just thinking the same thing. You are highlighting how you both think alike. Stay away from any direct confrontation. This will only fuel the desire to micro-manage you more. If you are flexible and the micro-manager's suggestions are technically sound then accept them and move on.

At the end of the first week of many meetings with the micro-manager, you should start to see them back off. This is a good sign, but you must continue the effort if you are going to be successful.

The second week is the time for a full-court press. The second week you increase the amount of time and make the decisions even seem more trivial but needing the micro-manager's approval.

You tie up the micro-manager as much as possible. If you are doing this correctly, the micro-manager will appear at your desk about halfway through the second week without being asked, and tell you to make more decisions on your own. The micro-manager will fess up that they no longer have the time for all your questions and discussions and they trust you to make the decisions.

At this point, if you did it right, they will be avoiding you! And the last thing they want to do is get tied up again meeting. Problem solved! You have the micro-manager's approval to take action and no longer need their input or approval on every little thing.

UNABLE TO MAKE A DECISION

"My boss is unable to make a decision; I cannot get anything approved. If only he or she could make a decision we could move on. The way it is now, we are stopped dead in limbo land." This is another common complaint. If you feel this way, you are not alone.

The basic root cause of this problem is that the boss does not want to make a bad decision and end up doing the wrong thing. Most likely the boss has made a decision in the past and it turned out to be the wrong choice. When it came time to place the blame, the boss was the person who ended up taking it. Consequently, the boss has learned to delay the decision until they are absolutely necessary and only make the decision if it is a sure, successful thing.

Employees also contribute to the boss's inability to make a decision. The employee is often the most knowledgeable about what has been going on and fails to realize that the boss is not at the same level of understanding as the employee. Consequently, the employee only brings the decision to the boss and not all the supporting data regarding the best choices. The employees feel this is an easy decision since they are the closest to the problem and know what to do and the boss should just trust them. Without having all the supporting data, instinct tells the boss that the best thing to do to minimize the chances for failure is delay the decision until further data can be collected.

A tactic some employees take in response to the delay by the boss is to go to the boss at the last minute and, hopefully, force the boss to make a quick decision and get on with it. This only compounds the problem since it makes the boss pressured into a decision they are not prepared to make.

You need to break this pattern. Here are simple steps you can take as an employee to get the boss to make decisions:

1. Forecast in advance and discuss with your boss (at least one week on big items) that a decision will be needed by a certain date. Make sure you highlight the consequences of delaying the decision in terms of extra cost or schedule delays. (This reduces the boss's fear of being rushed into a bad decision.)

2. Several days before the decision is required, educate your boss about the factors going into the decision. Bring them up to your level of understanding. Make sure you cover all the options and the pros and cons associated with the decision and generate a chart to support your findings. Writing it down in an organized fashion often helps. (In our boss's mind, this will reduce the risk associated with making the decision.)

3. Finally ask them point blank, can you make a decision on this date and if not, what other information do they need so the decision can be made?

4. Make sure you have backup plans identified if anything goes wrong so that you can recover quickly. (This is like an insurance policy to the boss and calms their nerves about making the decision.)

If you follow these steps, the boss should be able to make decisions in a timely manner and allow you to do your work and not delay the tasks.

▶ *Career Tip.* Involve your boss early and often in the decision-making process to get them to make decisions quickly.

Will you have to do this for every task? Only for a short period of time! Hopefully the boss will begin to develop trust in you after you have gone through this several times. Each time through it, the boss should be developing more trust and confidence in your ability to make the correct decision. With trust and confidence in you, the boss will start to delegate the decision process more to you. The boss can easily see all the effort you are putting into making the best decision and eliminating risk and chances of failure.

CANNOT STAY FOCUSED

"My boss cannot stay focused. I go to him or her about a single issue and before I know it, we are discussing six or seven other things that have nothing to do with the original issue!"

This boss has what is called in engineering terms, a diverging process or diverging series. They start out on one subject and bring in related adjacent subjects. Then they repeat the process on the adjacent subject and bring in other related adjacent subjects and before you know it, you are discussing items that have nothing to do with what you originally started out with. You do not converge to a single solution but quickly generate an indeterminate state with infinite solutions or no solutions.

The root cause of this problem stems from the boss wanting to cover every possible related subject to make sure all the bases are covered. However, the boss quickly loses track of what was the original purpose of the meeting.

You, as the employee, need to control this and bring your boss back and focus on the reason you are there. To control this, you first have to recognize when it is starting and take steps to refocus the emphasis of the discussion.

When you refocus the conversation back to the original subject at hand, you should do this in a professional and non-threatening manner. Remember, the boss does not realize he or she is moving off into other directions you do not want to go; they simply think they are helping you. Here are some simple non-threatening and professional questions you can ask or statements you can make to help them refocus.

1. I think we may be getting off the track a little; how does this relate to the original subject of . . . .
2. I do not see the relation between this and what I originally came to talk about. Can you help me understand?
3. Please clarify for me where we are going with this; I am lost.
4. This is a good subject and I would like to discuss with you after, but for now can we focus on . . . . and get through this and then come back to . . . .

These are not the only questions and clarifications you can ask. The point is to pick one you are comfortable with using and use it. Remember not to be abrasive but professional. People indicate to me that they are not sure they can do this. If they do not change and do something different, I point out that they are doing the exact same thing over and over and getting the same, but expecting a different, result. I believe a famous scientist (Albert Einstein) pointed out that this is a sign of stupidity.

FIGURE 31-1   Paralysis by analysis.

PARALYSIS BY ANALYSIS

In this scenario, the manager or lead engineer usually has to analyze every possible option only to come to the conclusion that nothing is going to work successfully. At this point the manager feels trapped or in a paralysis state — unable to move on anything. All work must come to a stop until the manager figures out what to do next (Figure 31-1).

The basic problem is the manager has put the emphasis of the analysis on what will not work instead of identifying what will work. He or she is looking at how empty the glass is instead of how full.

There are several actions you can take to get your manager through this and move on. First is to review the basic assumptions going into the analysis: what is considered unacceptable and what is considered just good enough. Maybe one of the potential solutions has a problem but the performance is acceptable for now. Review your assumptions and see if they are really necessary to meet your goal. If you can relax one of your assumptions that allow a solution to meet your objectives, you are out of the paralysis state. As a first step, try working through your assumption with your manager.

The second thing you can do is help your manager realize all solutions have a problem but the problems are manageable. Many managers want the perfect solution with no potential failures, so they think of all the possible bad scenarios that can happen.

Another action is to generate a comparison chart of the possible approaches and the good and bad points. Sometimes when you put it all down in writing in an organized fashion, it is easier to make a decision and move on. All the terrible problems do not seem that big, and often one solution will clearly stand out above the rest making the decision easy to make.

TAKES CREDIT FOR ALL MY WORK

"I do all the work and my boss takes all the credit!" This is another common complaint. There are several things you can do on this one. First, you must be direct and upfront with your boss to solve this problem. Simply sit down with your boss and express your concerns about lack of getting credit. Make sure you are professional and non-threatening in your discussion. Tell your boss you would like to get more credit for the work. Ask if they are highlighting in reports that you are the one responsible for the work. Give specific examples if you can of where you feel the boss is taking credit for your work. Typical examples are the boss taking credit in weekly reports for your work to upper management and never reporting that you are the one actually doing the work. Or another case is when demos are given and you are never involved. Your boss may not realize he or she is doing this, and simply bringing it to their attention, is enough to get the situation corrected.

▶ *Career Tip.* The first step to correcting the boss from taking credit for your work is to confront him or her directly and ask for more credit.

If this does not work then you will have to be more aggressive. Some of the more aggressive actions include sending a copy of your weekly activity reports to your boss's boss directly. The activity report should clearly highlight all your activities and the progress you are making.

If you are required to demonstrate things, make sure you are the only one that can run the demo. You hold the passwords and knowledge how to run things, so that way you are visibly running things when it is time to get credit.

If you issue reports or charts, put your name as the source in the footer. This way when people see the charts or read the report, your name is identified.

If you take pictures of the equipment or hardware, make sure you are in the picture, so your face gets associated with doing the work.

One final action is for you to share your feelings of disappointment and how you feel you have no option but to perhaps not work for them. However, this can quickly backfire since the boss may want you to leave and quickly arrange for this. This is only a safe card to play when there is absolutely not a single person in the whole wide world who can do what you are doing!

KNOW IT ALL

The "Know It All" boss simply believes their job is to share all their knowledge with employees since they already know what is the best thing to do. They have more knowledge than anyone on the team and by sharing and educating people, the whole team benefits. They feel a need to explain how everything works since they clearly have more knowledge than anyone they are talking with.

These types of bosses quickly drive employees crazy by constantly explaining every detail and why this works this way or that. They often have bad listening skills and seldom ask for your input. This leads to the tendency of people to quickly tune them out.

There are several methods for dealing with this type of boss. One action is to simply not respond and do not ask any questions. Sooner or later the boss is going to have to stop talking and involve you in the conversation so he or she can get your input on progress. By not talking, you are signaling to the boss that you are not taking part in this one-sided conversation. By asking questions and acknowledging their good points, you are only reinforcing to the "Know-It-All" boss that he or she is doing the right things and they need to continue to explain and educate you. They are looking for this type of behavior and interchange.

Another action is to constantly ask questions. Simply ask, "how does that work?" I have used the technique of asking "how does that work," with great success. One boss tried to keep up with my all questions just to show his knowledge and capability to explain everything. I asked question after question on everything he brought up until he was to the point where he realized he was far from the original question and needed to refocus back to the original question. He was worn out.

CONSTANTLY RE-PRIORITIZING MY WORK

Constantly re-ordering priorities is also a very common problem. The root cause of constantly re-ordering work priorities is the boss simply believes that every time he or she thinks of something that needs to be done, it is their job to assign it to you. They feel they are paid to think of things to be done, and since they rely on their memory rather than lists or schedules, they want to make sure it is in queue of things to do.

This can be bothersome to some people since they do not know whether to rush the rush they are working on or rush the new rush. The employee is continually put in the difficult situation of having to start a new task before they can complete the last task.

This is an easy one to change around. Keep a list of the tasks you are working and in the order you are working on them. Then, when the boss shows up with a new unplanned task, you simply go through the list and identify where it falls on the list. If it is so important you must start right away, then you ask the boss what task do you want to take off the list or which task should you stop working on? This should quickly stop the behavior and make the boss think before coming up with a new task. The boss will realize that they are constantly in the mode of telling you to stop working on the task they gave you yesterday and work on the new one. This highlights to the boss the delay in getting the task from yesterday done because they have just asked you to drop everything for today's project.

You should also be pointing out that you are unable to bring a task to completion since you have to shift your attention to a new task before the old one was closed. Point out that this is very inefficient and you are wasting time, resources, schedule, and increasing cost by doing this.

BOSS' WAY OR THE HIGHWAY

Many bosses firmly believe if people would just follow their direction, things would go better. Simply follow their instructions and we would not have these problems in the first place. This attitude can best be characterized by the saying "The boss' way or the highway." (Figure 31-2) This attitude by the boss creates mindless people who just follow directions and do whatever the boss wants, even if it is wrong. Trying to reason with this type of boss or change this behavior is an act in futility. This is like trying to reason with a tank driver. Tanks rollover everything in sight, take no enemies, crush people, and kill everything that stands in their way.

You only have two choices on this one: join the boss or flee. If you join the boss, you must be able to take orders and execute without question. Questioning an order is an act of insubordination! On the other hand, if you can execute orders flawlessly, you will be rewarded. You become a team member, and after time, will be allowed to give orders also. You must first prove yourself and your loyalty to the leader before becoming a commander in these ranks—capable of thinking and doing things for yourself.

The alternative is to flee and look for another position where more sanity exists. Some place where you are allowed to think for yourself and have a difference of opinion.

▶ *Career Tip.* Your way or the boss' way. You decide!

Obviously, I have overdramatized these potential actions to see if you identify with one more than the others. If you find yourself clearly aligning with either one of the oversimplified reactions, then it is a clear indication what you should do.

Let me make this perfectly clear so there are no misunderstandings about what I want you to do and not to do!

FIGURE 31-2 Boss' way or the highway.

If you are not sure what you want to do, then you are in trouble. Oftentimes, the work is great and it keeps you around under these circumstances. But the problem is, you cannot stand the "my way or the highway" attitude of the boss. If you find yourself in this position, then you have a very tough decision to make: stay and work at it or seek work elsewhere. There is no simple answer for this; you have to examine your position fully and decide what is really best for you!

MY BOSS ALWAYS SAYS THE OPPOSITE OF WHAT I SAY

"No matter what I say, my boss always says the opposite. If I say white, my boss says black. I say black, my boss says white." Does this sound familiar to you? If so, then you have a problem with your boss. Generally, this behavior is associated with you and your boss starting off on the wrong foot as they say. Another cause could be your boss' belief you are not competent. Employees contribute to this type of reaction by trying to tell the boss what to do or only offering a single solution to a problem when many solutions exist.

This behavior is easy to overcome. When presenting a solution or choice to your boss, always present at least two or three with no real indication of what your choice is. Next you must let the boss pick first. Never pick first; if you pick first or come with only one preferred solution, you are setting the boss up to pick the opposite of you. After the boss has made the final selection, then comment on how you were thinking the same and how you both think alike. This will diminish this difference between you and the boss over time. Also, it is very important to observe your boss' facial expression and body language when you inform him or her that you were thinking the same thing. If a smile occurs, it is an indication that the boss is pleased with your choice and you are moving the relationship to be more on the same side rather than opposing sides. If the boss' facial expressions are one of surprise or worry that you both selected the same solution, then you have more work ahead.

▶ *Career Tip.* Let your boss select the solution first and then agree!

Another method of getting the boss to pick your preferred solution is to select first the opposite solution that you wanted. Then relying on nature to take its course, the boss will select your preferred solution that is the one you really wanted.

All of the recommended solutions will work. You can alternate between them if you have to. The key is to stop setting yourself up for failure and take positive steps to change this behavior.

THE BOSS HAS FAVORITES

"My boss has favorites in our group. The boss always goes to them first or listens to only them." If this is your problem then you have a tough problem.

First you must look inward and examine what you want to do. If you have the attitude that you should not have to play this game and you do not really want to become a favorite, then your best career option is to make a job change, leave the group, and find another group. Having this attitude and staying in the group is only going to make the situation worse and eventually your attitude and resentment is going to show in your work. By not buying into the idea of becoming a favorite and working toward this, is basically saying that you do not support the boss and the boss needs to change. Trying to change the boss to conform to your ideals is a very career-limiting move.

▶ *Career Tip.* Trying to change the boss to conform to your ideals is a very career-limiting move. Spend the effort on improving yourself.

The problem you must solve is how to become one of the "favorites." The first step to solving this problem is to observe who the favorites are in the group and how they act. How are their actions different from yours? Are they complimenting the boss in front of others and supporting his ideas with all their efforts? Do they speak highly of the boss' actions and have a good mental attitude? Do they get the tough jobs done when it is critical? Basically, by studying the favorites and their actions you are identifying the potential reasons the boss likes them so much. Your goal is to emulate their actions and hopefully you will migrate to a "favorite" standing.

The second step is to have a conversation with the boss about who gets what assignments in the group and how does the boss select who he or she goes to? During this conversation you need to ask open-ended questions, like what traits are you looking for in an employee? How can you and the boss better work together? Answers to these questions should provide guidance on how you can change your behavior. Sometime during the conversation you need to be direct and forward and mention that you feel that the boss is assigning tasks in a way that suggests to the others in the group, who are certainly the favorites. The boss may not even realize this is occurring and correct the situation once it has been pointed out. Make sure you have concrete examples that you can cite at this point in the conversation. A perfect example of this might be when the boss leaves, the same person is always put in charge of the group. A good management strategy is to alternate this responsibility around to all the members of the group. Another concrete example is always picking the same person to present the results of the entire group to upper management. In any case, make sure you have specific examples that you can point to on why you feel the way you do.

One of the reasons the boss may be using favorites is because of their ability to perform the task. The boss may feel there are only certain people they can count on. If this is the case, then you want to point out to the boss that you would like to be considered for these assignments. If you need further training and development you are willing to take the training so that the boss can feel confident in your ability to perform the tasks. You would like the opportunity to show the boss you can perform the tasks and you will make sure the boss is not disappointed.

▶ *Career Tip.* If your attitude is one of learning, willing to make improvements, and willing to go the extra mile then you should be successful.

When addressing this problem like all the others in this chapter, calm open discussion without emotionally charged words is key to the resolution. If your attitude is one of learning, willing to make improvements, and willing to go the extra mile then you should be successful. If your attitude is fueled by resentment, blame, or anger, your chances for success are going to be very limited.

THE BOSS MAKES DECISIONS WITHOUT CONSULTING ME

Many times a boss will take action and make decisions in order to get things moving or break a dead-lock situation. The boss views this as a necessary part of the job to get things rolling again and on track. However, the employee may consider this meddling with their work and even feel insulted when they are not involved in the decision process.

This should be an easy one to fix unless you are the problem that the boss is trying to change. Simply sit down with your boss and discuss how you feel left out when he or she makes decisions about your work without being involved. Discuss how you feel disappointed and you feel as though your opinions are not valued. Simply letting the boss know you want to be consulted, or be involved in any of the decisions that involve your work, should rectify the situation.

INFLUENCING YOUR BOSS

If your boss is driving you crazy, then you need to make changes in your approach to dealing with the boss. Most likely your boss is not going to change his or her behavior for you. He or she is set in their ways and it is up to you to change your style or behavior if you want to be successful.

▶ *Career Tip.* Learning to control your boss and his or her perceptions about you is an essential ingredient for career success.

Remember the laws of physics and Brownian motion. Left alone, matter will randomly move about, tend to settle into an equilibrium state, and you can never really tell where matter is at any given moment.

Translating into career terms, your boss will randomly move about, things tend to settle into an equilibrium state defined by the boss and you can never really tell where your boss' head is at any given moment.

The obvious paradigm to this chapter is that you need to do a better job of influencing your boss. The best and most professional means of changing things is to simply sit down with your boss in a calm and professional manner, without a lot of emotion, staying away from blame, and discuss the things you can both do to improve the situation. Make sure you identify the things bothering you in a non-threatening manner and also hear the boss' side of the story. Maybe you are driving your boss nuts without realizing it?

Dealing with your boss on these difficult issues can be very nerve-racking and emotionally upsetting as well as very frustrating. Work force violence is on the rise and is not necessary. Please, if you feel you are losing control, you should remove yourself from the situation immediately and take the time to compose yourself and calm down. If necessary, seek professional counseling.

SUMMARY

Effective communication with the boss is best. The most professional thing you can do is simply have a talk about the things bothering you. In most cases, this can quickly solve the problem.

When you talk with your boss, you can set up the situation for success by getting your boss away from their office. Meet at some neutral place like the cafeteria or a conference room. Some place quiet and peaceful where you can both talk openly and yet not be afraid of being overheard by other employees.

If meeting with the boss does not change things, then it is up to you to take control of the situation and implement some changes. Making changes in your behavior may be the best way to influence the situation in your favor. One good rule to follow is to always use tact and polite manners when implementing change. Tact and polite manners coupled with the realization of changing behavior will take time and patience.

Have you identified any career actions you want to take as a result of reading this chapter? If so, please make sure to capture these ideas before you forget by recording them in the notes section at the back of the book.

ASSIGNMENTS AND DISCUSSION TOPICS

1 What is the best action to take if you are having problems with your boss?
2 How do you control a boss who is constantly re-prioritizing your work?
3 How can you influence your boss if you are bothered by their micro-management?
4 If your boss will not change is quitting the job the only answer?

REFERENCES

1. Willis, Gerri, "Dealing with an Abusive Boss," CNNMoney.com, http://moneycnncom/2004/10/15/pf/saving/willis_tips/indexhtm.
2. "How to Deal with a Difficult Boss," http://wwwbadbossologycom/book_menu. (Books on the subject).
3. Weinstein, Bob, *I Hate My Boss!: How to Survive and Get Ahead When Your Boss is A Tyrant, Control Freak, or Just Plain Nuts!*, McGraw-Hill, 1997.
4. Lubit, Roy H., *Coping with Toxic Managers, Subordinates . . . and Other Difficult People: Using Emotional Intelligence to Survive and Prosper*, Financial Times Prentice Hall Books, 2003.

CHAPTER 32

DIVERSITY—ENDORSE IT OR BE LEFT BEHIND

The engineering labs and offices of the 21st century are more diverse than ever before. Our global economy, international business, desires to have the best people, and ease of migration, contribute to having a diverse workforce. Any manager or company that endorses an attitude of hiring only people who look, think, and act similar to themselves, are dinosaurs who will not survive. In this chapter, we discuss what diversity is and how it is good for your company and your career.

WHAT IS DIVERSITY? WHAT ARE THE ADVANTAGES?

Workplace diversity refers to the differences in people with regard to their gender, religion, race, age, personality, ethnic background, education, physical handicaps, and many more factors. Diversity is not only recognizing these differences, but also accepting them and working with others who are different from us for the common good of the team, project, and your career.

Diversity has been around for many years but it is only recently that HR departments and employers have taken to making it a top priority. If we look back over time we can see some of the most significant contributions to science have been from people who would be considered diverse. One example is Charles Steinmetz who was severely crippled and had immigrated to the United States. He was almost not let in because of his physical disabilities. Steinmetz developed the theory for our modern day power systems in the early 1920s. One early pioneer of women in science was Marie Curie, who was responsible for developing the science of radioactivity. Steven Hawking, although physically handicapped, is an internationally recognized scientist with his theories on the Big Bang; the list goes on and on. I believe the best example of where diversity is alive and well is NASA Space Station and Shuttle program. The international participation on the space station and the diverse background of astronauts has allowed us to do things never thought possible.

An engineering team's success depends on its members' ability to accept people of different backgrounds and work together. Engineering teams that endorse diversity have advantages. These advantages include:

1. Increased versatility and better potential for solving a wider range of problems
2. More effective and efficient engineering efforts
3. Exposure to new and different ideas.

With a diverse team, members bring different experiences and methods to solving problems. This results in versatility and a wider range of solutions to choose from. Having good options is what it is all about when faced with difficult engineering problems. A diverse team offers the potential to be more effective since members can provide a wider variety of resources and solutions to the problem. With better solutions, the team is ultimately more efficient.

Diverse engineering teams where the members cannot work together will not be successful and will ultimately cost the company more money to develop products. Sondra Tiedemann in her book, *Making Diversity Work: 7 Steps for Defeating Bias in the Workplace*, identifies that consequences of bias can include lost business, discrimination, and even lawsuits [1].

WHY ENDORSING DIVERSITY IS GOOD FOR YOUR CAREER

Accepting the differences among your teammates and learning to work with a diverse team is a critical skill you will need throughout your career. Learning to judge people on their technical ability rather than their background or physical looks is a critical skill for advancement. When you find yourself opposed to a teammate's idea or work stop, ask yourself why you are objecting. Is it because the technical idea is not good or are you biased about the idea simply because of the person presenting the idea? If the reason you are objecting is solely because of the person and not based on the technical merits of the idea, are you really being fair? Could your bias and prejudices be coming out? One quick test of your bias is to consider how your reaction to the idea might change if another person on the team would have suggested it. Imagine for a moment that a teammate who you are very comfortable with, had suggested the idea? Would the idea suddenly seem more reasonable and acceptable? If so, it is time to rein in your bias and start acting more responsible and accepting of diversity.

Throughout your career you will be placed on teams containing diverse members. Your ability to perform on the job will directly depend on your capability to work in a diverse environment. Engineers who show they can perform in a diverse environment, are going to move up and get the more challenging assignments. Those engineers who hold on to their bias and prejudice and can't perform in a diverse environment, will be isolated from the group as a minimum, if not encouraged to find employment at other places.

▶ **Career Tip.** Judge people only on the technical content of their ideas and their ability to get work done successfully. Do not judge people's ideas and technical ability based on their background, race, or physical appearance.

As you move up the corporate ladder, you will be moving more into a leadership role where you will be directing diverse teams. Your ability to unite the team and bring out the best in them for the benefits of the project and ultimately for the company will depend on how well you deal with their diverse backgrounds and your ability to remove biases.

Endorsing diversity and capitalizing on the differences and strengths of people is a career accelerator. One way to encourage diversity is to celebrate diversity among your team members. When I found out that one of my team members was an immigrant from Somalia, I asked him to give a short presentation on their customs and dress. The engineer showed up to a noontime brown bag in traditional Somalia dress for men to give a presentation about his homeland. The engineer talked about his upbringing and living in a warlord society where people and parts of the country were segregated based on one's ancestry. People did not dare travel to an opposing warlord controlled city for fear of being killed. He discussed the life styles of his nomadic people and the dating rituals of the young people. He shared with us his story on how his family migrated to the United States and the problems they encountered.

The room was packed and it was one of the best-attended brown bag presentations of the year. Everyone left the room wiser and more accepting of the differences in our teammates. There are many well-written books to help guide people on diversity. To find additional information, search the Internet for "Diversity in the Workplace."

▶ **Career Tip.** Help facilitate a diversity event or meeting in your company. Diversity events can be great for people to share their backgrounds and life experiences with each other.

Have you identified any career actions you want to take as a result of reading this chapter? If so, please make sure to capture these ideas before you forget by recording them in the notes section at the back of the book.

REFERENCE

1. Tiedemann, Sondra, *Making Diversity Work: 7 Steps for Defeating Bias in the Workplace*, Kaplan Business, 2003.

PART 6

LOST YOUR JOB? LOOKING FOR A NEW JOB? STRATEGIES TO MAKE YOU SUCCESSFUL!

CHAPTER 33

TIME TO CHANGE JOBS OR STAY

Long before people are fired or laid off there are indicators that it is time to change jobs. Some of these indicators are obvious and some are very hard to recognize [1–3]. In this chapter, we explore what these indicators are so that you can assess your situation to determine if it is time to change jobs or stay put.

Changing jobs does not necessarily mean that you have to leave your present employer. Changing jobs can mean simply doing a different job in your department or moving to another department in the company with better opportunity. In fact, it is a good career move to stay with your current employer if better positions are available within the company. Let's start with some of the more obvious indicators that it's time to change jobs and then move to the less obvious ones.

INDICATORS THAT IT'S TIME TO CHANGE JOBS

Poor Ratings and/or on Probation. If you are receiving poor ratings and/or your performance has degraded to the point where the company has put you on probation or official notice, then it is time to change jobs. The next step in either case is normally termination of employment. It will take a super human amount of work to turn this situation around and at least three years of significant continuous improvement before your next raise or promotion. If this is the situation you are in, it is best if you immediately look for a change of employment.

Hate What You Are Doing. If you hate your work and even dread going to work in the morning, this is another clear indication that it's time to change jobs. Your work should be exciting for you, and getting up and going to work should not be something you dread doing. Left unchanged, your feelings and attitude are eventually going to start manifesting themselves in the quality of your work. It is only going to be a matter of time before your performance starts to deteriorate and ratings start to decline.

You Have Stopped Learning and Growing. When your work becomes very routine or monotonous, it is another indicator that it's time to change. If your job has become very routine to the point there is no more learning or improving your job skills, then you have hit a stagnation point. Without continuous training and improvement for your next job, you are setting yourself up to be eliminated once the job has been completed.

Your Role in the Organization Has Significantly Diminished. If your role in the organization has diminished to the point you are only kept abreast of information after the fact, whereas before you were part of the team making the decisions, then it is time to change jobs.

Others Want to Hire You. Do other departments or other companies want to hire you but your boss thinks you barely qualify for a raise? Does it seem like others are impressed with what you do, but your present department barely knows you exist? If this is the case, then it is time to talk seriously to these companies about better opportunities for you.

No Respect and Trust. Are others in your department distrustful of you and you feel like your work is not respected? There is clearly something wrong with this situation since the key ingredients for successful teamwork are respect and trust.

Always Disagreeing with Boss. If you continually find yourself disagreeing or on the opposite side of the table as the boss then it's time to change jobs. Disagreeing once in a while is good, but constantly disagreeing is a clear indicator that the relationship is not a good one for you or your career. Managers simply do not promote people they cannot get along with and are constantly fighting with.

Working for a Department that Can Be Easily Outsourced. Do you work for a department that is considered a support group to the main organization and management is considering budget cutting measures? If so, you are not on solid ground with your career since the whole department's work can be easily outsourced to contractors to save money.

Bad Economy Is Hurting Your Company. In a downturn economy companies are laying off, job anxiety significantly increases, and the common question asked: is it time to look for a new job? [4,5]. If your company has announced a round of cost cutting measures and workforce reductions, this is another good indicator it is a good time to evaluate whether to stay or not.

All these indicators clearly signal that trouble is ahead for you and maybe considering a change in jobs is a great career move for you.

INDICATORS WHEN TO STAY

Correspondingly, there are indicators when it is time to stay put at your present job. These indicators are signals that the present course you are on is a good one and to keep up the good work. It will be only a matter of time before career advancement happens for you.

Being Asked to Work on the More Challenging Assignments. If your boss and others are constantly seeking out your help for the new and challenging assignments, then you may want to stay put. Opportunity is seeking you out and this is exactly the spot you want to be in.

You Are Receiving Awards. Your manager is taking the time to submit you for awards and special recognition in the company. Your work is being appreciated and rewarded.

You Are Sent for Special Training. If your company or manager is sending you for specialized training to enhance your skills, this is very positive. Training costs are very high and managers like to send only the people who will benefit the company the most and be here in the future—hopefully in leadership roles.

Consistently High Performance Ratings. If you are consistently receiving high performance ratings review after review, then it is time to stay put. Your management obviously appreciates your work and is doing something about it.

Profits and Stock Prices Are Rising. The signs that your company is healthy and growing are increasing stock value and large profits. Management can easily justify promotions and raises when the company is doing great. Correspondingly, it is very difficult to justify promotions and raises when the company stock is dropping and profits are falling.

You and Your Boss Hit It Off. The boss agrees most of the time with you on how to accomplish work and is pleased with your results. The boss trusts you enough to completely turn the assignment over to you and let you run with it. This is a time to stay put.

All these indicators clearly signal good things are happening for your career so staying put is probably best.

SUMMARY

If you are wondering whether or not to change jobs or stay, there are indicators to help you in this decision. Some of these indicators are obvious and some are very hard to recognize. The indicators that it is time to change your job include such things as poor ratings, your role in the organization has significantly diminished, constantly disagreeing with the boss, and the business outlook is not good. If you feel it is time to change, then changing jobs does not necessarily mean you have to leave your present employer. Changing jobs, can mean simply doing a different job in your department or moving to another department with better opportunities. If these do not work, then the next option is to seek employment with a new company.

There are also indicators that clearly signal good things are happening for your career and staying put is probably best. These include such things as being asked to work on the more challenging assignments, you are receiving awards, consistently high performance ratings and profits and stock prices are rising.

Have you identified any career actions you want to take as a result of reading this chapter? If so, please make sure to capture these ideas before you forget by recording them in the notes section at the back of the book.

ASSIGNMENTS AND DISCUSSION TOPICS

1   How fast do the indicators change and how long should you monitor them before making a change or decision?
2   Name three indicators that it is time to change.
3   Name three indicators that it is better to stay.

REFERENCES

1. Fisher, Anne, "8 Signs It's Time to Change Jobs—It's Best to Move on Before You Get Fired, Lose your Sanity, or Both," CNNMoney.com, September 25, 2007.
2. JOBWERX, "Determining the Right Time to Change Jobs and Doing It," http://www.jobwerx.com/news/careers_biz-id=947712_102.html.
3. Taylor, Hunter, "Signs That it May Be Time to Change Jobs," Associated Content, http://www.associatedcontent.com/article/8044/signs_that_it_may_be_time_to_change.html?cat=31.
4. McKee, John, "Sit Tight or Jump Ship," *Computerworld Magazine*, November 10, 2008.
5. Hoffman, Thomas, "8 Signs it's Time to Look for a New Job in 08", *Computerworld Magazine*, January 21, 2008, p. 30.

CHAPTER 34

USING THE COMPANY'S JOB OPENING SYSTEM TO YOUR BENEFIT

Using your company's job opening system (JOS) to your benefit is one of the most important things you can do for career development. The JOS is the means by which companies inform their employees of the jobs that are available in the company. Are you aware of the JOS in your company and how to use it to your benefit? If not, you could already be missing an opportunity for advancement. In this chapter, we will explore how you can benefit from your company's JOS.

Typically, the personnel department in most companies post all job openings. The company does this because most often, the best candidate for the new position is someone already in the company, which could mean less time and effort than training a person from outside the company. So you, as an employee, often stand a better chance of advancement than an outsider.

Often the job openings are posted on bulletin boards, or accessible through the company computer network. Do you read these listings? It amazes me how many people will read the daily sports, financial, arts, or the comics sections, but hardly ever look at the company's job opening listing. To get ahead you must read the company's job opening listings continuously. You can never tell when a better opportunity will present itself. In addition to keeping abreast of the job market internally you should also be continuously looking outside the company. Study and read all the job ads you can. You may even want to contact some professional search firms to do some searching for you.

The Engineer's Career Guide. By John A. Hoschette
Copyright © 2010 John Wiley & Sons, Inc..

THE IMPORTANT THINGS YOU CAN LEARN FROM READING THE COMPANY JOB POSTINGS

Job ads will provide you with a wealth of knowledge about the job market and your chances for advancement at any given time. By studying the company's job ads you can quickly tell which divisions are hot and hiring. You can also tell which divisions may be having difficulties since they have no openings. How is your division doing? Is it hiring? Chances for promotion are significantly enhanced if you are working for a division that is hiring.

Study the job listings further. Do they identify the job level, type of engineer required, and the pay level? The job ad should identify the experience and background they are looking for. The background sought by the majority of the job ads indicates the background you may need to get ahead in your company. If all the openings are for chemical engineers and you are a mechanical engineer you may have a difficult time advancing. Compare the salaries being offered in the ads to your job level. How does the pay compare to yours? If you are getting underpaid and it's important to you to move to obtain more money, then it's time to move on. If the pay is comparable and you are satisfied, there is no reason to move.

Determine the salary being offered for the jobs one level above yours. Would you like a raise and a promotion? How will you find out if you qualify unless you go for an interview? Remember, your supervisor may be limited in the salary increase they can offer you. However, it may be easier for a different supervisor to give you the raise you want simply because you would be a new hire to the department.

Study the job ads to determine which divisions have openings and what products you would be working on. Which product line is hot and undergoing rapid development? What divisions and products are not growing? This is easy to tell by lack of job ads. Is there any chance of transferring to a new division or product and pick up a promotion in doing so?

▶ *Career Tip.* Read the job postings weekly, but be discreet and keep it to yourself.

A Word of Caution About Company JOS Ads. Most companies have internal policies regarding promotions. One of these policies is that if a supervisor intends to promote an employee to a new level, they must have a need and a position open for that grade level. The company wants to put the best person into that position, not just the one the supervisor wants. Therefore, they require that the supervisor post a listing for the position and interview all possible candidates. The intent is to get the best possible candidate for the position.

▶ *Career Tip.* When you find a job opening that interests you, find out everything you can about it before applying. Who posted it, the lead engineer on the project, the type of work, and other people in the group who might apply.

What this means is that often a JOS listing will be posted with someone already in mind. Most engineers are unaware of this policy and therefore think they stand a good chance of getting the job. If you are responding to a JOS listing, do some research first and try to find out if there is someone already in mind for the position. The obvious question to ask the supervisor when you are interviewing is, "is there anyone presently in your group that is interviewing?" Is that person more than likely the best candidate and is it a "done deal?"

In addition to reading your own company's job listings, you should constantly be checking what your competitors are doing through job listings. Are they hiring? What type of people are they looking for? What are they willing to pay for employees with a similar background? What are the competitors' growing and expanding divisions and product lines? Are they looking for your background and experience level or some other type? Most of these questions can be answered through studying the job listings. All this information is very important for career development.

The next question that usually arises is "What do I do if I find something?" The answer is straightforward—check it out. You should check at least once a year for other job opportunities through some type of job interview. This is important for three reasons. First, it will keep your interviewing skills sharp and your resume updated. Second, it may result in a better opportunity. Third, you may find out that your job is not so bad after all. In any case, you benefit from the experience and lose nothing.

MANAGING YOUR PRESENT SUPERVISOR WHILE INTERVIEWING

Most people react to the reasons for pursuing career advancement listed above with "That's great, but if my supervisor finds out, I'll lose my job." If this is really the case I highly recommend that you make a change immediately. You do not want to be working for this type of person. If your supervisor does find out that you are exploring new opportunities, there are several things you can say that should result in benefiting your career. First, you can say that you are happy to be working for him or her, but you heard of this wonderful opportunity and thought you would just check it out. You are happy with your present position but the opportunity sounded interesting and you were going to check to see what it involves.

Or you might respond with a compliment to your supervisor. You can compliment them by pointing out that the only reason you are looking is that they have done a great job developing you, and the only reason you stand a chance for a better opportunity is due to their excellent training and development!

▶ *Career Tip.* You have to have a strategy for communicating with your present supervisor during the time you are interviewing for a new position.

Still another approach is to be frank and open if you are unhappy with your present position and rate of growth. Tell your supervisor that you do not see much opportunity in the group and you feel that you must explore the options. They can react in one of two ways to this. One way might be by informing you that there is not a lot of opportunity in their group for you and you that should probably look elsewhere. In this case, you have found out some very valuable information about how your supervisor perceives you and what your real chances for advancement are by staying in the group. Therefore, you have made the right decision to search for work in other places (Figure 34-1).

One objection to your wanting to leave by your present supervisor is the fact that there may be no obvious replacement for you. You are leaving your present supervisor with a problem. There are several methods to overcome this objection. First, identify potential candidates who can replace you. Or offer to stay and train your replacement so the transition goes smoothly. Another action is to help write up a job description and duties for the next person.

The second way your supervisor might react is to express concern and describe how valuable you are and how they would like you to stay. You have now opened the door to further discussions about your chances for advancement. Use the opportunity to express what your future career objectives are, and hopefully, the two of you can work out a plan to get you what you want. In other words, use the opportunity to start planning your next promotion together. If you have a good rapport with your supervisor, you might point out that it's easier for them to give you the raise or promotion than hire a new person and spend time training them.

FIGURE 34-1   Managing your present supervisor while looking for jobs.

▶ *Career Tip.* Identify and train your replacement, if possible, to make the transition easier.

Regardless of what your supervisor says, continue to seek out new job opportunities. If the new opportunity is within your company, make sure you inform your supervisor before you talk to the person doing the hiring. It is professional courtesy to inform your supervisor about the opportunity. Don't let them find out from someone else. It will only end up hurting you.

There is a need for discretion when you are looking around. If you decide to go job shopping, keep it a very low profile activity. If you are too visible and start interviewing in too many places this can backfire. Your supervisor will quickly find out from other people what you are doing and want to know why. It does not help their career if you are highly visible while you are checking out other opportunities. Do not volunteer any information unless you are asked directly. Don't discuss your plans with coworkers; they may leak the information to others. For this activity, low profile is the way to go.

When you go for the interview, you should have some good answers to some very difficult questions. The first question is usually, "Does your present supervisor know that you are interviewing?" A very good answer to this question is that your supervisor does know and they do not want to stand in the way if this is really a good opportunity for advancement. The second is, "Why do you want to make a career change?" One way to respond to this is to say that you are looking for a better opportunity to grow. Whatever you do, do not speak negatively about your present position or supervisor. Anything that you say about your supervisor or group could be shared all over the company.

▶ *Career Tip.* Remember that soon after you leave the interview the hiring supervisor is going to call your present supervisor.

After you leave the interview the hiring supervisor is going to call your present supervisor. How your present supervisor responds can immediately make or break the opportunity. If you are on good terms with your present supervisor and they think highly of you, they may respond with "You are a great employee and I don't want to lose a person like this." If your supervisor says this, it will probably make the hiring supervisor want you all the more. If they respond with "You are a marginal performer," this can end the opportunity immediately. The point here is "Do you know how your supervisor will respond?" If you don't, you are taking a big risk. You should know how your supervisor is going to react before you spend time interviewing (Figure 34-2). Therefore, keeping on good terms with them during this whole process is essential.

FIGURE 34-2   Interviewing for the new job.

If you are already on poor terms with your supervisor this can be extremely good or bad for you. Sometimes I've seen supervisors go out of their way to help an employee get a new position when they are on poor terms. The supervisor wants to get rid of the employee, so rather than fire the employee, the supervisor finds new opportunities for them and may even pass along great recommendations about their performance. The hiring supervisor receives a glowing report about the applicant's past performance and hopefully will be encouraged to hire you.

On other occasions, I've seen the supervisor take a different approach. They use the opportunity to get even. Due to poor relations, the supervisor will purposely underrate the employee's past performance and spoil any opportunity for advancement. If you have poor relations with your supervisor, you might want to ask them how they would react before you go through all the effort. Watch their reaction. If they are giving you the "disgruntled' response you will have to deal with it and cannot ignore it.

HOW TO GET THE MOST FROM A JOB OFFER

Even after you get a job offer, your work is not done regardless of whether the offer comes from inside or outside the company. In either case, the ideal situation is to get your present supervisor to make a counteroffer in the hopes of keeping you. You should inform your supervisor of the new offer. Explain why it appeals to you, and why you intend on leaving if a counteroffer is not provided. The intent here is to try to get an even better offer out of your present supervisor, if what you really want is to stay in your current position.

It is surprising how fast some supervisors can move and what they can counteroffer when they know they are about to lose one of their key performers. Ask them to at least match it. Sometimes another offer is just what is needed for building a case for promotion with your supervisor. If someone thinks you are promotable, why shouldn't your present supervisor think so also? Sometimes this is all that is needed to push some supervisors into action. It hurts their ego to know that someone else in the company is trying to steal away their people.

If you get your supervisor to counter the offer, your efforts are still not finished. Take that offer to the new supervisor and find out if they can offer more. Your goal is to get both supervisors to offer the absolute most they can. You repeat this process until both supervisors can no longer offer any more, at which point you have gained the most out of the process.

Some engineers feel this unethical, to which I say look at other professions, athletes in particular. Athletes even go as far as hiring agents to negotiate job opportunities for them. Actors, actresses, lawyers, and nearly every other profession do the same thing. Knowing how to negotiate is essential for career advancement.

WHICH IS BETTER—INTERNAL OR EXTERNAL OFFERS?

When the best and final offers are in, you need to consider other aspects before you make a job change. If you change jobs for a promotion and stay within the company, this is the best of all possible worlds. By staying within the company you keep your vacation benefits, seniority, and retirement benefits. Studies have shown that an engineer who stays with one company until retirement, in most cases, will retire better than someone who has changed jobs. This is the case even if the person changing companies has had larger pay raises.

Remember, when you leave a company, in most cases, you lose your vacation benefits and retirement benefits. Leaving the company may also cause you unexpected expenses you never planned on. For instance, if you have to drive further to work, or move out of state, there are additional expenses you will incur. There will be costs to change your car insurance, place of residence, and real estate fees, just to name a few. Do you have any of the raise left after you pay all the new expenses? How will your retirement benefits be affected? For example, a 3% raise inside the company may be equivalent to a 6% increase outside the company.

Another point one must consider is that in the new company you will be starting all over and should not expect a large raise anytime soon. It will take you over a year to re-establish your position and your performance. If you get a raise when leaving, make sure it's large enough because it may be a long period of time before you get another one.

▶ *Career Tip.* Remember there is more than a simple pay raise involved when leaving the company. You have to decide for yourself if the move is worth it or not.

If you are lucky enough to be offered a new opportunity in or outside the company, make sure you give your supervisor at least a chance to match it—if you would still like to continue working in the same group.

WHAT TO DO IF YOUR SUPERVISOR TRIES TO BLOCK YOUR MOVE

If your present supervisor tries to block your move you have several options (Figure 34-3). First, try to get your supervisor to agree to a convenient transfer date some time in the future. Let your present supervisor know that you are not going to leave the group high and dry. You should be able to work out some gradual transfer plan. If your supervisor continues to fight what you can do is point out what your reaction will probably be over time. At first, you will continue to work hard, but after a while you will probably lose interest. It's hard for you to give it your all when you gave up a promotion and raise simply because your supervisor didn't think you deserved it. Try to make them see that holding you back is not good for them, you, or for the company in the long run. Another idea you might try is to volunteer your help in finding and training your replacement. This always makes the transition easier.

▶ *Career Tip.* In all discussions focus on what is best for the company and for your career.

Finally, if necessary and as a last resort, you can escalate the problem to higher levels of management or Human Resources. This has to be done with tact and careful attention on how this will reflect upon your present supervisor. Resentment and blame on your part will come off badly and other people may step in to block the move. You need to unite these resources behind you, and at the same time, cast your supervisor in a good light. Your intent is to establish and create a win–win situation and do what is best for the company, and your career. In all discussions, keep the focus on what is best for the company and for your career and not just what you want.

FIGURE 34-3 The blocking supervisor.

FIGURE 34-4   Always leave on good terms.

ALWAYS LEAVE ON GOOD TERMS

If you do accept a raise and promotion from elsewhere in the company, there are several things you must do. First, make sure you thank your present supervisor for all past help. Point out all the good things your new supervisor liked and how you qualified for the new position in part due to the fine effort your old supervisor did in developing you. Never leave on bad terms; it will only come back to haunt you later (Figure 34-4).

▶ *Career Tip.* Never leave on bad terms; it will only come back to haunt you! You may miss future opportunities.

SUMMARY

In summary, using your company's JOS to your benefit is important for your career development. The JOS allows you to compare what you are making to others in the company. The JOS is the first place new jobs are usually listed, and by constantly reading the ads you will be aware of the latest opportunities. In addition to reading your company's JOS, you should be reading the trade journals for job ads, Internet job sites, and salary surveys. This information will reveal what people outside the company are making as well as give you insight into what the competition is doing. When you consider taking jobs outside the company take into account all the new expenses you will incur and how your benefits are affected. And, finally, if you decide to make a move, leave on good terms; do not burn your bridges behind you.

Have you identified any career actions you want to take as a result of reading this chapter? If so, please make sure to capture these ideas before you forget by recording them in the notes section at the back of the book.

ASSIGNMENTS AND DISCUSSION TOPICS

1   Find out how your company JOS works.
2   Study the ads. Do you know all the codes? What do the ads tell you? Which divisions are hot, which are not?
3   Review some Internet Job Ads. Study them. What are they looking for and how much are they willing to pay?
4   Pick one JOS listing and go on an interview.

CHAPTER 35

SURVIVING LAYOFFS, CORPORATE TAKEOVERS, MERGERS, SHUTDOWNS, AND REDUCTIONS IN WORK FORCE

Corporate takeovers, mergers, shutdowns, and reductions in work force are hazards for engineers' careers. For successful career development you must know how to deal with and survive these career damaging events. Your goal is not only to minimize the damage to your career, but in fact to walk away with a career advancement opportunity.

In this chapter, we will explore how you can do just that. First, we discuss what happens during takeovers, mergers, shutdowns, and work force reductions. Next, we provide guidelines to help you assess how much at risk you are of losing your present job. Sometimes layoffs come without warning. Learn what to do immediately should this happen and what to continue to do to successfully survive.

UNDERSTANDING WHAT HAPPENS DURING TAKEOVERS, MERGERS, SHUTDOWNS, AND WORKFORCE REDUCTIONS

Lack of knowledge about future events during takeovers, mergers, shutdowns, and workforce reductions creates additional anxiety and fear. This additional anxiety and fear can work against you. It can cloud your judgment and paralyze you, preventing you from taking action at times. If you are aware of the sequence of events that normally occur during takeovers, mergers, and workforce reductions, it should help reduce the anxiety and fear, since you will know what is coming next.

The Engineer's Career Guide. By John A. Hoschette
Copyright © 2010 John Wiley & Sons, Inc.

With a lower anxiety level and knowledge of the sequence of events, you are better prepared to plan your next move and take appropriate action ahead of time. Let's first summarize some of the common events that occur in companies during a takeover, merger, shutdown, or work force reduction.

The first indicators that trouble is on the horizon are declining profits and stock value. If the profits continue to fall, most companies will not be able to stay in business, so management has to take some type of action to turn the trend around. In response to profit and stock value problems most companies will announce a series of predictable actions to be carried out. These actions are usually identified as cost cutting measures and reorganization.

▶ *Career Tip.* Monitor your company's stock price and earnings/profit statements as an indicator of possible troubles ahead.

The cost cutting measures will include phasing out old product lines that are no longer profitable, or reorganizing to help reduce costs and streamline the company. Often the corporate staff is cut as a cost saving move; a hiring freeze may be announced. The company will reorganize to eliminate duplication of effort. For example, combining stockrooms to have only one, or combining multiple data centers into a single one, one lab instead of two, and so on.

If profits still continue to fall, it is on to the next level. This may include reduction of workforce through early retirement incentives or elimination of part-time and temporary help where possible. Usually, at this stage the poorest performers are laid off.

If the profits still continue to fall or the company's markets decline, then it is on to the next level of trimming. The work force is rated and totem rankings are made. Those on the bottom of the totems become candidates for the next round of work force reductions. Managers will often protect those who have been loyal to them. This may not result in saving the best people—only the favorites. Upper management is aware of this practice and usually takes steps to counteract it. Such as, upper management will reorganize the departments and assign new supervisors who are not interested in loyalty but only in doing whatever it takes to turn the company around. If this means laying people off, then so be it. These hatchet men, as they are sometimes referred to, quickly cut the unwanted help with little attention to what has happened in the past. It is easier for them to do this since they have no personal ties with the employees they must supervise.

▶ *Career Tip.* Sudden changes in management may be an indication your company is headed for bad times.

If your supervisor should be suddenly replaced during one of the reorganizations, you have your work cut out for you. It is almost like starting a new job.

You must start all over to prove yourself to the new supervisor. This will take time and effort, so start immediately. If you recognize this early, and start immediately, you will be able to prove your worth to the new supervisor in a shorter time and hopefully, keep your name off the layoff list. Ignoring the fact that your supervisor has changed and assuming your past record is in good standing is not what is going to help you keep your position. When your supervisor changes, it is like the clock has been reset and everyone starts over.

The best thing you can do is react as though you just started a new job. You need to interview with your new supervisor. Take time to sit down, one-on-one, and discuss what your accomplishments were in the past and the value you bring to the group. Provide a copy of your resume and show your portfolio. Do not assume your supervisor will automatically do research and review your personnel file.

If the profits continue to fall, this process of reorganizing and cutting continues until the company either becomes profitable or folds. The company morale will cycle up and down with each wave of layoffs. Each time there is a layoff announcement the morale usually dips to another low. Everyone is worried and the efficiency of the organization can, at times, slow down. Workers are talking about who will be the next to get laid off and no one seems interested in their work. The best thing you can, do is keep your efficiency up and do not participate in the hallway discussions about who will be the next to go. Your time is better spent on your work and your backup plans.

▶ *Career Tip.* Use your time and energy wisely by developing backup plans.

Soon after the layoffs have been completed, and those laid off depart, the morale will start to climb again. The hopes and expectations are that the profits will increase and the layoffs will stop. Try not to cycle up and down with each wave of layoffs and reorganization. These are very tough conditions to work under. If you are going through this and feel you cannot handle it, then get professional help. Often companies will have counselors and psychologists on hand to help people deal with stress [1]. If your company has a psychologist available, seek out this help if you need it. Sometimes just being able to talk to someone can be a tremendous benefit. They are trained professionals who can help you. Use this company benefit if it is available to you.

During company mergers several different events may occur. The first step toward a merger is the preparation of a statement of net technical worth by the company executives. This is simply a sales brochure describing all the valuable assets and contracts the company possesses. Next action is the big announcement that company officials have decided to merge or sell controlling interests to other companies to remain profitable. At this point the company is literally up for sale. Those companies interested in purchasing your company now take tours of the facilities to see for themselves the assets of the company. The tour groups will include bankers, lawyers, executives and in some cases engineers from other companies to help assess the company's technical worth.

After the tours comes the big announcement: the company has been sold and there will be a merger if the Federal Trade Commission approves the sale. During this time many people from the purchasing company will tour your facilities and interview selected people to determine what stays and what doesn't. Then the big day arrives and the sale has been approved by the FTC, and an official takeover is announced.

From this point on, one of four things can occur. The first and least damaging is that the name on the building is changed, a few people are let go, and the operation stays pretty much intact. The second is that major reorganizations will occur. There is a major influx of people from the new parent company to teach everyone how things will be done in the future and there are a large number of layoffs. Those people getting laid off are usually the ones the parent company determined, through interviews, are not needed. They may not be needed because the parent company already has people doing their jobs and would not utilize their skills.

If you are selected for an interview make sure the interview goes both ways. Not only tell the new parent company about your skills but make sure you find out if they need your skills or already have a department doing the same operations. This will give you insight as to whether your department and you will survive the merger.

▶ *Career Tip.* Interview the new company to determine if you will fit into their business plans.

The third thing that can occur is the company is dissected. This means the company's operations will be cut up and separated. The new parent company will usually transfer parts or operations of the acquired company to other locations. Operations may be transferred to other states where they will be used to support or strengthen existing operations in the new parent company. If this occurs you may be asked to move to the new facility if you wish to keep your job.

The fourth and final thing that may occur is complete shutdown. In this case, the new parent company lays everyone off and closes the doors. This occurs usually for two main reasons. One reason is that the assets of the company are more valuable than the engineering work. For instance, the land may be more valuable than the building or the products it produces. The new company can make a quick profit by shutting everything down, letting everyone go, and selling the land. Another reason is that a competitor has bought out the company to get its contracts. The new owners can handle the contracts with its existing facilities and do not need the acquired facilities or people. Therefore they simply transfer the contracts and shut down the company, letting everyone go.

The steps I have described have been generalized and simplified. The actual events that occur will vary from company to company and situation to situation. Usually, it follows the sequence that has been described. Hopefully, these descriptions are enough to make you aware, allowing you to make better decisions about your future.

Now that you understand the steps that occur, let's look at some guidelines you can use to determine if you are at risk of losing your job.

ARE YOU AT RISK OF LOSING YOUR JOB?

The first step to successfully surviving is to determine if you are at risk of losing your present job [2]. Once you have determined the risk involved then you can develop a plan and react appropriately to ensure your continued employment. To help you in determining how much you may be at risk in your present job situation, I have come up with the following list of questions. Please keep count of the number of times you respond with a "yes" answer as you read the questions.

1. Does your company have high debt and low cash flow?
2. Has management already tried an early retirement workforce reduction effort?
3. Is there a hiring freeze on at your division?
4. Has your company merged with another that duplicates your work?
5. Has your supervisor or their supervisor been laid off?
6. Is the health of your industry poor? Are people ordering less of your company's products?
7. Has your management announced cost cutting measures to be implemented?
8. Is your supervisor constantly revising the department workload forecast for upper management to review?
9. Have there already been layoffs at your plant or division?
10. Are you on the lower portion of the employee totem?
11. Is your supervisor or program manager forecasting an end to a contract with no replacement?
12. Are more layoffs forecasted for the division?
13. Did you receive a below average rating on your last employee appraisal?
14. Has your supervisor announced there will be a layoff in the group?
15. Has the marketing department stopped advertising your product?
16. Is there a plan to phase out the product line you are working on?

17. Has the company announced a loss for the last quarter?

18. Is your supervisor meeting with the personnel department on a steady basis?

19. Is your supervisor's door always shut when meeting with the personnel manager?

20. Are other departments doing the work that you normally do?

21. Are you not getting invited to meetings that you normally used to attend?

22. Has the building maintenance department stopped working on your area?

23. Is your supervisor forecasting a workforce reduction?

24. Is your equipment being transferred to other groups or divisions?

25. Has your supervisor's secretary suddenly stopped talking about the workforce reduction to you or anyone else?

26. Are there several people in your group doing exactly the same work?

27. Are there several people in your group at exactly the same level?

28. Are you the highest paid senior person in your group and can lower-level employees perform the same work?

29. Are you the most junior person in the group with the least experience?

30. Do you have poor relations with your supervisor?

31. Are you doing "make work" assignments that do not really contribute to the company's profit line? (For example, writing procedures, manuals, or standards.)

32. Was your company bought out by another in the last six months?

33. Have you lost interest in your job?

If you have answered yes more than 20 times, then you probably are at great risk of losing your job. The chances are excellent that you may be part of the next workforce reduction. You should be on what the military calls red alert.

If you have answered yes 10 to 20 times, you should be on yellow alert. There is a fair-to-good chance you may be next in line for layoff. Both red and yellow alerts indicate you should be taking aggressive steps to ensure your employment in the future.

If you answered yes less than 10 times, then chances are you are not in immediate danger of being laid off, but you should still at the very least, continue to monitor the situation.

What should you be doing if you are on red or yellow alert? Get busy! The first thing you need to do is some investigating to determine how much you are at risk, and there is no better place to start than your supervisor's office. Simply reserve some time on their schedule and have a serious talk with them.

Start out by clearly identifying that you are concerned about your future with the company. If your supervisor stops you right there and lets you know that you are not in trouble, it is a good sign. Get them to expand on why they think you are not in trouble. Have they been told their department will not be affected by cutbacks? Does the workload indicate enough work? Have they spoken to their supervisor about the layoffs and who they will affect?

If your supervisor leads you to believe that workforce reductions are coming, ask straightforwardly, "am I on the layoff list?" Usually they will have one of two answers: "No" or "I am not allowed to share that information with you." Each answer tells you what you need to know. The first answer possibly indicates you are not in danger for now, the second answer says your name may be considered.

▶ *Career Tip.* Probe your supervisor if you are unsure about your future in the company.

Probe further; do not leave without a clear understanding of exactly how they feel things are going. However, sometimes supervisors themselves do not know because they will also be part of the workforce reduction. Upper management may not be sharing workforce reduction plans since they need supervisors to keep functioning until the end comes. In certain cases, supervisors may also fear that if they say anything, it could throw the entire group into chaos. They quickly conclude that it's best to be silent until the end comes. In any case, once you leave the office you still have further investigating to do.

Utilize your other contacts in the company to find out what might be going on. Check with your mentors. Do they know anything that they can share with you? Visit with the group accountant and pick up a copy of the workload forecast for your group. Is she forecasting fewer employees in the months ahead? What are the grade levels that will be cut? How about the last totem taken? Is there any chance of finding out where you stood on it? If you were on the bottom for your grade level, then it is likely that you are being considered for layoff. Again, visit the program manager on your project. Do they consider you critical to the program? If you have been identified as critical to the success of a project, you are usually not laid off.

Tap your social connections in the company for any information they have. Can you tap any other social contacts you made in the company? Perhaps people on the committees you served on? Often they will share things with you. In any case, if you hear something, check the source of it and do not take it for granted. Rumors get started and circulate around the company very easily.

▶ *Career Tip.* Tap into your social network at work. Sometimes other employees in different departments hear and share information.

One upper-level manager started the latest rumor sign-up sheet outside his office. He simply put up a blank sheet of paper and told the employees to write down the latest rumor they heard and he would tell them if it was true. He did this during a merger that was going on with another division. Within days he had over 30 rumors that ranged the spectrum from everyone getting fired to raises for everyone since the merger was going well. The truth of the matter was that nearly every rumor was just that, a rumor not based on any fact. For something as serious as your employment make sure you double check and get verification from multiple sources before you act on rumors.

The ultimate objective of all this research is to come to a conclusion about your future employment with the company. To help you better understand where you stand, gather all the information together and write it down. Diagram what you know; list the positives and list the negatives. If the negatives clearly outnumber the positives this will tell you something and aid you with your decision. The conclusion must be based on facts and then you must decide to either stay and take your chances or start looking for a new job. Only you can make this decision.

If making a decision like this frightens you, you can take some comfort in knowing that you are not alone. Everyone is frightened when it comes to making a decision as big as this. Most of us do not want to leave a company we have worked years for or to change jobs and start all over. The thought of it is very frightening. However, you must decide to do one or the other. If you ignore the situation, you are setting yourself up for a worse failure by letting others control your career. By choosing one course or the other you are taking control of your career and you are determining your outcome. You will fare significantly better if you are in control. Let us now look at some actions you should take when you choose one versus the other.

MINIMIZING YOUR CHANCES FOR LAYOFF

If you feel the risk of being laid off is small, and you decide to stay with your present employer, don't think you can sit back and relax. You must work toward minimizing your chances for being laid off in the future. There are actions you can take to minimize your chances for being laid off [3,4]. Taking these actions will not automatically guarantee you will be immune from future layoffs. They should, however, help reduce the chances.

The first action is to critically evaluate your recent performance. Are you meeting deadlines? Has your work been above average? Are you getting good visibility with your supervisor and the managers? If you are not getting good visibility then you should change your work habits. If it requires that you put in extra effort, commit yourself and just do it. It doesn't hurt to put in extra effort if it results in job security. Showing up for work early and leaving late will also help. Volunteering for extra assignments so that you become critical to the company is another action.

▶ *Career Tip.* Critically evaluate your performance and get feedback from your supervisor.

During workforce reduction everyone is on edge due to the uncertainty. As a result, people are nervous and tend to easily end up in arguments. Don't start any shouting matches. The last thing you need is your coworker or supervisor thinking how difficult it is to work with you and how you will not listen to reason.

Make sure your work is the best it can be. Remember, it will be compared with others in your group. Spending a few extra minutes making sure your work is neat, organized, and clearly communicated, is absolutely essential. Do not leave anything half done—carry it to completion. Before you present it to your supervisor, try to think of questions or criticism that he or she may have, based on other works you have done. Anticipate these things and have answers ready. Don't present problems without solutions. Supervisors are looking for engineers who are willing to work on the problems, not complain about them.

Make sure you get credit for all your work. Often engineers will help other engineers at the expense of not getting their own work done. This is not a bad thing to do, but I recommend at this point you don't help others out unless your supervisor knows about it. To help others at the expense of not getting your own work done, is career limiting to start with. Doing this during a merger or workforce reduction can be fatal to your career. I'm not saying not to help other engineers out; everyone needs help from time to time. What I am saying is, if you are going to put in the extra effort to help someone, make sure you also put in the extra effort to tell your supervisor. During workforce reductions your supervisor must decide who to keep and who to let go. They will make this decision based on your visible and demonstrated performance. It is essential to make sure all your work is visible to them.

▶ *Career Tip.* Don't present problems without solutions and get credit for all your work.

If your group is going through a workforce reduction but other parts of the company are strong, try to transfer to another group. Get busy reading the company job ads. Maybe, there is another group that could use your talents. To help you find out about other jobs, talk to your contacts, your mentors, and anyone else who might know about openings in other divisions. Typically, the personnel department is doing this before you even learn about it. Most companies try to transfer people to other groups rather than lay them off. Often you will be notified by your supervisor that you are being transferred to another group. Cooperate and make the move if it is a good one. But make sure you check into everything that is involved before you commit to moving.

MAKING BACKUP PLANS JUST IN CASE YOU LOSE YOUR JOB

Even if you decide to stay, you should develop some backup plans just in case things quickly take a turn for the worst [5]. This should not be foreign to you since all good engineers make backup plans in case their work encounters unexpected problems. This is commonly called *contingency planning*.

The first thing your contingency plan should include is updating your resume. Obtain several books on how to write a resume. Start reading and start updating your resume. Refer to Chapters 39–43 in the book on how to write resumes and conducting interviews resulting in job offers.

Second, start looking through job ads outside your company. You do not need to respond to them but clip and save them in a folder for future reference. If you are considering making a move to a new location, contact the major newspapers in that area and order the Sunday paper with all the job listings and job fairs. Clip the job ads from trade journals and go online to search for jobs.

Third, identify any coworkers or previous supervisors in your company you might get a letter of recommendation from. These may be people with whom you worked in the past who have complimented your work and might be willing to write a good recommendation letter. You need not contact them but at least make a list. In the future, if you suddenly need a letter of recommendation you will know who to talk to. If you feel comfortable enough about contacting them, you might ask, "In the future, if I ever need a letter of recommendation would you write one for me?" If they respond with a yes, you're all set. If they respond with " no," then you know who to stay away from.

Fourth, start a portfolio of your work and awards. Put into a binder any photographs of significant projects you have worked on. Awards and certificates are also good to include. You can think of this as a scrapbook of your accomplishments. My personal experience has shown this to be one of the most valuable interviewing tools. Most people I've shown my portfolio to have told me that I was the best prepared candidate they ever interviewed. The portfolio clearly showed all my experience, something that I found extremely difficult to do with words. To quote a very old saying, "One picture is worth a thousand words." See Chapter 41 on how to assemble your portfolio.

▶ *Career Tip.* Being well prepared is the best contingency planning you can do.

If you make the decision to leave, since you feel the risk is too great in staying, then you must start immediately preparing yourself to leave. This will not be easy, since you may be constantly fighting back the temptation to stay. I found myself minimizing the risk and telling myself, "They would never lay me off; I'm too important to the project." I quickly found out how wrong I was when the company downsized by 50% and I was part of it. Making the decision to leave will be frightening, but let's look at how you can make the transition easier.

WHAT TO DO IF YOU DECIDE TO LEAVE THE COMPANY

The first steps to follow once you have decided to leave are very similar to the ones recommended for backup plans. First, start your portfolio. Gather all the photographs, impressive diagrams, and awards that you have. Organize them into an impressive portfolio that tells a story of your work and your accomplishments. Next, search out people who might provide you with good letters of recommendation. Approach these people if you can and get them to write a letter. A suggestion I have is to write down what you feel would be important for them to say in the letter. Give them a copy of the list and ask if they might highlight the things you have come up with, since you will be stressing those skills in your resume. It's also good to give them the freedom to add others if they think of any.

Before we move on, I would like to share two very interesting observations with you. The first has to do with your ability to find a new job. It is a well-known fact that it is easier to find a job if you already have a job. This would indicate that if you feel that you are going to be laid off, the best chances for finding a new job would be while you are still working. So if you find yourself in this position, take immediate action!

The second observation has to do with the timing of your departure. If you leave before you are laid off, you give up any severance package that you are entitled to. These severance packages can be very significant. For instance, some companies offer up to two weeks severance pay for each year of service. If you have been at a company for 15 years, this means you should be entitled to six months of pay when you leave. Based on this, the optimum time to interview is before you are laid off and the optimum time to start the new position is within a few days after you have been laid off. If you can time everything just right you can walk away from one company, receive a bonus, and start at the new company with a raise.

One engineer I know started interviewing before he was laid off and found a new job. He timed it so that he started the new job just two weeks after he was laid off. Since he was laid off, he received a severance package of several months of pay. With the new position, he received a slight pay increase. To summarize, that year he received an extra paid two-week vacation and banked a two-month severance pay package as a bonus for going out and finding a new job. Not only that, the new job was closer to his home and less of a commute for him. Now that is really surviving a layoff! If you have trouble getting started once you have made the decision to leave, simply think of it as getting yourself a well-deserved bonus. This should help you keep going!

LAID OFF, WHAT'S NEXT?

If you are laid off, the question becomes what should you do. Before we can discuss what your best actions will be, you must first understand what you are going through. When you are laid off, fired, or let go, workforce reductions and downsizing become personal and overwhelming. Normally, no one is prepared for the shock. I was totally numb and felt paralyzed with shock when it happened to me. Luckily, a company counselor was at hand and she sat with us the first few hours after we were notified to help us get through it [6].

Losing your job will result in a grief reaction. This reaction will be similar to the grief reaction that a person experiences when losing a loved one. Your loss may be the loss of self worth, the loss of coworkers, the loss of security, or the loss of responsibility. You will be experiencing some or all of the following emotions:

Anger	Guilt
Depression	Fear
Bargaining	Denial

There is no set order in which you will experience these emotions. You may move from one emotion to another with no set pattern. The goal is to recognize them, accept the situation, and move on to more constructive activities as soon as possible. Even after you have reached acceptance, you will continue to re-experience these emotions from time to time.

The important thing to recognize is that you will be under great stress and you need to take action to effectively deal with the stress if you are to successfully survive. Some effective ways in dealing with the stress include:

- Re-establish a routine in your life (such as the time you wake up and go to bed)
- Develop and utilize a support system (family, friends, counselors)
- Exercise regularly
- Eat regular, balanced meals
- Reward yourself with enjoyable activities
- Get plenty of rest
- Talk to others; share your feelings with family, friends, and counselors

Similarly, there are things you should not be doing that will hurt your progress and ability to accept reality and move on. These include:

- Excessive use of alcohol
- Blaming people (your coworkers or family)
- Running away from or ignoring the situation
- Working on something to the point of exhaustion

You now need to stay alert and deal with problems as they occur in order for you to survive. More than likely there will be other stresses occurring in your life at the same time. These may include the death of a loved one, divorce or separation, injury, marriage or financial problems, children leaving home or change in residence, just to name a few. Dealing with a layoff is extremely tough; dealing with a layoff as well as other stress can be devastating. Your goal is to reach acceptance and move on to constructive actions that will lead to a new job opportunity [7].

If you are laid off you need to keep a level head and immediately take action. Normally, when you are laid off you are given a termination date that may range anywhere from one day to two weeks or more. Immediately upon learning of your termination, you need to start taking the following actions before you leave the company for good. These actions will be hard to do, but to successfully survive, you must. They usually involve your supervisor or personnel department.

First, make sure you will get your vacation pay. You are entitled to it, you have earned it, you deserve it [8,9]. Next, find out if there is a severance package. Most companies offer a severance package based on years of service. If nothing is offered, then start bargaining. You will need this money to support yourself until you have found a new job.

Next, you need to make sure your vested rights in profit sharing and/or pension still stand and the money you have accumulated is not affected by the layoff. Contact your personnel department immediately and discuss with them how to handle your account. Do not bypass this; it could end up costing you money if you mistakenly withdraw from your IRA or other plans.

▶ *Career Tip.* If you are laid off, obtain extensions on your benefits and apply for state benefits if you have too.

Get an extension on your medical insurance. Most companies will offer you the opportunity to continue your medical and dental benefits, but you will have to pay for it personally. Take the coverage; it is usually the best coverage available. If the company does not offer an extension on your benefits, then contact your health insurance carrier directly. They often have plans available for people who have lost their jobs due to layoffs, but, again, you will have to pay for it personally. The monthly premiums may seem expensive compared to what you were paying, since you are without an income. But with no insurance coverage you will be wiped out if a major medical emergency occurs to you or one of your family members.

You will be experiencing an overwhelming amount of self-doubt, pain, and anger. Put those feelings aside; you still have more work to do before you leave. Because of this self-doubt and anger you may not want to be around your coworkers or even talk with them. The best thing you can do, however, is just the opposite. Tell everyone of your situation. Now is the time to utilize your contacts—every contact you have made over the years, if possible. Leave no stone unturned, as they say. Contact everyone and anyone in the company who may have a job lead for you. Do not let your ego get in the way and become a roadblock to an even more successful job. Don't be afraid to ask for names; it is surprising how people will open up and help you once they are aware of your situation.

You may find that some people will help you simply to bolster their own egos. It provides them with an opportunity to show off who they know and how important they are in the industry. Others may help you because they have been in the same situation themselves and know how hard it is. Whatever the case, let them help you. If it gets you a job, isn't it worth it?

When someone gives you a lead make sure you get all the information you can about the opportunity. Who is the person doing the hiring? What kind of help are they looking for? What is the salary? You need to ask these questions for two reasons. First, you have to determine if this is a good lead for you. Often people will provide you with leads that are not what you are really interested in. If you blindly start chasing every lead without finding out how good they are, you will be wasting precious time. Second, if it is a good lead you will want to go to the interview as well prepared as you can. You will know in advance what they are looking for. This allows you to emphasize the skills you have that match the job requirements. This is always a plus in your favor during an interview.

When someone provides you with a lead, always check with the person giving you the lead to see if it is all right to mention their name. Their answer to this question will help you qualify the lead. If they only heard about it through the grapevine and have no direct contact with the person doing the hiring, then you know this is a long shot. However, if they personally know and are good friends with the person doing the hiring, then you know this is a good lead to follow up on immediately.

Who should be the first person you start asking for leads? Perhaps the same person who just laid you off. He or she may even offer to furnish you with some job leads; if not, ask. From your supervisor it is on to your close coworkers, mentors, and any other contacts you have within the company. You do not know where the lead that lands you a new job will come from. Therefore you have to contact everyone.

As you search around the company, be sure to ask key people, including your former supervisor, for letters of recommendation. You should ask several people for letters; not everyone you ask will get around to writing one. In any case, do not walk out of the company without at least three letters of recommendation. Sit in their office and wait for them to write it if you have to. You need to be persistent because the chances of getting a letter of recommendation decrease significantly once you walk out the door.

If you remain calm and follow these steps you could leave the company with several hot job leads and several letters of recommendation; this is very important. The leads will provide you with hope that other opportunities do exist and provide something for you to look forward to. If you are lucky the job leads could result in a new job opportunity, but do not pin all your hopes on them. The letters of recommendation will help bolster your sagging ego when you read how much people thought of your work. This prep work will establish a foundation from which to start your new job search.

▶ **Career Tip.** Don't burn bridges when you leave. Leave on a positive and upbeat note.

If you feel up to it, visit with old coworkers and friends to say goodbye one last time. Spending some time saying goodbye is good for you. With everyone wishing you good luck, it helps you to start moving on. It also prepares you mentally for the changes that are to occur. Take the time to say goodbye. Please remember these may be the same people your future employers will be contacting. It is of no value to leave on a bad note.

Coping with coworkers or friends who have been laid off can also be very difficult if you are not the one laid off. Show care and respect to those who are loosing their job. Look for ways to help and be supportive by being there even if it's just to be a listening ear. Stay with them while they pack up their desk and office. Don't say things like "Don't Worry" as this minimizes their feelings. Be positive and up about how talented they are and surely will find something else, it is only going to be a matter of time.

Cleaning out your desk will be very hard to do. Before you leave make sure you take any material that could be of use to you at a future job. If you are unsure as to what you can take, check with your supervisor. Do not take any proprietary information with you; this is industrial espionage. It is punishable by fines and prison terms.

Where should you be headed when you walk out the door? To either the company sponsored job placement center, an employment agency, or some-place where you can work on your resume and make phone calls on the leads you have. You have now successfully made the transition to the job search phase.

PROTECTING YOURSELF WHILE JOB SEARCHING

The job search phase can last anywhere from a month to over a year. During this time you must take steps to protect yourself [7–9]. The first step is to establish a new financial budget. Determine the minimum amount you can live on and start cutting your expenses immediately. Determine how long you can survive before you will deplete your savings. This will give you a deadline by which you must have a new job. Sometimes a deadline is an excellent motivator.

Lock up your credit cards immediately and start to make minimum payments wherever possible. Eliminate or delay any payments you can. Sometimes you can write a company in advance and notify them that you cannot make the monthly payments but you plan to make a certain amount. Conserve energy whenever possible. Delay purchases on clothing and anything that is not absolutely essential. You may even consider selling a few things if you have to.

After being gainfully employed you are entitled to unemployment benefits. Go immediately to the unemployment office and register for your benefits. It may take several weeks before you receive anything, so do it immediately.

Is there any part-time consulting or part-time jobs that you might be able to obtain? This will provide you with some income and still give you time to do some job searching.

A final note with regard to discussing the bad news of losing your job with family members. The best thing you can do is tell them immediately. You will need your friends and family for support in the days ahead. Give them the opportunity to help you. Let them know exactly what happened; explain that it may be a while before you get another job, so you are counting on their support and cooperation. You cannot effectively look for another job and hide what you are doing at the same time. Other companies will be calling you at home, so it will only be a matter of time before the family finds out.

SUMMARY

Successfully surviving takeovers, mergers, and workforce reductions requires you to be alert and actively assessing the situation. You must first determine if you are in danger of being laid off. Assess the warning signs. If there is danger, at what stage are events and how much of a danger do they pose to you?

If you are in danger you must make the decision to stay or move on. If you decide to stay, then you must take action immediately to secure your position as well as develop backup plans, just in case. If you decide to leave then you need to get busy looking for new job opportunities.

If you are laid off you will experience the grief reaction. This reaction will be similar to the experience of losing a loved one. You will be experiencing anger, guilt, depression, denial, and fear. To successfully survive you must recognize these feelings, deal with them, and move on to acceptance and more constructive activities. Before you leave the company, you must go into a high energy state of making sure you get all your benefits, any letters of recommendations you can, and a list of potential job leads.

Once you leave, you must continue in your high energy state. You need to take an inventory of your skills and interests. After completing these inventories it is on to resume writing, job searching, and interviewing. To successfully survive you will need to make use of all your resources. These include friends, family members, professional contacts, and employment agencies, to name a few. You will not know where your next job will be coming from. Therefore, you need to leave no stone unturned in your job search. If you are doing things right, your job search will hopefully require you to sort and prioritize all your opportunities.

Have you identified any career actions you want to take as a result of reading this chapter? If so, please make sure to capture these ideas before you forget by recording them in the notes section at the back of the book.

ASSIGNMENTS AND DISCUSSION TOPICS

1 What are some of the warning signs that your job may be in danger?
2 What is the sequence of events during a takeover?
3 When is it better to stay than to leave? When is it better to leave than stay?
4 What is the grief reaction?
5 What makes a good letter of recommendation?
6 Does your supervisor always know what is going on?
7 Name two things you must do before leaving the company for the last time.

REFERENCES

1. Razzi, Elizabeth, "Surviving the Storm," *The Washington Post,* September 7, 2008.
2. CareerBright, "Am I Going To Be in a Layoff? What Are the Signs of a Layoff?," January 31, 2008, http://careerbright.blogspot.com/2008/01/am-i-going-to-be-in-layoff.html.
3. "Surviving a Layoff," Job-Hunt.org, http://www.job-hunt.org/layoffs/surviving-a-layoff.shtml.
4. Levinson, Meridith, "7 Secrets for Surviving a Layoff in a Down Economy," *NetworkWorld Magazine,* Nov. 6, 2008.
5. Society of Professional Engineers, *Layoff Handbook,* http://www.speea.org/news/files/layoff_handbook.pdf.
6. Nutter, Ron, *Twenty Ways to Survive A Layoff,* Network World, August 25, 2008. http://www.infoworld.com/article/08/08/25/Twenty_ways_to_survive_a_layoff_1.html.

7. Viscusi, Stephen, *Bullet Proof Your Job, 4 Simple Strategies*, Collins Business, 2008.

8. CareerBright, "Accepting and Dealing with a Layoff," February 1, 2008. http://careerbright.blogspot.com/2008/02/accepting-and-dealing-with-layoff. html.

9. Westbury, Maurisa, "Dealing With a Layoff in an Economic Downturn," http://ezinearticles.com/?Dealing-With-a-Layoff-in-an-Economic-Downturn&id=12930ZZ.

CHAPTER 36

UNDERSTANDING THE JOB SEARCH PROCESS

Finding a new job is no easy task. It requires one to successfully carryout a series of actions that require multiple skills with finesse and all directed toward marketing yourself as the best candidate for the job. In this chapter, an overview of the job search process is presented which identifies the sequence of steps and actions leading to successfully finding a new job.

To be successful you must know the skills required at each step in the job search process and since you will not know what opportunities will finally result in a job, you more than likely will find yourself taking actions in all steps of the process simultaneously. Your ultimate goal of accepting a new job offer is dependent upon you knowing what the optimum actions are for each step, and how to quickly advance the opportunity to the next step in the process.

JOB SEARCH ACTIONS LEADING TO SUCCESS

In this chapter, an overview of the job search process is presented and the key actions are identified. Along with the overview of the process, time-proven and helpful motivational tips are discussed to help you overcome the very difficult and demoralizing reactions you may experience from getting rejected. In the chapters following this one, the detailed actions and tips for making your job search more successful at each step in the process are greatly expanded upon.

A general overview of the job search process is shown in Figure 36-1. The process starts with you making the decision to look for a new job. Once you have decided to look for a new job, the recommended first step is to take a self-inventory.

Self-Inventory. A self-inventory is identifying your job skills and interests. It is very important for you to know what your strengths and weaknesses are.

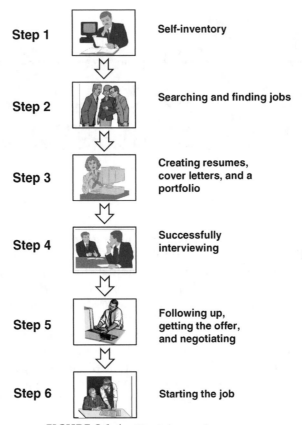

FIGURE 36-1   The job search process.

Make an inventory of your abilities, interests, and significant past accomplishments. This inventory will make the job search easier.

When you are done, you will have identified those skills that you consider your strengths and significant past accomplishments you can discuss during an interview. This list becomes the starting point from which you will generate your resume. Details on how to do a self-assessment is given in Chapter 37.

Searching and Finding Jobs. In this step you will search for job leads. Where do the leads come from? Everywhere! Get busy and start looking. To find job opportunities you will be using your networking skills, Internet search skills, and social skills. In Chapter 38 the details of how to find jobs are presented, and tips so you can even find multiple job opportunities.

Creating Resumes, Cover Letters, and Portfolios. Writing a generic resume is not the best approach. Closely reading the job ad and tailoring your resume, cover letter, and portfolio to the skills identified in the job ad is the best approach. Tips are discussed on how to best do this in Chapters 39–42.

Interviewing. Your biggest challenge once you have identified a potential job opportunity is how to get an interview. How do you best submit your resume so that you get selected for an interview? If you are called for an interview, what is the best way to prepare? Most people are very frightened and anxious about interviewing. They feel very uncomfortable about how they should act during the interview and what is best to discuss and what they should not say. Interviewing tips and proven techniques for impressing people during the interview are presented in Chapter 43.

Following Up, Getting the Offer, and Negotiating. Sometimes an offer does not come immediately after the interview. You may be left hanging and not know what to do next. Should you call or just wait? If you get an offer but it is not acceptable should you reject it or try negotiating? What is fair in negotiations? These issues are addressed in Chapter 44 with guidelines and tips for you to follow to resolve the situation in your favor.

Starting the Job. Leaving the old company on a good note and handling all the challenges associated with quitting one job and starting the next can be very difficult. In Chapter 45, tips are provided on how to write a resignation letter, making the relocation easier, and what to do the first days on the job to make a good first impression.

YOUR NEW JOB IS FINDING A JOB

Once you start looking for a new job whether you have been laid off or still employed, your new job becomes finding a job. You are working for your own company at this time. This is a very important mindset to have, especially if you have been laid off. If you are unemployed, you may be feeling lost and not sure what you should be doing next. If you follow the job search process described, and have the mindset that your new job is to find a job, you now have a purpose to your actions and a strategic plan to accomplish your goal. This combination of purpose and mindset is what you need to be successful and you should find your feelings of being lost and without purpose greatly diminished.

▶ *Career Tip.* Your job is to get a job.

WHAT ARE THE ODDS FOR FINDING A JOB

One of the most common questions asked is what are the odds for finding a job? Data provided by Drake Beam Morin, Inc. indicates on the average that it takes 10 resumes submitted before you obtain an interview. Again on average, it takes about three interviews to receive an offer. And most people would like to have at least two offers before they make a final decision and accept one. If you work the numbers backward, this means the average person has about six interviews. To get the six interviews the person needs to submit about 60 resumes.

There are several immediate conclusions you can draw from the odds. One is that it is a game of numbers and if you submit enough resumes, you will find a job. Second, you can expect a lot of NOs, or rejection letters. Finally, to cut the odds down, you need to better qualify the leads and determine if it is worth submitting a resume. Your resumes need to be specifically tailored to the ad to improve your chances and to conduct a successful interview. This is so critical, but worth the effort. Now let me show you how you can successfully handle all the NOs.

EASILY OVERCOMING ALL THE NOs AND REJECTION LETTERS

Since getting NOs is all part of the process, expect to get them. All people interviewing get rejection letters and NOs. It is all part of the process. So don't take the NOs and rejection letters personally. Just accept them and move on. Some people thank the people for saying no. Why? Because it means they are one step closer to a job. When I was laid off I kept the rejection postcards and pinned them up over my desk at home for motivation. As the number of rejection postcards grew, my hopes increased since I knew it was a numbers game and my chances of finding a new job were significantly increasing.

At our networking placement center the people would make a sport out of it and share how many rejection notes they had received which meant the closer they were to finding a new job.

▶ *Career Tip.* Look at your rejection notices as one step closer to a job.

When you receive your first couple of NOs, be prepared for the shock. For most people it takes the wind out of their sails and they are temporarily stopped dead in their tracks. When I received my first NOs, I was totally devastated. I could not believe that people would not hire me. Didn't they know I was the best engineer and candidate? Be prepared for this reaction and realize it is all a numbers game and do not take it personally; just move on. I recommend you go out and get the first NOs as quickly as possible and get them out of the way. Realize you have to move on and keep the job search going. It is all part of the game. You don't have time to sit and be upset about your rejection; it is on to the next opportunity.

▶ *Career Tip.* Go out and get your first NOs right away and get them over with.

In the ideal case, it would be nice to know why you were rejected, but this rarely happens. However, if you feel comfortable and on good enough terms with the people you interviewed with, you might ask them for some feedback on what you can improve. A simple email asking them to let you know what you could improve may yield valuable insight into what to change for the next opportunity.

▶ *Career Tip.* If possible, ask for feedback on improvement areas.

EARN $1,000 A DAY SEARCHING FOR A NEW JOB

For some people the thought of earning money really motivates them to action during the job search process. Would you like to earn $1,000 a day while doing a job search? Well, you can. Think of it this way. When, and note, I said "when" and not "if" your job search ends in success, you will be receiving a new salary. Let's say, for example, your job pays you $60,000 dollars. If you are an average person and had to send out 60 resumes, then you earned approximately $1,000 per resume submittal.

$60,000 Salary/60 resumes = $1,000 per resume submittal

If you consider on the average it takes six interviews to get a job, then for each interview you earn $10,000.

$60,000 Salary/6 interviews = $10,000 per interview

When you think of it this way for each interview, it can really be motivating. So I tell people to get out tomorrow and earn $1,000, send out a resume. I would come home after a day at the placement center and brag to my wife that I made another $3,000 today; I sent out three resumes. Others at the center would also announce it was a $10,000 day for them; they had an interview. If money motivates you, then think about how much you are earning each time you send out a resume or have an interview. Having a positive attitude during the job search process is one critical aspect to successfully finding a new job. Which gets us to our next point—it's all about attitude.

IT'S ALL ABOUT ATTITUDE

One of the critical components to conducting a successful job search is having a positive and upbeat attitude all the while you are searching. Your attitude comes out in your writing, networking, and interviewing. If you have a poor attitude and walk around with a chip on your shoulder about being laid off, or talk yourself into believing you have nothing to offer, then you are headed for failure. This comes out in your lack of energy to network and find jobs. It comes out in your unappealing resume and your body language during the interview. This poor attitude manifests itself in subtle ways to hurt your chances.

However if, on the other hand, you firmly believe you have great skills to offer a new employer and enjoy meeting and talking with people about your worth, then you are going to be more successful and sooner. Your positive attitude will manifest itself in your networking, resume writing, and interviewing. One of the best things you can do during this process is to remember to smile. Your smile indicates how you are doing and sends a big message to the people around you that things are good.

▶ *Career Tip.* If you think you can, you will be successful.

SUMMARY

Finding a new job is no easy task. It requires one to successfully carry out a series of actions that require multiple skills with finesse and all directed toward marketing yourself as the best candidate for the job. To be successful you must know the skills required at each step in the job search process, and be capable of taking actions in all steps of the process simultaneously. Your ultimate goal of accepting a new job offer is dependent upon you knowing what the optimum actions are for each step and how to quickly advance the opportunity to the next step in the process.

The major steps in the job search process are completing a self-inventory to identify your job skills and interests, searching and finding jobs, creating resumes, cover letters and portfolios, interviewing, following up, getting the offer, negotiating, and starting the job. Expect to receive NOs and remember for each no you receive, you are one step closer to finding a job. Go out and earn yourself an $1,000 today.

Have you identified any career actions you want to take as a result of reading this chapter? If so, please make sure to capture these ideas before you forget by recording them in the notes section at the back of the book.

ASSIGNMENTS AND DISCUSSION TOPICS

1   Why is attitude so important?
2   How can you increase the odds in your favor?

CHAPTER 37

SELF-INVENTORY

The first step to conducting a successful job search is to start with a self-inventory as shown in Figure 37-1. Simply put, you need to make a personal assessment of your job skills and interests. It is very important for you to know what your strengths and weaknesses are. Make an inventory of your abilities, interests, and significant past accomplishments. This inventory will make the job search easier. To help with the inventory processes refer to Figures 37–2 through 37–4.

Shown in Figure 37-2 are three different lists of abilities or skills that engineers utilize on the job. These abilities or skills have been divided into technical, managerial, and interpersonal skills. You should first review the lists and mark the skills that you feel are your strengths and those that are your weaknesses. Next identify, if you can, a specific past accomplishment that clearly demonstrates that you have that skill.

When you are done, you will have identified those skills that you consider your strengths and significant past accomplishments you can discuss during an interview. This list becomes the starting point from which you will generate your resume.

The next step is to try to visualize the type of job you wish to move into. To help with this visualization complete the interest inventory given in Figure 37-3. This inventory will help you identify those things that are important to you in a job. It will also help you to identify those things you wish to avoid in a job. Sometimes, when you are unsure of what you would like in a job, just listing the things you don't want in a job will help.

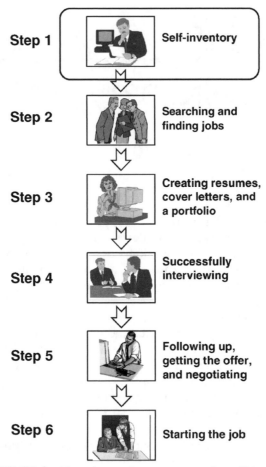

FIGURE 37-1 The job search process step 1—self-inventory.

You will use the interest inventory to help you sort through the job ads. By looking at your interest inventory and closely reading the job ads, you should be able to quickly eliminate those jobs that you are not interested in.

After you have completed the interest inventory, it is on to the significant accomplishment inventory. This inventory sheet is shown in Figure 37-4. It is simply a listing of your most significant accomplishments and the skills that were demonstrated by the accomplishments. This inventory is very important. This information is what you should be putting on your resume and discussing during interviews.

Type of skill	Strength or weakness	Past accomplishment demonstrating strength
Technical		
Product design		
Product build		
Laboratory test		
Laboratory research		
Technical publications		
Computer modeling		
CAD design and modeling		
Analysis and modeling		
Experimental research		
Patents		
Technical awards		
Programming		
Producibiliy		
Manufacturing		
Project Management		
Planning		
Budgeting		
Organizing		
Developing policies		
Developing procedures		
Cost tracking		
Schedule planning		
Customer interface		
Team formation		
Salary administration		
Department budgeting		
Capital planning		
Presentation skills		
Interpersonal Skills		
Motivating		
Team leadership		
Conflict resolution		
Work relationships		
Meeting skills		
Versatility		
Team dynamics		
Communication style		
Customer relationships		
Social abilities		
Mentoring		

FIGURE 37-2   Job abilities and skills inventory.

Visualization of your ideal job

1. Job title: _____

2. Job salary: _____

3. Company location: _____

4. Home location: _____

5. Commute time: _____

6. Company size: _____

7. Company products: _____

8. Size of engineering group: _____

9. Job functions: _____

10. Office size: _____

11. Company benefits: _____

12. Freedom to work on: _____

13. Travel: _____

14. Laboratories look like: _____

15. Personal computer: _____

16. Supervisor who: _____

17. Coworkers who: _____

18. Career advancement paths leading to: _____

19. In five years I'll be doing: _____

Visualization of your worst job

Things I absolutely will not put up with on my next job:

_____

FIGURE 37-3 Visualization of your ideal job.

My significant accomplishments were	Abilities and skills demonstrated
1._____	1. _____

FIGURE 37-4   Inventory of significant accomplishments.

A blank copy of these inventory sheets is given in the *Self-Inventory Workbook* located in Appendix B. Use these blank sheets to conduct a record of your thoughts. Once you have completed your inventories, it is time to start writing your resume and looking through the job ads.

CHAPTER 38

SEARCHING AND FINDING JOBS

NETWORKING IS KEY TO FINDING JOBS

The next step in the job search process, as shown in Figure 38-1, is finding job opportunities. Research conducted by Drake Beam Morin, Inc. has shown that networking is key to the job search process. Their research has shown the following statistics on how people find jobs.

Face-to-face people networking	70%
Using search firms	15%
Sending targeted mailings	10%
Answering ads	5%

This data indicates a person is nearly five times more likely to find a job through face-to-face people networking than any other means. So with this in mind, where should you be looking and networking? *Everywhere.*

DEVELOP A 30-SECOND COMMERCIAL AND CONTACT CARD

Before you start networking you need two essential items: a 30-second commercial about yourself and some type of business card you can leave with a potential contact so that they contact you. You need to have a 30-second commercial that highlights your skills, background, and experience that you can share with people. It should be short and directly to the point highlighting the most impressive features about you. Developing this commercial takes time and a good tip is to prepare and practice many, many, times so that it comes off natural and like second nature to you.

The Engineer's Career Guide. By John A. Hoschette
Copyright © 2010 John Wiley & Sons, Inc.

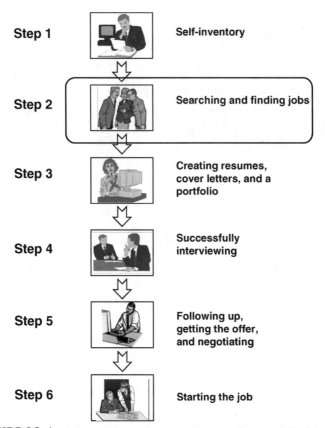

FIGURE 38-1   Job search process step 2—searching and finding jobs.

In addition, you will need a business card with all your contact information. If you are unemployed, create your own business card with your contact information on it. The cards should list your name, title, areas of expertise, and contact information including phone number and email address.

BEST PLACES TO PEOPLE NETWORK TO FIND JOBS

Business Associates. This is the first place you should consider contacting when you start looking for a new job. These people already know you and you have an established line of communication with them. Business associates include your suppliers, vendors, subcontractors, and customers you have worked with. You more than likely already have all their phone numbers and email addresses. It is easy to approach them. All you have to say is that you are looking for a new job and wondering if they have an opening at their company. If they do not know of any openings, then ask who they know that might know of job openings.

▶ **Career Tip.** Every contact leads to another contact.

Technical Societies Members. Contact any professional societies that you belong to or attend one as a guest to see if you would have interest in joining. Most technical societies allow you to attend their local meeting even if you are not a member. Once at the meeting network and explore the opportunities. Some technical societies even have free job placement services that are available to members and have a job posting website.

College Alumni Association. If you are still living in the community where you attended college then contact the alumni association and, if possible, attend meetings.

Attend Trade Shows. If you live near a major convention center, go online and look up the planned trade shows for the year. Most convention centers will list their planned events and give you contact information about the show and exhibitors. You can use this information in advance to contact companies about jobs even before the show. Some trade shows also have a job posting area. Post your name, technical expertise, and contact information. If companies in the show have job postings, then make contact during the show. Go prepared to discuss your qualifications and bring a complete resume package and portfolio.

Attend Job Fairs. Many technical societies and search firms sponsor job fairs to fill open positions. There is no better place to go than a job fair where employers with jobs are easy to approach and located all in one spot at one time.

Job Placement Center. If the company that laid you off has a job placement center, utilize it. These centers are excellent for picking up leads. They often have a large number of free trade journals and newspapers for you to review. In addition, other people using the center will often share information about who is hiring and who is not. Some job placement centers even provide staff to help you with resume writing and free use of phones. Make sure you take full advantage of this service if it is available to you and network with others using the center.

Contact Federal and State Employment Agencies. During economic downturns this is where many companies send their job listings first. Most agencies have a resume posting capability and update the job posting list weekly. In addition, they have many free career services available for you including counseling.

Your Friends and Neighbors. Don't forget about your friends and neighbors; they are another means to find jobs. Your neighbors and friends may not have the same technical background as you, but their company may need people like you. It doesn't hurt to casually mention that you are looking and ask if their company is hiring people with your skills. Once again, if they do not know, then ask if they could put you in contact with someone who might know.

HOW TO UTILIZE SEARCH FIRMS TO FIND JOBS

Using search firms has the next highest success rate for finding jobs. Your best and safest way to approach selecting a search firm is to find someone who has recently found a job utilizing a search firm. Ask them who they would recommend for you.

If you are left to select a job search firm on your own then you should consider several important aspects prior to selecting one. Here are some questions that are recommended for you to ask prior to selecting a job search firm.

1. What area of technical expertise do they represent? Do they have expertise and clientele in your technical area of expertise?
2. What companies do they have personal contacts with?
3. How many people does the agent represent at any given time?
4. How many people have they placed in the last year?
5. Are they local only or a national firm?
6. Do they charge you a fee? If so how do they charge?
7. How are resumes handled? Online? Can you update your resume for each submittal job ad? How often do they allow you to update your resume?
8. Do they have any job postings they are trying to fill immediately that are directly related to your technical area?
9. Does their company provide any support for resume writing or interviewing?
10. Will they disclose who they are submitting your resume to?

Running through these questions with a job search firm prior to selecting one will give you a much better chance of being successful. Those employment agencies that specialize in your area are the best. You will usually find these agencies advertised in technical trade journals or by searching online. Selecting and posting your resume with two or more job search firms can be a good idea, if the firms service totally different industries. However, you should be careful if you elect to post your resume with several firms since this can backfire for several different reasons. One example is when a company gets your resume from several different head-hunters. This can complicate the situation. The second is that if other head-hunters find out that someone else is pushing your resume around they may drop you.

Once you have selected a job search firm and posted your resume with them, you will need to follow up weekly to keep your resume on the top of the recruiter's pile. Simply submitting your resume and sitting back to wait for someone to call does not work.

SENDING TARGETED MAILINGS

The best response to a hot lead you discovered through your networking is usually to send a targeted email directly to the individual. The target email means that you refer to the person in the email and attach a cover letter and resume that has been updated directly for the position available. This rapid response shows that you are definitely interested and it does not hurt to follow up with a phone call just to make sure the individual's SPAM blocker did not intercept your email or you had a bad email address. The key here is rapid response, custom tailoring your email to the person, and updating your resume for the specific job opening. If you feel comfortable enough, you might want to contact the individual in advance of sending your package and get their input on what to emphasize on your resume.

ANSWERING NEWSPAPER ADS

In a down economy the number of job ads significantly drops. However, newspapers are still a good source to checkout. Contact the major newspapers in the cities you are interested in working in and get a subscription to the Sunday Want Ads only. You can have them mail you a copy or go online and sign up for an electronic subscription. In addition to newspapers, trade journals are another good source for finding jobs but usually these are only published monthly.

INTERNET SEARCHING

There are many venues to finding job postings online. You can post your resume to such websites as monster.com or explore individual company websites for job posting. These methods, according to the data, have the lowest chances of success but are still considered worth doing. When you post your resume to a website, you can be assured that no less than 10,000 others are on the website or in the data bank. When a corporation runs an ad in a newspaper or has a listing on their website, they often receive in excess of 500 resumes for the posting. So your chances of being selected are usually very small unless you have the outstanding qualifications.

In addition, many companies post job openings that are considered "bait" job openings on their company website. This means that companies have these generic positions posted continually on their website and wait for the "golden nugget" resume to be submitted. This is a resume from an engineer that is exactly what they are looking for and has all the highly desirable skills and capabilities plus more.

These are just a few of the avenues available to you to find job opportunities. I'm sure that when you put your mind to it you can think of at least a half dozen more ways to generate job leads. I'm also confident that if you really put the effort into it, you will quickly find there are more job opportunities than you thought. You will probably have to sort through the listings and leads and prioritize which ones you consider the best. If you get to this stage you are doing all the right things, which brings us to the next tip in doing job searches.

MAINTAIN A CONTACT NOTEBOOK

Once you start your job search in earnest, you will be talking to many different people hopefully every day. You will receive many tips or contacts that you will need to follow up on. To do this you will need to create a contact notebook. A contact notebook contains all the vital information you have about your leads. It should contain the following information as a minimum:

1. Name of contact
2. Company
3. Phone number, email, and website
4. Position(s) you are considering
5. Date you submitted cover letter and resume
6. Adhere their business card to a page
7. Journal of notes on these conversations that transpired with dates and times
8. List of next actions with this contact.

If you are disciplined to record conversations and organized so that you can easily retrieve the information, this is going to save you time and go a long way in shortening your time to your next job. The contact notebook is extremely valuable in the event you are suddenly called and you need to refresh your memory on what has transpired in the past. One final reason for keeping a contact notebook is to share your contact information with others who may be looking. If you have a contact that is looking to hire but your qualification turns out not to be a match, why not share it with someone else whose qualifications may be a better match. You never know when you share a contact with someone they, in turn, might do you a favor and share a contact they have that could be your next job.

KEY JOB SEARCH STRATEGIES YOU MUST FOLLOW

Now that you know how to locate jobs, you will need to decide upon a strategy on how to rank and prioritize the opportunities. The priority ranking strategy will be up to each individual. For instance, some people will want to stay in the local area, so their priorities when deciding will be job location. So they will rank and sort their job posting on the basis of the following order:

Job Ranking by Location

1. Job in local area only, no relocation required (other technology jobs ok to accept)
2. Job in close-by city, relocation to another city nearby
3. Job is 1 to 2 days driving distance from present home
4. Job in nearby states
5. Across the nation opportunities (relocation required to another state)

This type of strategy will limit your job searches to local companies and, more than likely, local recruiting firms and Internet searches on the basis of company location.

Another ranking criteria may be salary with location not considered important. Or another ranking criteria may be technology and job scope has to be the same as the present job, which means the person does not care about location and is willing to relocate across the nation or even to another country.

If you do not select a strategy for conducting your job search, you will be all over when you consider potential jobs, and more than likely this shot-gun approach will not be successful. The following line describes characteristics involved when considering a new job. Please review the list and rank them from the most important to least important characteristics when you consider finding a new job. This will help you prioritize how to conduct your job search.

Job Title, Salary, Location, Technology, Office/Lab Facilities, Commute to Work, Family

Many engineers start looking for a new job that is exactly identical to the old job, in the same technology, within 5 miles of their home, with a salary increase, and no loss of benefits. This is the ideal case and if you find this opportunity you are very lucky. However the reality is, this rarely happens and we are all forced to compromise. I have counseled engineers who have held tight to the belief they could find a new job that was exactly the same as the old one, only to be unemployed for years. Although quite a noble goal to strive for, the reality is you are going to have make changes and compromises to get a new job. Having a prioritized list of characteristics for selecting your new job helps the process move quicker.

BEST ACTIONS FOR GETTING A JOB

Up to this point in the chapter, we discussed the tools available to you to help with your job search. However, these are only tools and tools do not get you a job, action does. You need to make sure your job search actions are focused on what is going to land you your next job. Here is a simple list of the best actions you can take to find a new job.

Networking and Work Your Relationships. Smart job searchers realize the best actions are networking and working your relationships. You should be spending most of your time meeting with people—either having coffee, lunch, talking to them on the phone, or emailing them. This also includes attending engineering society meetings at night and other meetings two or three nights a week, reconnecting with people and re-establishing old relationships, as well as developing new ones. Pick up the phone and call someone every chance you have. If they do not know of any job openings, ask if they can give you name of a couple of people who might know of potential job openings. When you network, make sure you have your 30-second commercial, a contact card, and even extra copies of your resume handy.

Job Search Firms. Job search firms are in daily contact with many companies about their hiring needs and help them fill positions. Contact job search firms that specialize in your area of expertise and that have clients. Work with their recruiters to quickly get your resume to the companies that are hiring.

Responding Quickly and Following Up. In the ideal case, a job searcher should be able to respond to finding a job opening and submit a cover letter and resume within hours. Waiting days and weeks to get the perfect resume and cover letter written is not going to land you a job. When ads come out in Sunday papers, or are first posted on websites, the smart ones have their resume on the desk of hiring managers the following morning.

Do Your Homework First. Responding quickly will not get you the job if you send in a cover letter and resume that misses the mark. Do your homework, identify keywords in the ad and skills they are looking for, search online to find out more about the company, if possible, probe your contacts for further information and then update your resume and cover letter.

Appear Desirable. All the experts agree that the best time to look for a new job is when you have a job. Looking while you are employed sends a message that you are a very desirable person to hire since another company has already hired you. If you are unemployed, find a temporary position and work as a contractor or for a temporary job agency. This shows flexibility and initiative. Once you are in a company people will be watching and evaluating your performance and this is the perfect time to excel. With everyone watching it is the time to sell you and your skills, and make them want to hire you permanently.

During the interview, prospective employers will want to know if you understand their business and how you can help them with their problems. It is not only about you being technically capable but also having the potential for making greater contributions to the company. This is the time to highlight how you understand the bigger picture and all the extras you bring to the job that will benefit the company. It is time to strut your stuff as they say.

Polish and Practice. Anytime you are not networking and have a few spare minutes you should be polishing, refining, and practicing your 30-second commercial and interviewing skills. Practice, practice, and practice is the best thing during your non-networking and spare time. Make sure your resume is clear of mistakes and power-packed with good information. Prepare for the interview by practicing with others and answering the tough questions. If you can, take classes on networking, interviewing, and resume writing.

Get Organized. You should have a plan for each day and keep a calendar of your planned actions just like you would do on the job. Have a notebook to keep track of contacts and actions. Allocate time each day for your different job search activities. You should have a routine each day to follow and should include time for such activities as planning, making calls, networking, attending events, and job searching. Prioritize your tasks each day to make sure the most important ones are always at the top of the list.

Always Be Ready. You never know when a potential employer might call you to discuss your resume prior to setting up an interview or to conduct a phone interview. Therefore you should always be ready. The best way to be ready is to have a copy of their job ad and copies of the resume and cover letter that you sent so you can look at them as you talk to the interviewer. Keeping a copy of the job ad, cover letter, and resume all together in a notebook that you can instantly grab, sets you up for success. Make sure you also have a business suit ready to go. Dress appropriately for the interview.

SUMMARY

The data indicates that a person is nearly five times more likely to find a job through face-to-face people networking than any other means. Every contact leads to another contact. Before you start networking you need two essential items: a 30-second commercial about yourself and some type of business card you can leave with a potential contact so that they can contact you.

Excellent places to search for jobs are with business associates, technical societies members, at industry trade shows, job fairs, federal and state employment agencies, friends and neighbors, job search firms, company websites, newspapers, and journals. Once you start obtaining job leads the best way to get organized is by creating and keeping a contact notebook. Finally, determine what your search strategy will be and what will be the priority ranking for the characteristics of your new job.

Have you identified any career actions you want to take as a result of reading this chapter? If so, please make sure to capture these ideas before you forget by recording them in the notes section at the back of the book.

ASSIGNMENTS AND DISCUSSION TOPICS

1 Why is face-to-face networking so effective for finding a job?
2 What are the best questions to ask when contacts come up empty-handed?
3 What are some of the questions you should ask when selecting a job search firm?
4 Write down your prioritized list of characteristics for your new job.

CHAPTER 39

CREATING RESUMES THAT GET YOU INTERVIEWS

In today's competitive job market, a well-written resume is critical to landing that new job. Initially, your resume is often the only means you have to influence hiring managers to select you for an interview. Once selected, your resume becomes the critical piece of paper that is reviewed by all during the interviewing process and provides you with a coherent and structured means to communicate your skills and experience. And when you leave the company, it is the only piece of paper that remains behind that the decision makers can refer back to, during the selection process, to refresh their memories and compare data on candidates. All these reasons only underscore the importance of having an outstanding resume. Engineering managers and Human Resource personnel probably review anywhere from 100 to 200 resumes in a year in response to job postings. To impress these people you will need a resume that is well-structured, easy to read, clear of typos, and full of vital career information. In this chapter, we will review the general structure of resumes, how to write impressive resumes in less time, and how you can power-charge the information you present in your resume so that you stand out from the rest of the applicants. As shown in Figure 39-1 creating resumes is the next step in the process.

GENERAL FORMAT AND LAYOUT OF A RESUME

The general format of a resume is shown in Figure 39-2. At the top center in large print is your name and contact information. Directly below your contact information is your objectives statement or career highlights section. This area on the resume is the single most important area. It is the areas that interviewers go to first, and after reading, usually form an opinion about the candidate.

The Engineer's Career Guide. By John A. Hoschette
Copyright © 2010 John Wiley & Sons, Inc.

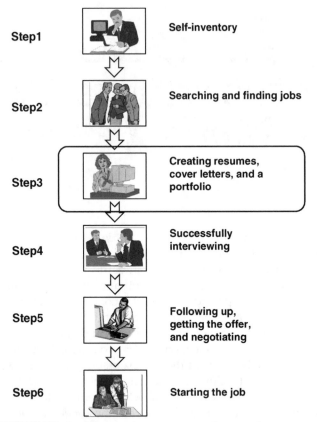

Step1 — Self-inventory

Step2 — Searching and finding jobs

Step3 — Creating resumes, cover letters, and a portfolio

Step4 — Successfully interviewing

Step5 — Following up, getting the offer, and negotiating

Step6 — Starting the job

FIGURE 39-1   Job search process step 3—creating resumes.

General Format of a Resume

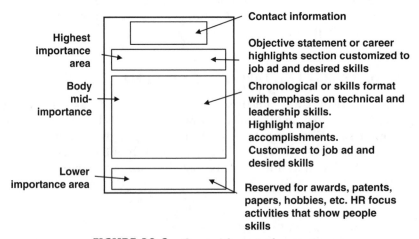

Contact information

Highest importance area — Objective statement or career highlights section customized to job ad and desired skills

Body mid-importance — Chronological or skills format with emphasis on technical and leadership skills. Highlight major accomplishments. Customized to job ad and desired skills

Lower importance area — Reserved for awards, patents, papers, hobbies, etc. HR focus activities that show people skills

FIGURE 39-2   General format of a resume.

The Objectives or Career Highlights Section. This should be a clear concise statement of your objectives with emphasis on how they relate to the company and the job position you are interviewing for. Please do not write "seeking a position that will allow me to utilize my skills for the benefit of the company." Instead, mention the company and specific skills as identified in the job ad. For instance, if you are writing an objectives statement for a job position with Microsoft you might write something like:

> Seeking a software engineering manager position in Microsoft's Seattle Research group developing advanced Internet search engines. A position that can benefit from my 20 years experience of conducting research on the theory of search engines development for MIT, Stanford, and DARPA labs, along with my award-winning managerial skills.

If you are writing a career highlights statement rather than an objectives statement, you might write something like:

> Award-winning software department manager with 20 years experience in the research and development of advanced software search engines for MIT, Stanford, and DARPA. Research lead to $150M worth of grants and receiving DARPA's highest award for innovative research.

The purpose of the objectives statement is to get the reader excited and interested in reading the rest of the resume.

Body of the Resume. The next section below the objectives statement is the body of the resume. The body of the resume can be arranged in a chronological or skills format. The emphasis of this section should be technical skills and leadership skills if applicable. Here is where you highlight your major accomplishments that you identified when doing your self-inventory. The technical skills and capabilities should reflect the same terminology as used in the job ad.

As shown in Figure 39-3, the body of the resume can be either arranged in a chronological format or skills format. It is best to use a chronological format when you are interviewing with people who are familiar with your present company, its products, or projects. List your jobs starting with the most recent and working back in time. A chronological format resume should emphasize the growth and development in your career over the years.

For each significant job assignment or employer clearly indicate the company name, your position, and the dates of your employment. Provide a brief overview statement that tells information about what the company does, its products, and customers. This will help the reviewer assess your technical skills and experience. Follow this by a concise power-packed statement on exactly what you did for the company. Don't make the reviewer look for the information or read between the lines. State the obvious. If the reviewer has to interpret or read between the lines, more than likely your resume will go to the reject pile.

Chronological Vs Skills Format

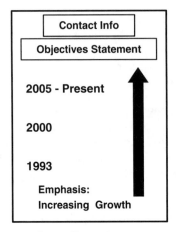
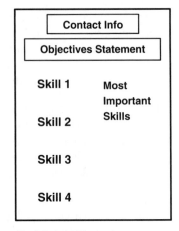

FIGURE 39-3   General resume formats for chronological versus skills format.

The skills format is best when you are interviewing with people who are NOT familiar with your present company or its products. The strategy is to list skills that are directly transferable to the new job rather than discuss projects or programs you worked on. For this format, you list the most important skill first followed by skills that are decreasing in value but necessary for the job.

Bottom Portion of the Resume. The bottom of the resume is usually the last thing read and the lowest value material. This area is reserved for other information like awards, patents, papers, hobbies, and other nontechnical accomplishments that reveal interesting and pertinent items about you.

One Page or Two Pages. The general rule is, for recently graduated engineers or junior engineers, a one-page resume should suffice. For more senior engineers a two-page resume is normally expected.

HOW TO WRITE AN IMPRESSIVE RESUME IN LESS TIME

The process for writing an impressive resume in less time is shown in Figure 39-4. The process starts with the job ad. Your objective is to write a resume direct to the job ad. Your goal is to finish with a resume that sounds like it is the perfect fit for the job. Take the job ad and read it several times. Then circle all the keywords in the description. The more times you use these keywords in your resume the better.

In fact, some companies will run your resume through special software to count the number of matches between the ad and your resume. Based on the

Creating Your Resume

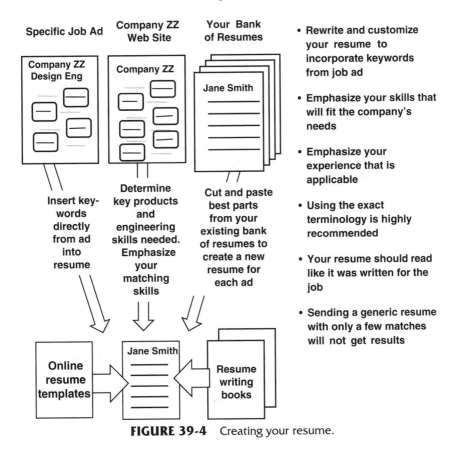

FIGURE 39-4 Creating your resume.

number of matches they count for your resume, they will either accept or reject your resume.

▶ *Career Tip.* Read the job ad and identify keywords. Then make sure you use these keywords multiple times in your resume.

Next, visit the company website and learn everything you can about their products, technology, and engineering skills required. If you have worked on similar products and technology then emphasis this on resume.

Next look at your inventory and previous resumes you have written. You should be able, in a matter of minutes, to cut and paste sections of previous resumes into your new resume and edit these to make sure your best skills from the inventory are included. The draft resume should only take a few minutes to complete. Go online and review the templates that are free for your use [1–3]. Finally, visit a library or bookstore and read the excellent books and articles on writing resumes [4–7]. When you have done all this, writing a resume should be a whole lot easier. Now onto power-packing your resume.

QUANTIFY, RELATE, AND SELL YOURSELF

You can create a power-charged resume that should get you interviews and hopefully a job offer by quantifying, relating, and selling yourself on the resume. The best way to do this is to sell your benefits and not list your features. This is the old story that people buy benefits and not features. To illustrate the point let's consider a new car salesperson who simply informs you about a car that you are interested in by going through a list of its features. For instance, the salesperson would say the car has four wheels of 16-inch diameter and tire pressure of 32 PSI. The wheel base is 75 inch by 108 inch and the inside compartment has 32 cubic feet of space. The car has a six-cylinder engine and runs on regular gas. This does nothing to motivate the customer to buy.

Now contrast this when the salesperson quantifies, relates, and sells the benefits of all these features. To sell benefits, the salesperson might say something like, "our car has the largest wheel base made giving it the smoothest ride on the road, the tires are specially designed for a smooth ride, better mileage, and improved traction in rain and snow. The interior is very roomy and will make it comfortable for a 6-foot tall man to drive. The engine is the highest efficiency six-cylinder made and also provides the power needed to safely pass other cars on the freeway. You will look awesome driving, it and be happy with our guaranteed, no-hassle warranty for 100,000 miles."

Which sales pitch are you more likely to want to buy? People buy benefits and that's what the company you are interviewing with wants to buy. The benefits for them in hiring you are most on their mind, not the features.

Now let's illustrate how to power-charge your resume. Here is a sample sentence taken from an engineer's resume from a chronological resume.

> Lead engineer on the XYZ Project working on digital electronics for our products.

And here is what you can to do to power-charge this by quantifying, relating, and selling yourself.

> Lead engineer on the successful XYZ Project overseeing and directing the efforts of a 40-person team including mechanical, electrical, and software engineers. Reported directly to the Vice President of Engineering and lead customer interface. Produced state-of-the-art digital controller in 6 months, 1 month ahead of plan, and 20% under cost. Project resulted in highest company award.

Here is a sample sentence taken from an engineer's resume from a skills resume.

> Skills—digital designer for FPGA and ASIC devices.

And here is what you can to do to power-charge this by quantifying, relating, and selling your skills.

> Superior Digital Design Skills. Lead engineer for digital design of state-of-the-art 4 million gate Actel FPGA. The design was completed in 4 months and successfully tested in 2 months. FPGA design ported to ASIC and presently in production of 10 million units. Utilized latest Mentor Graphic tools for modeling and Lab View Software for testing. Design has built-in testability and error correction capabilities. Customer presented highest company award to team.

For each of these examples you can see the power-charged second statement significantly improves the saleability of the engineer for the job. One of the keys to power-charge your resume is using action words and verbs. This brings us to our next section.

GREAT ACTION WORDS TO INCLUDE ON YOUR RESUME

Accelerated	Collected	Demonstrated
Accomplished	Completed	Designed
Acquired	Computed	Determined
Achieved	Conceptualized	Developed
Adapted	Concluded	Devised
Advised	Condensed	Diagnosed
Analyzed	Conducted	Directed
Apply	Conferred	Enforced
Appraised	Configured	Engineer
Approved	Connected	Enhanced
Arranged	Conserved	Ensured
Assemble	Consolidated	Established
Assess	Constructed	Estimated
Assign	Consulted	Evaluated
Assisted	Contributed	Examined
Attained	Controlled	Executed
Authorized	Converted	Expanded
Awarded	Coordinated	Expedited
Budget	Counseled	Fabricated
Built	Cultivated	Facilitated
Certified	Customize	Finance
Chaired	Dedicated	Focus
Changed	Defined	Forecast
Coached	Delegated	Formulated
Collaborated	Delivered	Funds

Furnish	Outperformed	Restructured
Gained	Participated	Retrieved
Generated	Performed	Reviewed
Graduated	Plan	Revised
Handled	Prepared	Revitalized
Helped	Presented	Schedule
Identified	Prevented	Screened
Illustrate	Process	Served
Implemented	Produced	Showcase
Improved	Program	Simplified
Increased	Promoted	Solved
Influenced	Proposed	Specialize
Informed	Prospect	Specify
Initiated	Proved	Sponsored
Inspired	Provided	Staff
Installed	Publicized	Standardize
Integrated	Purchased	Started
Interacted	Pursued	Succeeded
Introduced	Qualified	Supervised
Investigated	Received	Supported
Launched	Recommended	Surpassed
Led	Reconciled	Sustained
Maintained	Record	Tested
Managed	Recruited	Tracked
Mediated	Reduced	Transformed
Merged	Refocus	Translated
Mobilized	Regulated	United
Modified	Reorganized	Updated
Motivated	Repaired	Upgraded
Negotiated	Replaced	Used
Obtained	Report	Utilized
Operated	Represented	Validated
Ordered	Research	Value
Organized	Resolved	Verified
Outpaced	Restored	Volunteered

RESUME MISTAKES TO AVOID

Typos. You will likely grow tired of hearing this, but correct spelling, appropriate grammar, no missing words, and no typing mistakes, make your resume an employer-pleaser right out of the starting gate. An error-free resume is rare. Indeed, some hiring managers will not consider your candidacy further if they find even one mistake. Every mistake makes me pause and think. Every mistake makes me question your carefulness, care, and attention to detail. Don't make me pause; don't make me think.

Large Gaps in Employment. The first scan of a chronological resume will reveal gaps in the job searcher's employment history and these are red flags indicating potential problems with the candidate. If you have significant periods of unemployment usually greater than 3 months, then you better be prepared to discuss this.

Problems with Grammar and Sentence Construction. If the resume is poorly written with mistakes, this is a bad indication especially since most computers can now instantly check spelling and even recommend grammar changes. Have others review your resume before you send it to any companies.

Wild Formats. Some people are tempted to use wild formats and fonts to attract attention to their resume in hopes they will be noticed. For most engineering hiring managers this is a danger sign and not looked at favorably. The safest thing to do is stick with the standard format and commonly used fonts.

Special Note on Resume Formats. Some people will work for hours and days trying to get the best format and font for their resume. This can be wasted effort in some cases. With online resume submittal and new software programs, the resumes are automatically reformatted when loaded into a company website and often converted to unformatted text to run through a comparator program. So before you spend all the time formatting your resume to make it look perfect, check and see what the requirements are for submittal.

Also, the type of software you use to create your resume may be not be compatible with the hiring company's software; when they open your resume file all the special formatting is lost. Sometimes, the resume format is changed to the point it is almost unreadable. The best way to prevent this is submitting your resume in PDF format. However, not all companies can handle PDF files, so again, the best thing to do is check with the company on the file format they require before submitting your resume.

WHAT THE INTERVIEWERS WILL BE LOOKING FOR IN YOUR RESUME

In most corporations the Human Resource department will do a first pass of the resumes received and eliminate those that do not meet the minimum requirements for the job. They will be checking to see if the candidate has the number of years of experience and correct engineering degree. Next, they may do a comparison of the resume to the job description by doing a word search to find out how many matches there are between the job description and the resume. If your resume passes this, then it is usually forwarded to the engineering manager doing the hiring.

The engineering manager may sort through the resumes or distribute them to a team of engineers and have the team give their opinion as to which engineers they should ask to come in for an interview. Once the final selections are made, the candidates are contacted and interviews set up. Those not selected usually receive a postcard or email rejection notice.

FREQUENTLY ASKED QUESTIONS ABOUT RESUMES

Should I Include a Picture? The jury is still out on this one. Some hiring managers and personnel like to see a picture of the person and others couldn't care less. A picture can help and can hurt, so use caution when considering putting your picture on your resume.

Should I Put My Education and Degrees First? Yes, for junior engineers since this is usually one of the first criteria hiring managers look for.

How Do I De-Emphasize My Age? There are two problems with age. Being too old and being too young for the position. If you feel you are too old you can de-emphasize your age in two ways. One, do not put the year you received your degree and also do not put the years you worked at each job.

If they are looking for a younger person since it is a lower level job, and you might be overqualified, you can play up your experience as a real benefit since you will not be getting paid any more but they will be gaining a person with vastly more experience. It is thought that older workers are more dependable and are not likely to leave on a moment's notice. Older workers are generally more settled down.

If you are younger than they expected then you need to emphasize all the energy and enthusiasm you bring to the job. You have boundless energy and willing to put in all the extra effort needed to make up for being young.

Should It Be One or Two Pages Long? For junior engineers or recent graduates one page is best. For senior engineers, two pages is a must.

Should I Put in Hobbies? Yes, it looks good on the resume but keep it to one line maximum at the end of the resume.

What Word Processing Program Should I Use? The program and version used by the hiring company is usually the best or submit your resume in PDF format.

What If They Want you to Fill Out an Online Resume Form? This is the way most companies are going with their efforts to go paperless and return to green. The best thing to do is to always have an electronic copy of your resume at all times, and simply cut and paste it into the companies' online system if possible.

Should I Use a Resume Writing Service? Only if you feel you are totally incapable of writing it yourself. There are excellent resume writing companies easily found online [8–11]. You might want to contact local companies just in case you need to visit them for in-person help. Remember, for best results you need to modify your resume to match each job ad. If the resume writing service charges you for modifications it could end up costing you a lot of money.

SUMMARY

In today's competitive job market, a well-written resume is critical to landing that new job. To impress people you will need a resume that is well-structured, easy to read, clear of typos, and full of vital career information. There is a generally accepted format for resumes, with your name and contact information at the top and your objective statement directly below. The main body of the resume follows this. The most commonly used formats for resumes are chronological and skills. You can create a power-charged resume that should get you interviews and hopefully a job offer by quantifying, relating, and selling yourself on the resume. The best way to do this is to sell your benefits. Action words are the best way to power-charge your resume, so, use them often. Always triple check your resume for spelling and grammatical errors before submitting since these are so easily avoidable. Finally, create your resume while looking at the job ad and use the keywords of the ad in your resume as much as you can. Tailor each resume to the specific job you are applying for.

If you would like further information on resume writing the best place to go is your local bookstore, library, or conduct an online search using the words "resume writing."

Have you identified any career actions you want to take as a result of reading this chapter? If so, please make sure to capture these ideas before you forget by recording them in the notes section at the back of the book.

ASSIGNMENTS AND DISCUSSION TOPICS

1 Why is having an impressive resume so important?
2 Describe the general format of a resume.
3 What are the differences between a chronological and skills format resumes?
4 What is the difference between benefits and features?
5 What precautions should you take when submitting a resume online?
6 Write your objective statement.
7 How do you quantify, relate, and sell your skills? Give examples.

REFERENCES

1. "Free Resume Templates for Engineers," http://www.resumelogic.com/rl.nsf/resumetemplates?openform.
2. "Engineering Resume Examples," http://www.bestsampleresume.com/sample-engineering-resume/sample-engineering-resume-2.html.
3. "Technical Resume Template," http://www.job-application-and-interview-advice.com/technical-resume-template.html.
4. Lewis, Adele, and Moore, David J., *Best Resumes for Scientists and Engineers*, John Wiley & Sons, Inc., 1998.

5. Enelow, Wendy, and Kursmark, Louise M. S., *Expert Resumes for Engineers*, JIST Works, 2008, 288 pp.

6. Isaacs, Kim, "Resume Tips for Engineers," http://career-advice.monster.com/resumes-cover-letters/Resume-Writing-Tips/Engineering-Resume-Tips/article.aspx.

7. Bolles, Richard Nelson, *What Color Is Your Parachute? 2009: A Practical Manual for Job-Hunters and Career-Changers*, Ten Speed Press, 2009.

8. "Best Ten Resume Writers," http://www.best10resumewriters.com/?gclid=CLXVjaSrvpgCFQw9GgodoEbzZw.

9. Certified Engineering Resume Writing Service, http://www.employment911.com/resumes/engineering-resume-writing-2.asp.

10. "How to Choose A Resume Writing Service," http://www.gotthejob.com/selectingaresumewritingservice.html.

11. Chapman Services Group, http://www.chapmanservices.com/Services.php.

CHAPTER 40

COVER LETTERS THAT WILL IMPRESS EMPLOYERS

People submit a cover letter with their resume as a means to introduce themselves to prospective employers. Your cover letter must capture the reviewer's interest; make them want to find out more about you, just like your 30-second commercial. The cover letter should make the reviewer want to read your resume to find out more about you. Your cover letter is like the cover of a great book; once you scan it, it makes you want to pick the book up and start reading it.

Your cover letter assists in selling you by highlighting the most key or desirable skills a company would want to buy. Your cover letter should be easy to read, direct and to the point, positive, and project a confidence by you in your talents and skills.

The reward of a great cover letter will be a resume that is read in detail instead of being passed over. Great cover letters, followed by great resumes, get you an invitation to interview.

Writing a cover letter to accompany your resume submittal is part of the job search process as shown in Figure 40-1. Generally one writes a cover letter after they have completed their resume, since the cover letter is really a summary or just the highlights of your resume that are meant to impress the reviewer.

GENERAL FORMAT AND LAYOUT OF A COVER LETTER

The general format of a cover letter is shown in Figure 40-2. In the upper right-hand corner is your contact information. Next should come the company contact or who the letter is addressed to. Following this is the greeting and introduction section. The objective statement is next followed by the best three reasons why they should hire you. And at the bottom of the cover letter is a strong closing and a request for an interview.

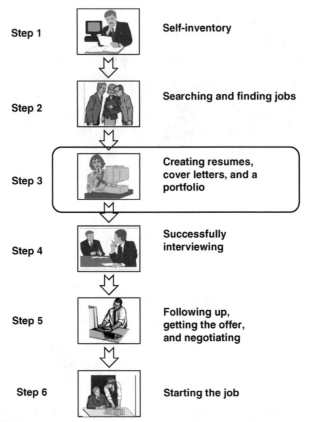

Step 1		Self-inventory
Step 2		Searching and finding jobs
Step 3		Creating resumes, cover letters, and a portfolio
Step 4		Successfully interviewing
Step 5		Following up, getting the offer, and negotiating
Step 6		Starting the job

FIGURE 40-1   Job search process step 3—creating cover letters.

Greeting and Introduction Section. For the opening greetings it is best to be traditional and use "Dear" followed by the person's full name. Directly below the greeting is a very short paragraph that tells them why you are writing to them; this should immediately grab their attention. There may be multiple job openings, so a good rule of thumb is to mention the job ad or opening that you are responding to. Make it short and directly to the point.

Objective Statement. This should be a short paragraph describing your objectives or highlights. This should be similar to, but not identical to, your objective statement on your resume.

Reasons to Hire You. Next should follow the three best reasons why they should hire you. These reasons should be clear, concise, and easy to read. They should jump out from the page and not be buried in the middle of a paragraph that makes it hard to read. One can use bullet points to highlight your superior skills.

Strong Closing. You should end your letter with a strong closing that directs the reader to your attached resume and requests an interview. And naturally, at the bottom of the closing, is your hand-written signature. All

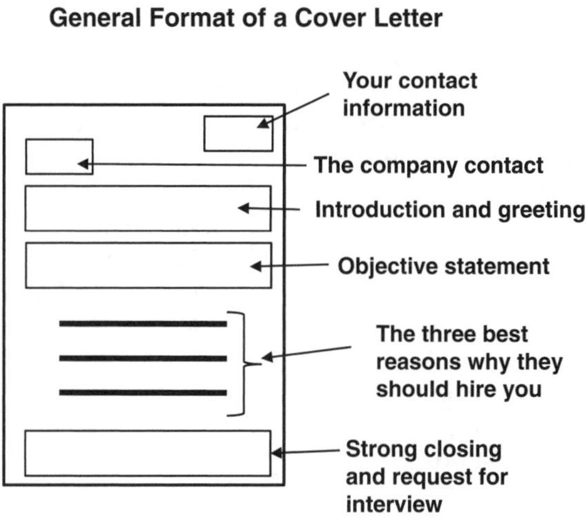

General Format of a Cover Letter

FIGURE 40-2   General format of a cover letter.

cover letters submitted should be originals. The standard closings include: Regards, Sincerely.

HOW TO WRITE IMPRESSIVE COVER LETTERS

The process for writing a cover letter is very similar to writing a resume. You should have the job ad, your resume, cover letter templates, and reference material close at hand to look at as you draft your cover letter. You start by asking yourself, "what do I have to offer that would make them choose me?" You determine this by finding the matches between their job ad and your resume. These are the items you want to highlight in your cover letter. Once you identify these matches, then you are ready to start writing. The following are great tips to consider when drafting your cover letter.

Use the Job Ad Terminology. Again, like when you wrote your resume, you tailor the cover letter directly to the job ad and use the same terminology and keywords the company put in the job ad. Your cover letter should highlight your superior qualifications for the job in the exact words used in the job ad.

Quantify, Relate, and Sell Yourself. Power-charge your cover letter by quantifying, relating, and selling yourself. The cover letter is all about high-lighting the great benefits you can offer to the company. Your cover letter should answer the question, "Why should they hire you?"

Use Action Words. Use powerful action words throughout the cover letter like you did when describing your superior skills in your resume. Please refer to the previous section for a list of action words.

Use Simple Language and Sentence Structure. Don't overcomplicate your letter by using industry jargon and acronyms. Stick to the terminology of the job ad and not your terminology. Keep the sentences short and easy to read.

There should be more white space than text on the page; your cover letter is not a PhD thesis or long narrative on why they should hire you. QuintCareers.com suggests your cover letter should be in the range of 200 to 250 words.

Proofread Many Times. Proofread your cover letter many times before you submit it and even have others review and get their comments.

> John Smith
> 123 First St
> Maple City, CA 76532
> Date

Ms. Mary Jones
National Engineering Company
465 Park Avenue
Trenton, New Jersey 12345-6789

Dear Ms Smith

I am interested in the position of senior engineering manager #34567 that your company has posted on your website. My skills and experience match your needs and, in some instances, offer more than the requirements of the position.

I am presently an engineering manager with over 25 years experience in the design and development of products utilizing Field Programmable Gate Arrays (FPGA) devices. I have a Master's Degree in Digital Design and a Master's Degree in Business from Stanford. Here is what I have to offer your company:

- Highly successful engineering manager responsible for 150-person group generating $110M business base with eight different products. Revenues and profits from my group have exceeded expectations the last three years.
- Demonstrated outstanding customer skills as evident by customer selecting our group as Supplier of the Year.
- Responsible for leading state-of-the-art digital designs for extremely large three-million gate FPGA utilizing advanced tools to complete design ahead of schedule and under cost.

I would appreciate the opportunity to talk to you about your exciting opportunity so that I can share more of my background and skills with you. I will be contacting you to confirm that you received my attached resume and if you have any questions. The best way to contact me for an interview is on my cell phone at 234-567-7890.

Sincerely,

John Smith
Senior Manager
Enclosed: resume

Sample cover letter

Now after reading this cover letter let's summarize the key points. John Smith has an advanced education, 25 years experience, manages a large group that produces very profitable products. He exceeds expectations, his customers select him, and he utilizes the latest technology to keep costs down and delivers products on time. Yes, I would say this applicant is definitely a good candidate. The cover letter is written to present the image that John Smith is an ideal manager to have working for your company. Make your cover letter read like you are the ideal candidate whom any company would hire.

Email Cover Letters and Electronic Copies. Some companies no longer accept paper cover letters and resumes. They inform applicants to email their resume to the company email address. In this case your email is actually the cover letter. If you find yourself submitting resumes in this fashion, make sure the structure, font, a and formatting of your email cover letter will remain intact once it is opened up at the company. The best way to do this is by selecting a common font, a simple paragraph structure, and staying away from special alignment tabs. A good test is to send it to a friend and have them open it up and see if the format goes unchanged. If the company wants an electronic copy of your cover letter, another good thing to do is convert it to PDF format before sending.

SUMMARY

Your cover letter should make the reviewer want to turn immediately to your resume to find out more about you. A great cover letter is like the cover of a book that once you scan it, it makes you want to pick the book up and start reading it.

Cover letters are there to sell you. Sell you by highlighting the most key or desirable skills you have that the company would want to buy. Your cover letter should be easy to read, direct and to the point, positive, and project a confidence by you in your talents and skills.

The process for writing a cover letter is very similar to writing a resume. You should have the job ad, your resume, cover letter templates, and reference material close at hand to look at as you draft up your cover letter. You start by asking yourself what do I have to offer that would make them choose me. You determine this by finding the matches between their job ad and your resume. These are the items you want to highlight in your cover letter.

The format of cover letters varies slightly and templates are available for you to start from [1–3]. In addition, there are excellent reference books available to guide you when writing cover letters [4–8]. If you are interested in more information, search online using "cover letters for engineers."

Have you identified any career actions you want to take as a result of reading this chapter? If so, please make sure to capture these ideas before you forget by recording them in the notes section at the back of the book.

ASSIGNMENTS AND DISCUSSION TOPICS

1 How does a cover letter differ from a resume?
2 Describe the general format of a cover letter.
3 How do you power-charge your cover letter?
4 What precautions should you take when submitting a cover letter online?
5 How do you quantify, relate, and sell your skills? Give examples.

REFERENCES

1. "Free Engineering Cover Letters," http://www.bestcoverletters.com/cover-letters/Engineering.
2. Cover letters for engineers, http://www.bestcoverletters.com/cover-letters/Engineering.
3. JobBank USA, http://www.jobbankusa.com/resumes/cover_letters_free_samples/examples_templates_formats/.
4. Borchardt, John K., Science Careers, "Writing a Winning Cover Letter," http://sciencecareers.sciencemag.org/career_magazine/previous_issues/articles/2006_03_10/noDOI.4819437018278975029.
5. Huffman, Libby, "How to Write a Cover Letter," http://www.officearrow.com/home/articles/the_officearrow_career_center/human_resources_and_job_search/p2_articleid/464/p142_id/464/p142_dis/3.
6. Tyler, Mary Anne, "Welcome to My Writing Cover Letters Website," http://www.writing-cover-letters.com/.
7. Lions, Elizabeth, "Writing Cover Letters That People Will Read," *IEEE USA Today's Engineer Online*, August 2006.
8. Engineer Resume Cover Letter Truths, http://www.resumelogic.com/cover-letter.htm.

CHAPTER 41

WHY PORTFOLIOS ARE IMPORTANT FOR ENGINEERS

From my informal surveys of engineers, I have discovered a very high percentage who do not have a portfolio of their work. When I ask why not, most engineers respond with the same line: engineers don't need portfolios. Then I challenge their thinking by pointing out all the other professional careers that necessitate the use of a portfolio. Lawyers have portfolios, as do architects, doctors, accountants, investment brokers, and politicians, and I am sure there are many more. I point out these professions find it extremely important for their career to have an impressive portfolio, so why do engineers think it is not necessary and never bother to assemble a portfolio? Are engineers missing something?

Yes, engineers without a portfolio are missing an extremely important career aid. As shown in Figure 41-1, having a great portfolio is part of the job search process. Having an impressive portfolio can make all the difference in the world when you are interviewing and is often a key discriminator between you and others interviewing for a job. In this chapter, I will show how you can quickly and easily assemble a portfolio that gets you hired.

WHAT CONSTITUTES AN EXCELLENT ENGINEERING PORTFOLIO?

An overview of a portfolio with some recommended items to use is shown in Figure 41-2. The first task in assembling your portfolio is to visit several office supply stores and look at the presentation booklets. Select one with a high quality cover and which has multiple clear pockets in the inside.

I recommend not buying a cheap book report binder kids use for homework or three ring binder since you are trying to impress people. It should have a sturdy cover and be only about $\frac{1}{2}''$ to $1''$ thick when completed so that it can be easily carried around during the interview. The clear plastic pockets should be hardbound into the cover and capable of holding $8\frac{1}{2}'' \times 11''$ sheets of

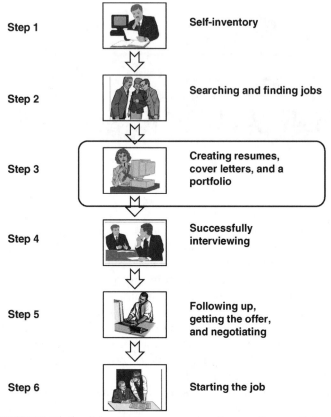

Step 1 — Self-inventory

Step 2 — Searching and finding jobs

Step 3 — Creating resumes, cover letters, and a portfolio

Step 4 — Successfully interviewing

Step 5 — Following up, getting the offer, and negotiating

Step 6 — Starting the job

FIGURE 41-1   Job search process step 3—creating portfolios.

paper. I recommend you stay away from 11″ × 17″ foldouts unless absolutely necessary. You want everything to fit neatly into a pocket and not fall out when you are paging through or if you accidentally drop it. Make sure the portfolio is well-constructed and will hold everything tightly in place.

▶ **Career Tip.** Buy a high-quality portfolio.

Once you have a portfolio selected, your next task is to collect the material you want to put in. As shown in Figure 41-2, the following are some of the recommended items you may want to place in your portfolio.

Resume. Place your resume in the pocket directly under the cover, so it is the first thing you see when opening up your portfolio. The resume acts like the table of contents to the portfolio. When you are discussing your resume and come to a point you would like to emphasize, you simply flip back a few pages to what you want to show. When you are done showing the supporting material, simply return to the resume and continue. Since the resume acts

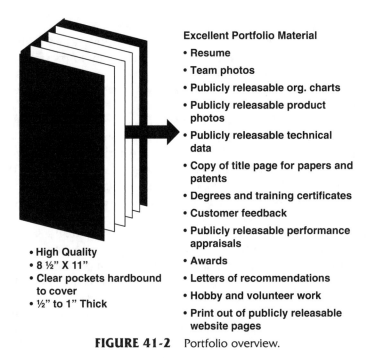

Excellent Portfolio Material
• **Resume**
• **Team photos**
• **Publicly releasable org. charts**
• **Publicly releasable product photos**
• **Publicly releasable technical data**
• **Copy of title page for papers and patents**
• **Degrees and training certificates**
• **Customer feedback**
• **Publicly releasable performance appraisals**
• **Awards**
• **Letters of recommendations**
• **Hobby and volunteer work**
• **Print out of publicly releasable website pages**

• **High Quality**
• **8 ½" X 11"**
• **Clear pockets hardbound to cover**
• **½" to 1" Thick**

FIGURE 41-2 Portfolio overview.

like a table of contents, the supporting material following the resume should be in the same sequence as you have written your resume. The following are recommended items to put in your resume and are in no particular order.

▶ *Career Tip.* Arrange the material in your portfolio to follow the order of your resume.

Team Photos. Team photos are excellent for portfolios since they allow people to see the size of the team and at the same convey a very positive team player and people-oriented image. Team photos are great to emphasize especially when talking to Human Resources personnel.

Publicly Releasable Product Photos. Visit your marketing department and pick up a few product brochures or print-out copies from your company's public website.

▶ *Career Tip.* Make sure all materials in the portfolio are approved for public release by your company. DO NOT show any proprietary information. Showing proprietary information without prior approval can get you fired.

Publicly Releasable Organization Charts. Any publicly releasable organization charts with your name on the chart showing how you fit into the organization. This is beneficial since it lets people see the size of your organization and your role. Interviewing managers are most interested in this.

Publicly Releasable Technical Data. Do you have any impressive publicly releasable technical data to show? 3-D graphs are really impressive and interviewers with a technical background like to see the technical depth of your work.

Papers and Patents. Put a copy of the cover page for any significant papers or patents you have published or received. Only putting a copy of the title page keeps down the volume of paper you have in the portfolio. If you have published multiple papers or received multiple patents, do not put them all in, but create a list and put the list in a pocket. Only put copies of title pages for the most important one or two in the portfolio. Again, people with a technical background will be impressed by this.

Degrees and Training Certificates. Put a copy of your degrees above and beyond your basic engineering degree, like an MBA or second technical degree. Did you take any special training and receive a certificate for successfully completing the training? These certificates make excellent portfolio items since they show how you are willing to update and improve yourself.

Customer Feedback. Do you have any letters, notes, or emails from customers who were extremely happy with your work? These are also excellent portfolio items.

Awards. Include any awards you have won from the company. Also include technical awards from outside the company. These are great portfolio items since they show your technical expertise and your work has been recognized not only in the company, but outside.

Performance Review. Do you have a copy of a very recent and publicly releasable performance review where you receive an outstanding rating? These are clear evidence that you are a superior performer.

Letters of Recommendation. If you have on your resume, letters of recommendation available upon request, there is no better place to store them than your portfolio. It's mighty impressive to simply flip to them in your portfolio and briefly show they exist and the excellent recommendations about you.

Community Service and Volunteer Work. A single page showing your community involvement and volunteer work is always good. Human Resources personnel will relate to this.

Only One Page of Personal/Hobbies. Please keep this to one page and make this the very last thing you talk about in an interview since it does not relate to your job. A simple montage of photos of you enjoying your hobbies or family goes a long way in showing your personality type.

STRATEGIES FOR INTERVIEWING WITH A PORTFOLIO TO ASSIST IN GETTING YOU HIRED

There is a strategy to interviewing with a portfolio. This strategy goes along the line of letting you take control of the interview and slowly walk through your resume and the supporting material in the portfolio. Some people doing the interview will not be receptive to this and want to control every aspect. They have a set way of doing interviews and do not want to deviate from it. If this is the case, then you need to follow their lead and simply fill in the blanks as they march through their process. Trying to get them to do it your way will only make them frustrated and want to move your resume to the bottom of the pile. To effectively respond to this style of interviewing, you will have to jump around your resume and portfolio as they bring up each point.

People who are open to letting you talk will ask questions like, "tell me about yourself." This is an indication that the interviewer wants to see how well-organized you are and how well you discuss your qualifications. At this point in the interview, ask if you can show your resume and portfolio. If they respond positively you are good to go.

Make sure there is a table or desk you can place your portfolio on where both of you can easily read it. The best is if the interviewer can sit next to you, side-by-side, to review your portfolio. If at all possible, move to the interviewer's side of the table as you go through the portfolio. If you cannot be alongside the interviewer, then you better be able to point to things and read upside down. Do not be constantly turning the portfolio around so you can read it and in doing so, make the interviewer read it upside down.

Do not go through all the material in your portfolio with every interviewer. You should know the background of the person you are interviewing with and flip only to the pages that you think are of most interest. For example, if you are interviewing with a purely technical person chances are they are going to be most interested in your technical expertise and accomplishments and not your team pictures, hobbies, and volunteer activities; a technical interviewer will be impressed with your papers and patents. However, if you are interviewing with a person from Human Resources, more than likely they are not interested in your technical background as much as people skills, which means they will relate more to your personal items such as team pictures, hobbies, and volunteer activities.

The key to a successful interview using a portfolio is to make sure you emphasize the material that best fits the person sitting across the table from you asking the questions. Correspondingly, an interview killer is to spend too much time on areas in your portfolio that do not relate to the interviewer's interest.

▶ *Career Tip.* Tailor the use of your portfolio to match the needs of the person interviewing you and be prepared to review upside down.

Limit the amount of time you spend on the support material in the portfolio. It is too easy to get sidetracked and start telling stories about your experience when showing team photos and products. Limit your discussion to a 1 to 2 minute review per support item and make sure you emphasize why it is important they see this. You may feel very tempted to discuss what fun it was and the great team party at the end of the project; however this will not get you hired. You can use a team photo to lead into a discussion of how the multifunctional team was comprised of software, electrical, mechanical, chemical, and nuclear engineers all located in separate locations. Your job was to lead the team and they successfully completed the project on time and on cost in spite of being located in three separate sites across the nation. This type of discussion about the team photo will get you hired.

▶ *Career Tip.* When showing portfolio materials limit the amount of time on any one page and make sure you quickly get to the point.

Portfolios need updating at least once a year and absolutely before a new interview. Your portfolio, like your resume, needs to be specifically tailored or focused to the specific qualifications of the job. Use the job ad to guide you when you update or modify your portfolio.

SUMMARY

Creating a portfolio is easy to do and can be fun. Purchase a high-quality binder at your local office supply store. Identify the items you want to place in your portfolio and start collecting them. Keep your eyes open during your normal work day for items you may want to place in your portfolio. The more pictures, graphs, and awards, the better. Organize your portfolio to follow your resume. Practice interviewing with your portfolio and make sure you can do it right side up and upside down. Tailor your use of the portfolio to match the interviewer's background.

Have you identified any career actions you want to take as a result of reading this chapter? If so, please make sure to capture these ideas before you forget by recording them in the notes section at the back of the book.

ASSIGNMENTS AND DISCUSSION TOPICS

1   Describe how you would use your portfolio during the interview.
2   Why should you practice talking with your portfolio upside down?
3   How often should you update your portfolio?

CHAPTER 42

INTERVIEWING

The job market is rapidly becoming more and more competitive due to virtual teams, economic changes, internet job posting, and globalization. Your competition for a job can originate from local, national, or even global sources. The result is employers often have many engineers applying for a single opening. This overabundance of candidates provides more choices for employers and allows them to be highly selective when making a job offer which results in significantly increasing your competition.

All this competition comes to a peak at the interviewing stage that is the main event in the job search process shown in Figure 42-1. All the engineers selected for an interview are considered very good choices to hire and pretty much equal. The purpose of interviewing is to determine which one of these engineers is the best one to hire. At this point, it all comes down to how well you present and sell yourself or how good your interviewing skills are.

Your interviewing skills are your means to stand out from all the other engineers selected for interviews. In this chapter, we will discuss how to develop and enhance your interviewing skills so that you clearly stand out from all the other candidates during the interview process. Putting significant effort into developing and refining your interviewing skills is well worth the effort and will pay great dividends.

PREPARING FOR THE INTERVIEW

The interview is the main event of the job search process and all eyes will be on you at this point. Consequently, you must spend time preparing and practicing your interviewing [1–9]. The first step is to collect the tools you will use for the interview: your resume, portfolio, job ad, and any other information you have about the company or people interviewing with you. With these items in front of you, your next step is to develop your personal theme statements.

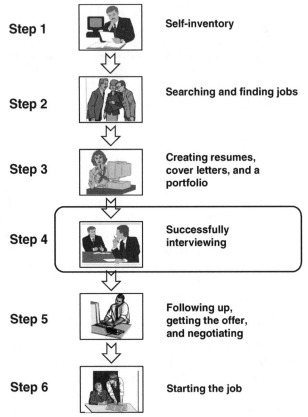

FIGURE 42-1 Job search process step 4—successfully interviewing.

Personal Theme Statements. Your personal themes are highlights of your skills, experience, and accomplishments you think the company will be most interested in and make you stand out [6]. These could also be your cover letter highlights.

Write up your themes in simple sentences. Your theme statements are how you summarize and verbally communicate to the reviewer what is special about you. It is what you want them to remember when you leave and they think of you when they go through the selection process. Here are some examples to help you develop and refine your theme statements.

- Consistently rated a high performer with a proven track record of on-time delivery and within costs.
- Superior technical designs that meet or exceed all requirements while implementing the latest technology.
- Team player with excellent communication skills who is always willing to take on those difficult assignments.

- Innovative and always looking for new methods to improve and enhance products.
- Problem solver with high energy who can drive until the project is a success.

To help generate your theme statements, look at each major assignment or skill you have listed on your resume and think of the benefits this offers to the company. Choose to state your benefit or personal theme in terms of key-words from the job ad. You are literally translating the meaning of these parts of your resume into terms the interviewer will understand and can easily identify as excellent reasons why you should be hired.

Once you have developed your theme statements it is on to the next step in preparing by practicing.

Practice, Practice, Practice. You will need to practice your interviewing at least 10 times before you go on your first interview. Practice by yourself first, just like you are on a real interview, walk into room, practice your greeting and then sit down just like you would during a real interview and start presenting. Use your resume as the talking document and start from the top and go through the parts you want to talk about.

After you have done this 10 times, it is time to practice with another person. Find someone who will help you practice and have them play the role of the interviewer. Give them a list of questions to ask you and see how well you can answer them.

You may want to videotape yourself during practice and then play back afterward to see how you did. Watch for body language, tones, facial expression, and how well you discussed your qualifications. You will not believe how you look and all the things you are doing without even knowing it. The first time I videotaped myself I was appalled at all the facial expressions I made without realizing it. Especially the squint and closing of my eyes reaction when answering a difficult question. I did not realize I was doing this and it looked terrible. I would have never even hired myself after reviewing the video.

Don't erase the videos, save them. The reason is at the end, when you think you have practiced enough, go back and play the very first one you recorded and then immediately watch the last one again. You are going to see a world of difference and this is going to give you more confidence than you can imagine. If you do not have a video recorder, then sit in front of a mirror and practice.

Timing. The next very important reason for practicing is to get a sense of timing. How long does it take to make all your points? Are you taking too long? Your goal is to be able to run through your resume and make all your key points and theme statements within the time allotted. Your time with people can range from 30 minutes on the short end to 2 hours on the long end. In the short interviews you need to make your points and move on. Practicing will give you a sense of timing and help you determine the pace you need to get everything in.

Dressing for the Interview. The next thing to prepare is your physical appearance. Haircut and dress in a professional manner. Dress up for the interview. For the men it is easy to remove ties and open your shirt if you come overdressed. But if you are underdressed you have no options. You have to project the right image for job interviews; the aim is to look very professional. If you feel uncomfortable wearing a suit and tie, try wearing a suit for a couple of days before the interview so that you get used to it. And remember to polish those shoes, carry an attractive briefcase or professional-looking portfolio.

DIFFICULT QUESTIONS YOU WILL BE ASKED

Now for the hardest part of interviewing: skillfully answering difficult questions [9,11–13]. Since all the candidates' resumes indicate they are all excellent candidates, the company must come up with a way to discover which one is the best. They do this by asking difficult questions to see how the candidates respond and who gives the best answers.

> So here are some typical questions you might expect interviewers to ask you. Tell me about yourself?

The answer to this one is very simple; you just start going down your resume. You start at your highlights or 30-second commercial and then move into your present position or most important skill. You can transition and talk with your resume once you get through your highlights/30-second commercial by saying, "as shown on my resume I am presently at" and then pull out your resume and point to what you are talking about.

Here are some more questions you should be prepared to discuss:

> Why are you looking?
> Can I contact your present employer?
> What are your likes and dislikes?
> What are your career goals?
> What are your weaknesses?
> Tell me about a project that went bad and what you did about it
> Why do you think you would make a good candidate?
> What are your strengths?
> Why are you interviewing with our company?
> Why should I give you the job?
> How soon can you start?
> What salary were you looking for?

This is only a very short list of questions that can be asked but some of the more common ones. Peter Veruki in his book, *The 250 Job Interview Questions*

You'll Most Likely Be Asked and the Answers That Will Get You Hired, is a good resource [11]. Also the book by Vicky Oliver, *301 Smart Answers to Tough Interview Questions,* is another excellent reference for handling the difficult questions [9].

Where do company interview questions come from? Most engineering managers are not in the interviewing business and therefore rely on the Human Resources department to come up with tough interview questions [10]. More than likely, if you visit a local bookstore and research interviewing questions, you will probably be looking at the same books as the Human Resource department is looking when they research interviewing questions. Therefore, by researching the possible questions you will have a high probability of knowing what they are going to ask long before you ever get to your interview. Like in engineering school, it sure was easier when you knew what questions were going to be on the final exam before you took it.

QUESTIONS YOU SHOULD ASK

Sometimes you can stand out from the crowd by having a couple of questions of your own to ask that show you are really thinking about the position. Here are some you might consider asking if the opportunity presents itself.

> Why did you choose this company and what do you like about it?
> What is the last person who had this job doing?
> What are the job responsibilities?
> What skills are most important for the job?
> What kind of training is provided?
> What are your company benefits?
> Who will I be reporting to and in what organization?
> What opportunities of advancement are there?

These are all good questions to ask but in order for you to make a decision whether you to work for the company you are going to need answers to some very basic questions about the job and the company so that you can compare it to other opportunities. These questions are usually answered best by the hiring manager or Human Resources representative.

> What is the salary range for the job?
> How soon before you can make me an offer? (always a good
> question to end with if you want the job!)
> How soon after the offer do you need an answer?
> When would you like me to start?
> What are the benefits? Do you have a package I can take with me?
> How much vacation do I get?

Does the company have a savings plan?
What are the pension or 401k benefits?
If necessary, do you pay for relocation?
If they flew you in to interview, how do you get reimbursement of expenses?

One good tip is to write up the questions you have and check the list to make sure you have all the answers you need before you leave for the day. That night after the interview fill in the answers before you forget.

CONTROLLING THE INTERVIEW IN YOUR FAVOR

The interviewers you will encounter all have their own style of interrogating you. At one end of the spectrum are those who believe it is up to you to impress them and they will have no set agenda other than to say "Tell me about yourself." They want to see how well you are organized and how good you can present your case. At the other end of the spectrum, there are those who want to control every moment of the interview and will ask you questions in their order and expect you to simply fill in the blanks. If you try and deviate from their agenda, they see this as negative and more than likely will score you low if you try to take control and change anything. For these interviewers you have no choice but to follow along and let them control the interview.

In the event you have the opportunity to control the interview, there is an order to how best to exchange information that favors you and here is what I recommend you do.

Greeting and First Impression. When you first meet the person, introduce yourself (name and title) and shake hands. (American tradition). Exchange business cards and listen to the interviewer's name and title. Repeat it three times quietly to yourself before you sit down. From the title, determine the interviewer's background. This is extremely important for reasons I will point out later in the chapter.

Company First. At this point, if possible, ask if the interviewer could give you a brief (2 to 3 minute) overview of the job opening and its requirements. I tell them it is best if they first share information about the position so I can then show them how my qualifications match their needs. Pay close attention to the keywords they are using. It is amazing when you do this; hidden information is often revealed which was not mentioned in the job ad.

If you talk first and go on about qualifications that they are not interested in, you are missing the mark and killing your chances. By letting them talk first you will know exactly what they are looking for and you can highlight the exact qualifications that you have that match their needs.

Next You. After they are done talking you immediately and enthusiastically start to review your resumes and highlight assignments and qualifications that match their needs. Remember to flash your portfolio as you go through your resume. Be prepared for the tough questions at this time. Make sure to pace yourself and look at your watch if possible, so that you know how much time you have left.

Company. After you have covered your resume it is time to regroup and ask a few questions of your own. Get the information you need to make a sound decision as to whether you want to work for this company.

Ending the Interview. You must end the interview on a positive note. Summarize why you would make a good candidate by highlighting how your qualifications are such a good match to their needs. You can repeat your theme statements. Thank them for their time. Express how you are looking forward to hearing from them. Shake hands and make sure they have a copy of your resume with a means of contacting you. Finally, ask for the job if you like it.

HOW TO IMPRESS ALL THE INTERVIEWERS

The most impressive interviewing skill is interweaving your resume, theme statements, keywords from the job ad, and your portfolio, to answer a question or present your point. For example, if the ad is looking for a lead engineer to oversee a project and have excellent communication skills, you might decide your theme statement is

"Proven record of success as a lead engineer with excellent team skills."

You might discuss your present assignment as listed on your resume by weaving your theme in. Your verbal response on the description of your present assignment could follow along the lines of:

In my present assignment I am a lead engineer on the $12M design project overseeing 24 engineers. I have learned, as a leader over this very diverse group, to be successful requires excellent communications skills. I have taken special training in leader communication and consider myself successful since my team received the company's highest award last year for their efforts. My excellent communication skills helped get the team through some very difficult times.

For this example the theme statement is interwoven with the description, and keywords are inserted from the ad to help the interviewer understand how your present work demonstrates that you have the skills needed for their job.

Then, flip your portfolio to the page that shows a photograph of you accepting the award from the president of your company, and on the page next to it is your award certificate with your name in large print.

This is the optimum combination of personal theme, keywords, and pictures, coming together. You relate to their needs, you weave in your theme so you stand out, and then you show physical evidence by a photograph of you, and an award certificate. It is almost like the interviewer is there at the award ceremony watching you receive the award. Your ability to combine all these clearly shows you have excellent communication skills.

The benefits of interweaving your theme statement, keywords, and job description, manifest later on during the selection process. Interviewers will be able to look at your resume and point to your present assignment and say, "this person has clearly demonstrated the key skills we need on their $12M design project. The candidate received an award for project leadership and has excellent communication skills."

Tailoring Your Interview to the Person and Their Background. To assist managers in selecting the best candidate, a team of people with different backgrounds generally conduct the interviews. This team may contain Human Resources people, managers, technical staff, and other engineers working in the group who will have to work alongside you. You will interview with all these people and afterward they will fill out evaluations and provide input on which candidate was the best from their perspective. So now let's look at how you need to tailor your interview to best fit these people.

Hiring Engineering/Business Manager. This person is strictly business-oriented. They usually have a managerial background and are not that interested in the technical details. They leave the technical details up to the technical experts in the group. Their consideration is how well you will fit into the group. How soon you are available and what salary are you looking for? Can they afford to hire you? At the top level they are concerned about your technical background and work experience. Does it show that you are dependable and able to produce good results?

For this person you may want to stress your team skills, communication skills, and ability to get along with other people in the organization. You also want to stress how your people and technical skills can contribute to the organization and to its bottom line. You probably do not want to talk about social items, such as company picnics and hobbies you enjoy unless they bring it up. Think strictly business, controlling cost, and generating profits.

Human Resources. The Human Resources personnel are nontechnical and have a psychology background. They mostly do not understand the technical jargon. You might as well be speaking a foreign language in most cases. They are on the team to assess your people and social skills. They are looking to see if your personality fits the company and will you fit in socially. They want to know: Do you have good social skills or are you a potential problem?

For this person you may want to stay away from all the technical jargon and focus on your social skills, hobbies, and company volunteer activities you participated in. Stress your ability to get along with people and how you like the people side of engineering. If you only talk technical to these people they more than likely will rate you low in the people and social skills area.

Technical Experts. You may be asked to interview with a staff engineer or technical expert. These are the most technically knowledgeable people in the organization and are only worried about the engineering or science of their work. They are similar to an university professor whose interest is with technical issues and who couldn't care less about people and business issues. These people are interested in what you have to offer technically and want to measure the breadth and depth of your technical knowledge. Do you have any advanced degrees and from what institutions? They are all about educational credentials.

For this person you may want to stay away from social, hobby, and business discussions. Stick to your technical background and discuss how you can contribute technically to the group. Discussion showing how you understand the science involved in their group, and how your technical background might help them along with any advanced technical training you have received, is excellent. Discuss any difficult technical problems you encountered and how you solved them. Do not shoot from the hip with these people and try to wing it; stick to what you know technically.

Future Coworkers. More than likely you will be asked to interview with other engineers in the group who could be above you, at the same level, or below you. They may also have different technical backgrounds than yours. For these people you want to focus on your technical background and your experiences—what you have learned which you feel makes you more valuable. They are considering if you are the right type of person to come into the group and really help out or are you going to be a problem to work with. For this reason, you want to emphasize how you are easy to get along with. For the people below your level, they are trying to determine if they would like you as their new lead engineer. For the people above you, they are wondering how well you take direction and what you bring to make their job easier. For the people at your level, they may possibly be wondering if you are going to be more competition for the next promotion, as well as how you can contribute to help out and make their job easier.

By now it should be intuitively obvious how important it is to tailor your interviewing to the person. The engineer who gets hired is rated the highest by all the evaluators. If only one interviewer rates you low it could knock you out of the running.

How do you get rated high by everyone? You determine the person's background, emphasize and highlight your skills, capabilities, and performance that they are most interested in and relate to. You do not interview exactly the same way for each person but tailor your presentation to match their background. Everyone you are interviewing with is evaluating you for different reasons; you must be a versatile presenter and change your presentation for each. And last, make sure you get every interviewer's business card so that you can send them a "thank you" note.

Body Language, Tone of Your Voice, and Attitude. The next area where you can stand out from the others when interviewing is through body language, tone of your voice, and attitude. Body language says many things about you. If you sit up, lean forward toward the person with open arms, a smile on your face, looking them directly in the eyes, and nodding occasionally in a positive manner when they speak, you are sending the best body language.

Contrast this with a person that leans back, folds their arms, and stares at the ceiling when talking; this is totally the wrong body language.

Next is the tone of your voice. Your tone should be upbeat, confident, positive, energetic, and show enthusiasm for the job. You should be clearly pronouncing your words and not mumbling. This is the best tone. Examples of this are:

> Yes, I can easily handle that. . . .
>
> Absolutely, I have done that before and know exactly what to do. . . .
>
> No problem. I would welcome the opportunity to show all I can do about. . . .
>
> That sounds exciting and something I would love to work on. . . .
>
> I am always willing to try something new. . . .

If you are mumbling your words and casting doubt, these are the wrong tones. Stay away from saying:

> I am not sure if I . . . .
>
> Never did that before. I wouldn't know what to do, but I think I could figure it out if. . . .
>
> Gee, I am not really sure. I never considered. . . .
>
> I wouldn't have to do . . . . if I accepted the job . . . .
>
> I am not going to do that if I accept the job because. . . .
>
> It all depends; I guess I could do that if you want me to. . . . . .

The next way to clearly stand out from the rest of the interviewees is your attitude. You need a great attitude. People will be watching your attitude about interviewing and your attitude about the job. Your attitude is being assessed from the time you start interviewing and keeps on going until the final handshake goodbye. Your attitude comes out in the way you act and talk about things. Here are some great things to say which indicate you have a great attitude:

> Glad to meet you, thanks for the opportunity to interview, I really appreciate your taking time out of your busy schedule to talk with me.
>
> I look forward to hearing from you, I can't wait to start, everything sounds so exciting. . .
>
> Your company and the job are very impressive and I would like very much to join the team.
>
> I find this kind of work exciting and enjoyable; I can't wait to start.

Now let's look at some bad attitudes that you do not want to convey during job interviewing:

> It was a miserable commute getting here and traffic was terrible. Is it always that bad?
>
> Fill out the forms? Can't you take the information from my resume?
>
> Do I have to interview with all these people. Can't you just make the decision?
>
> So what do you want to know about me?

Can you see the differences in attitudes and how your attitude can be sabotaging your interview?

TELEPHONE INTERVIEWS

Some companies do phone interviews of candidates prior to actually inviting them to the office for interviews. If this happens, you will need to be prepared and set yourself up up for success when interviewing over the phone [6,7,8]. Here are some great tips to follow:

1. If they call at a bad time when you are not prepared, it is best to ask if you can call them back in a few minutes so you can get to a place that you can talk. Most companies will understand this and gladly let you call them back. If you do this make sure to have their number and call back exactly when you said you would.
2. Go to a place where you can talk openly and without being interrupted. A room or office where you can shut the door and other people will not walk in on you.
3. It is best to have a copy of your resume, cover letter, and the company job ad in front of you. Also any other information you may have found out about the company.
4. Turn off your computer and any other items that may distract you. If you get an incoming call during the phone interview ignore it and let it roll over into voice mail.
5. Most often the person calling is the hiring engineering manager and they have a stack of resumes in front of them. Their job is to determine the 2 or 3 they should invite for an office interview. Therefore, your interviewing highlights should be focused to a manager—strictly business.
6. Check to make sure the phone line is clear and they can hear you ok. If it is not, hang up and try another line.
7. You start out the interview just like you would a face-to-face interview with introductions. Get their name and title. Write it down immediately. Once you have completed the introduction it is like a normal interview, and you should conduct yourself accordingly as previously mentioned.

You should practice phone interviewing many times just like you practiced face-to-face interviews, only over the phone. Have someone call and practice phone interviewing. Make sure you get feedback on how you sounded and your tone. One excellent way to practice a phone interview is with a mirror so that you can see your facial expressions. This visual feedback will let you know if you are frowning, rolling your eyes, or wincing. These facial expressions are associated with bad tones in your voice. Smiling and nodding are good facial expressions associated with good tones in your voice.

HOW COMPANIES DECIDE WHICH PERSON TO HIRE

Once you leave, the interviewers are required to fill out an interview summary sheet on you. A sample is shown in Figure 42-2. The summary

Acme Engineering   **Interview Summary Sheet**

Candidate Name: _____

Date Interviewed: _____

Person Conducting Interview: _____

Phone: _____

	Poor	Fair	Good	Very Good	Outstanding
Theoretical Understanding					
Analytical Ability					
Technical Knowledge					
Practical Experience					
Judgment					
Ability to Express Her/Himself					
Initiative Indications					
Attitude					
Appearance					
Personality					
Leadership					
Capacity to Grow					
Estimate of Potential					

Remarks
_____
_____

Recommended for Employment   Yes ☐   No ☐   Other ☐
If "No" or "Other" Explain
_____
_____

This form must be completed on the interview date and returned to personnel

FIGURE 42-2   Sample interview rating sheet.

sheet is used as an aid to help the interviewer summarize their thoughts about you and give you a rating. Let's look closely at the form and what they are asking the interviewer to rank you on.

You will notice in this sample form that there are 13 criteria to rate you on. The first 4 of 13 are concerned with your technical ability to perform the job. You will be ranked on how well you communicated your technical skills and how well they fit the company's needs. Most engineers think this should be the only criteria. However, the remaining nine criteria are concerned with your people skills. For these criteria you will be ranked on how well you communicated your people skills. It is interesting that there are more people skills than technical skills. Can you guess why? Most forms are made up by Human Resources departments whose primary focus is on people skills. The message here is, during the interview you need to clearly exhibit your technical and people skills to stand out from the others.

Following this is the remarks section and then comes the bottom line. Recommended for hire or not. And anything marked other than "yes" requires an explanation. This is only a sample summary and not meant to reflect that all companies use this exact form. Each company will have their own interview summary sheet and they are constantly changing.

After all the forms are filled out, several different things can happen. The hiring manager can collect them all and make the decision on who to hire. Or the manager may call a meeting and go through all the summaries with the team and hopefully reach agreement with the team who to hire. A second candidate may be identified in case the first candidate rejects the offer. Once the selection is made, the offer is usually extended to the candidate.

SUMMARY

The job market is rapidly becoming more and more competitive due to virtual teams, economic changes, Internet job posting, and globalization. Your competition for a job can originate from local, national, or even global sources. All this competition comes to a peak at the interviewing stage that is the main event in the job search process. At this point, it all comes down to how well you present and sell yourself, and how good your interviewing skills are. You must spend time preparing and practicing your interviewing. Develop your themes and practice, practice, practice.

To stand out from others during the interview process interweave your themes while you discuss your resume, remember to describe your skills in terms from the job ad, and show physical evidence by using your portfolio. Tailor you interview to match the background of the interviewer. Do research and be ready for those difficult questions. Your body language, attitude, and tone are all being judged as you interview; let them see you at your best.

Have you identified any career actions you want to take as a result of reading this chapter? If so, please make sure to capture these ideas before you forget by recording them in the notes section at the back of the book.

ASSIGNMENTS AND DISCUSSION TOPICS

1 How do you prepare themes?
2 How long should you spend preparing your themes?
3 Where is the best place to find answers to difficult interview questions?
4 What questions do you consider most important to ask interviewers?
5 What do the body language and actions of the interviewer tell you?

REFERENCES

1. Yeager, Neal, and Hough, Lee, *Power Interviews*, John Wiley & Sons, 1998.

2. Bolles, Richard Nelson, *What Color Is Your Parachute? 2009: A Practical Manual for Job-Hunters and Career-Changers*, Ten Speed Press, 2009.

3. "Guide for Successful Job Interviews," Engineers International web site, http:// www.engineers-international.com/careersinterview.html.

4. "10 Interviewing Tips for Engineers," Engineersalary.com.

5. Beatty, Richard H., *The Interview Kit*, John Wiley & Sons, 2007.

6. Gottesman, Deb, and Mauro, Buzz, *The Interview Rehearsal Book*, Berkley Trade, 1999.

7. Kennedy, Joyce L., *Job Interviews for Dummies*, John Wiley & Sons, 2000.

8. Stein, Marky, *Fearless Interviewing: How to Win the Job by Communicating with Confidence*, McGraw-Hill, 2002.

9. Oliver, Vicky, *301 Smart Answers to Tough Interview Questions*, Sourcebooks One World, 2005.

10. Beatty, Richard, *Interviewing and Selecting High Performers*, John Wiley & Sons, 1994.

11. Veruki, Peter, *The 250 Job Interview Questions You'll Most Likely Be Asked and the Answers that Will Get You Hired*, Adams Media, 1999.

12. Allen, Jeffrey G., *The Complete Q & A Job Interview Book*, John Wiley & Sons, 2000.

13. Yeung, Rob, *Answering Tough Interview Questions for Dummies*, John Wiley & Sons, 2007.

CHAPTER 43

FOLLOWING UP, GETTING AN OFFER, AND NEGOTIATING

Your work is not done once you have completed the interview; there is still more to do. As shown in Figure 43-1, once you complete the interview you move on to the follow-up and negotiate the offer stage. In this chapter, we will discuss follow-up actions to increase your chances of getting an offer and how to negotiate for more.

SENDING THANK YOU NOTES

Within 1 or 2 days after the interview, send all the people you met a simple thank you note. The thank you note can be a handwritten note or email. If you obtained their business card or contact information during the interview, you know exactly where to send it. If you did not, as sometimes happens, you can always contact the Human Resources person and ask for the contact information.

The thank you note should be addressed directly to the person and express your gratitude for taking the time to interview with you. It is also a very good practice to mention one or two specific items you discussed during the interview. This will help them remember who you are and it sends a very positive message that you were indeed listening and paying attention. Here is a sample thank you to guide you when writing yours.

The Engineer's Career Guide. By John A. Hoschette
Copyright © 2010 John Wiley & Sons, Inc.

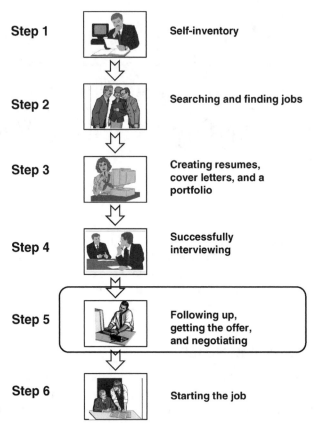

FIGURE 43-1 Job search process step 5—following up, getting an offer, and negotiating.

Mary,

Thank you so much for taking the time to interview with me, I enjoyed hearing about the company and the position. Your software development efforts on Project XYZ and the database challenges with user authentication sound exciting. I look forward to working with you and it was nice to meet another alumnus from Tufts University, I enjoyed sharing stories.

I look forward to hearing from you so I can be part of your organization.

Sincerely,
John Jones
Senior Software Engineer
Phone Number

Sample thank you note

In this example, I mentioned two specific items. One was related to the assignment and the other was a personal item. Mentioning a job-related item and a personal item are powerful ways to reconnect with the interviewer and allows you to be more personable to the interviewer, rather than just another candidate.

The timing of the thank you note is also critical. The best timing is 1 to 2 days after the interview. It sends a message that you are on top of things when responding so quickly. Some interview coaches recommend you to fill out the thank you note immediately after leaving the company, since everything is fresh in your mind, and mail it that night so that it shows up the next morning; an email thank you will also accomplish this.

WHAT TO DO IF YOU HEAR NOTHING

If the company is really serious about hiring you and desperately needs help they oftentimes give you an offer in writing before you leave, or tell you they want to hire you and should have an offer out within a day. They may even ask if they can email you the offer when it is ready.

Getting an offer this quickly is not the norm and usually the company needs several days or even up to a couple of weeks before they can process an offer of employment to you. There are a multitude of reasons, some of which include the need to obtain upper-management approval of the selection, more candidates are being interviewed, or the hiring manager just has gotten busy with normal business and deadlines. The decision is put off until later when they have more time to make the decision. So the question becomes—what to do if you hear nothing and how long should you wait before contacting the company.

The answers to these questions depend on several things. If during the interview they told you it would take time and don't expect anything before so many days, then you should do nothing until the date they mentioned comes and goes without a word from them. If the date has passed and still no word, then contact them. If they did not indicate a date when they would get back to you during the interview then contact them within a week of your interview and inquire when you might expect a response.

The two people to contact are the Human Resources manager and the engineering manager to find out if they will be making you an offer. In the event you are not selected, you may ask for the reasons why. This will/may help you on the next interview.

GET THE OFFER IN WRITING

Make sure you get an offer in writing that identifies the job title, position, salary, and start date. If a company cannot put the offer in writing, then something is wrong. Putting it in writing protects you if anything goes wrong. You have legal documentation to support your case in the event things fall apart. I know of cases where companies made offers of employment to people because they thought they were going to get a huge contract. Well, the contract never came, people had already quit their jobs only to be notified before they started that there was no job, and they were unemployed. Having written documentation is your only means of recovering from a situation like this.

HOW TO NEGOTIATE MORE OUT OF THE OFFER

When you get the offer you do not have to accept it as is, especially if you are unhappy about the terms of your employment or the offer falls short of your expectations. Everything is negotiable and job offers are no exception. Contact the person who is making the offer and start to go through it in detail. Discuss the position, salary, benefits, and start date.

As you go through the offer and you agree with each particular item, let the hiring manager know the areas you are pleased with. However don't stop there. If there is something you want to ask more for then discuss this item further. Explore the options for getting more.

If you are disappointed with the salary, ask if this is negotiable. Normally, there is a salary range that a hiring manager is allowed to offer and naturally all good managers do not offer the highest salary to start with. State you would like to request an increase, counter their offer with the salary you were expecting. Ask if they can come up with any higher offer to meet your needs. Stress that this is a hard issue for you which you want to resolve before accepting the offer. The hiring manager may go back and try to get you more. Or they may say take it or leave it. If they have no more they can offer you, then discuss the possibility of potential salary adjustments within the first year. Some hiring manager's hands are tied and they cannot offer any more than was approved. However, they have the means to adjust your salary upward after you have been with the company for a period of time. In this case you may be able to get the salary you wanted but it will take a little longer.

Another option is to request if they can offer you a signing bonus. When companies desperately need people they often offer a one-time signing bonus to entice people to come to work for them.

Next option to consider is regional salary cost of living adjustments. Companies know the cost of living in their area as compared to the area you are presently in. If the new job is causing you to relocate to a higher cost of living area, the company might offer a slightly higher salary to help keep your standard of living up as incentive to make the move.

Negotiate for title or rank in the company. If the new company asks if you want to be called a system engineer or electronics engineer, ask which one has a higher salary range. Select the title with the higher salary. Remember, the company is probably going to offer you a higher salary based on including next year's raise already in it. So do not expect a raise anytime soon once you are in the company.

If the salary offer is too low and they are not going to offer you any more after negotiating, you can feel assured that at least you tried to obtain the most you could; its your decision at that point to accept it or turn it down.

Another item up for negotiation is vacation benefits. If you are enjoying 3 or 4 weeks vacation a year, then accepting a new job where you only get 2 weeks vacation can be tough. You can discuss this with the hiring supervisor and see if you can have an exception to this rule. You may not be able to get your full 4 weeks but maybe they can offer more than the standard 2 weeks.

As long as you negotiate your items of concern in good faith and discuss your concerns in a professional and nonthreatening manner, you will be fine. Remember, it never hurts to ask and after all is said and done, you are no worse off if they say no to your request; the job offer still holds.

There is one more thing to consider before you finally accept the new offer. Would your present company be willing to make a counteroffer [1]? This negotiation option is also available to you. Your boss may ask you to remain with the company and make a better offer. In advance of your decision to try a counteroffer out of your present company, you should ask yourself if you receive a counteroffer, does it address all your concerns and will it move you toward your career goals? Would your relationship with your present company be damaged because you had wanted to leave? Would your employer now be suspicious of your loyalty if you stayed? Thinking through these questions and answering them in advance provides a valuable exercise that can prevent you from being caught offguard and enable you to decide in advance if you would accept or reject a possible counteroffer.

Your final step to finding a new job is to sign the job offer. Make a copy for yourself and return the original to the company. Congratulations, you now have a new job.

Keep all your information to yourself and do not talk to anyone at your present company about your new opportunity during the negotiation process.

RESIGNATION LETTERS

After you have signed your new employment agreement and have a starting date, it is time to turn in your resignation letter. Resignation letters are short and to the point—you are leaving [2,3]. Resignation letters are not for blaming or explaining all your thoughts. Here is a sample resignation letter to help you get started with yours.

Dear Mr./Ms. Manager:

This is my notification that I am resigning from (company name) as (title). (Date) will be my last day of employment. I have accepted a new position outside the company.

I appreciate the opportunities I have been given here, and wish you much success in the future.

Sincerely,
(your signature and date it)
Your Name
cc: (names of those being copied on the letter)

Sample resignation letter

The best professional action upon your part is to give your present company a minimum of two weeks notice prior to your last day. This will give them time to find a replacement and not give the impression you are leaving them high and dry.

LEAVING YOUR OLD JOB ON A GOOD NOTE

Once you have turned in your resignation letter, the word will spread quickly and everyone will want to know why you are quitting. You have to behave in your best professional manner and your goal is to leave a positive image of yourself; this will help you retain important professional contacts that could be helpful in the future [3]. Your resignation letter and how you conduct yourself from the time of your official announcement, to the time you leave, will form lasting impressions.

Be prepared for emotional reactions. Your employer may be surprised by your announcement. They may react emotionally or even take your leaving personally. Be cooperative and cheerful with your supervisor and coworkers. These people may experience anxiety concerning your unfinished projects and the sudden need to train someone to replace you. Try to anticipate these situations and calm their fears with positive communication as well as your cooperative nature and desire to help with the transition.

If you are up to it, you can also volunteer to train your replacement. In your final days, if you are comfortable with it, you might share how people can contact you once you leave in case there is anything they may need assistance with.

Here is a list of things to do before you walk out the door for the last time:

1. Go to Human Resources and find out how to extend your medical and life insurance benefits to cover you when you are in-between jobs.
2. Start taking home a box of your items each day so the last day you have very little to remove.

3. Check to make sure your vacation is going to be paid in your last paycheck.

4. Find out when your last paycheck will be coming.

5. Work until the end of the pay period so you get a full pay check.

6. Collect any magazines or books that you have loaned to other people.

7. Return all material that is charged out to you: computers, cell phones, and so on.

8. Make a list of all the key people you know, their phone numbers, and email addresses.

9. For those coworkers you want to stay in touch with provide you new contact information.

10. Discuss with Human Resources how to best handle your 401k, savings plan, or retirement pension plan. Record all the points of contacts for these should you need to contact them when you start at the new company.

11. The last day send an email to all your good friends at work saying goodbye.

12. Stop by the boss' office one last time and say goodbye. Shake hands, thank them, and wish them good luck.

13. Make a list of key business associates outside of work and let them know you are taking a new position and how to get in touch with you in the future.

If your coworkers host a going-away party, you should be the gracious recipient thanking everyone for all their help over the years and wishing them well. If you are asked to talk, you should be positive and upbeat and mention all you have learned from them and how much you enjoyed working with them. And if they give you a going-away gift you should send a thank note so that they can post it for all to see.

▶ *Career Tip.* Leaving your old company on a positive note is great for your career.

SUMMARY

Your work is not done once you have completed the interview there is still more to do. Within 1 or 2 days after the interview, send all the people you met a simple thank you note. If you have not heard from the company within a week (5 working days) from your interview, call them to see what is happening. All offers are negotiable and you can negotiate for more if you are unhappy with the offer. Get the offer in writing. After you accept it, then turn in your resignation letter. Resignation letters are very short and to the point. Give at least two weeks notice and leave on a good note. Be very professional and leave your company on a positive note. Don't burn the bridges behind you. Thank people for the opportunity to work with them and wish them good luck in the future.

Have you identified any career actions you want to take as a result of reading this chapter? If so, please make sure to capture these ideas before you forget by recording them in the notes section at the back of the book.

ASSIGNMENTS AND DISCUSSION TOPICS

1 Why send thank you notes to people?
2 Why is it important to get the offer in writing?
3 What terms of the job offer are negotiable?
4 Explain how you leave on a good note.

REFERENCES

1. Barkdull, Larry, *How to Write the Perfect Resignation Letter*, WriteExpress, http://www.writeexpress.com/resignation-letter.html.
2. About.com Tech Careers, http://jobsearchtech.about.com/od/resumesandletters/a/letrofresign.htm.
3. Rudloff, Alex, Emurse web site, 2007, http://www.emurse.com/blog/2007/05/23/sample-resignation-letters/.

CHAPTER 44

RELOCATING AND STARTING THE NEW JOB
TIPS TO MAKE IT EASIER

Congratulations, you are now ready to make your move to the new company. As shown in Figure 44-1, you are ready to relocate and start the new job. In this chapter, we will share proven tips with you that should make your move easier.

I would like to share some of my experiences that I have learned after six relocations over the years. I have moved from the Midwest to the East and West coast. I have moved my spouse who also had to re-establish her career as well as my children who experienced loss of friends and family and relocation to new schools.

I also know how exciting it can be to obtain positions in different companies and how it enhanced my career. New experiences of moving around the United States proved to be surprisingly wonderful experiences for the entire family.

There are so many things to think about and prepare for when job relocations are necessary. Hopefully you can learn from my experiences and make your move less stressful and easier on your family.

PREPARING, PLANNING, AND MAKING THE MOVE

There are so many things to think about and prepare for when job relocations are necessary. For some, the relocation effort is one that is a career choice and a welcomed position to be in; for others, you may have been faced with a job loss and are possibly moving reluctantly. In either case, it is going to take a significant amount of preparing and planning to have your move come off successfully.

This takes a great deal of careful work and coordination. Possibly the largest of tasks you will take on to coordinate is the move. It's time consuming

The Engineer's Career Guide. By John A. Hoschette
Copyright © 2010 John Wiley & Sons, Inc.

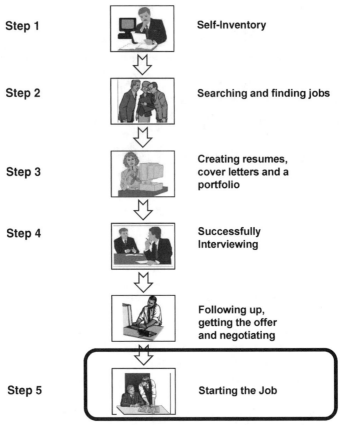

FIGURE 44-1   Job search process step 6—starting the job.

and can be stressful just because of the sheer nature of the multiple items associated with relocation. The first step to planning your move is to understand your relocation package.

UNDERSTANDING YOUR RELOCATION PACKAGE

Make sure you understand the terms and conditions of your relocation package from your new employer. Don't be afraid to ask questions and negotiate for what you need. Review your contract with your spouse or partner before you sign-off—make sure you have thought through the details of the contract and identify the impact and limitations to your move. Here are some very important terms and conditions normally contained in the relocation package that you should consider.

1. What is the total budget you have for your move? The budget should include costs for paid trips to find a new home, real estate fees at both

ends, days allowed in a hotel and the daily per diem. It also should include a budget for moving your household goods; often this has a weight limit in conjunction with it. What budget is allowed for your cars to be moved?

2. What is the length of job commitment?
3. Will they pay real estate fees for the sale of your home?
4. Do they plus-up your salary to cover any extra taxes you will incur?
5. How long do you have to purchase a new home?
6. Will the company buy your old home if you cannot sell it?
7. Will the company help your spouse find a new job?
8. Do they have a preferred real estate company to work with?
9. Is car rental part of the relocation package?

If you have any uncertainties contact your Human Resources department immediately. Once you understand the terms and conditions of the relocation package you should be able to generate your plan.

GENERATING A PLAN

Start your plan by preparing a timeline of tasks to accomplish for your relocation. This timeline will keep you focused and will enable you to include others in the planning and provide a good look at the big picture. First, begin by sitting down with a calendar and lay out the events that will occur once you accept the position. Lay in the key activities. Here is a list of activities to help you create your timeline.

1. Meeting with realtor and listing your home
2. Last day of school for kids; meeting with school counselors
3. Movers and packing up current house
4. Sale and closing on home
5. House hunting at new location
6. Packing and final day at old house
7. Travel to new location—two visits (usually for house hunting and area visits)
8. Visit school and local areas
9. House hunting time at new location
10. First day of new job
11. First two weeks: lodging, meals, car rental, and so on.
12. Purchase and close on new home
13. Moving van arrives at new house
14. Unpacking and first night in new home

Make use of all the resources available to you when generating your plan: company Human Resources representative, babysitters, relatives, realtors, friends, whomever you think will be available when you need help. This will help you to quickly draw upon those resources when you need them. Lining up the resources ahead of time will also provide comfort to you by knowing that you have many people to assist you in your transition. Once you layout all these events on your timeline you will quickly see there is no way you can accomplish all this and be settled in your new job and home all within a few weeks. Here are some tips to help you with this.

- Have your family stay in the old house and live there until you sell it. This will take the pressure off you having to sell it so quickly. Also, the old home will show better for sale purposes with all the furniture in it. Empty homes are harder to sell. Start your new job and live in an apartment until the family arrives. This will save money just in case your old house does not sell immediately and also help if you need the proceeds from the sale of the old house as downpayment for the new house.

- Schedule the house hunting trip after you have your house up for sale or even wait until after you have an offer. Knowing what you can get for your old house will help you determine what you can afford at the new location. Include your kids in the second house hunting and make a mini-vacation out of it. Do something fun for the kids so that they get excited about the new location.

- Set up meetings with realtors at your new job location. Usually realtors are referred through company Human Resource departments that are taking care of your move to your new job. However, you can do a great deal of checking online, or as my wife did, she worked with the realtor we had our existing home for sale with, to refer her to a realtor specialist in the new job location. Allow yourselves at least two house hunting visits. One where you get to shop and then another trip to revisit everything once you have it narrowed down to some must sees and visit homes together. Of course, you have many things to consider: location, price, schools, proximity to new job and spouse's job, as well as medical facilities, airports, and so on. Define your goals for the right area to settle in.

- Once you start work, ask your boss if you can have a little extra time off and you will make up for it later. Explain all you have going on. Normally they understand. This will free up some time for you without having to take unpaid time off since usually your vacation will not start until after you have been at the new company for a minimum of 6 months.

SPOUSE'S CAREER

Oftentimes the company you are working for can provide great information, leads, and direction to the spouse looking for a job too, so seek out advice if the company you are joining provides such services. Otherwise, help your spouse get an updated resume, find headhunters that will help in the job search, and be supportive. It is a difficult time for the spouse as well.

TIPS TO MAKE IT EASIER FOR THE KIDS

Review area schools and meet with counselors ahead of time to make sure you have all the information you need to share with your children. Pick a time when you can all visit schools. Some say that it is good to move at the end of the school year, but I have found that if they can jump in while school is in session it makes for an easier transition. Moving during the summer makes it hard for kids to meet schoolmates and they start off sometimes missing out on making friends faster. This, of course, does depend on your children's ages. If your children are school-age, you will also need to bring or send a copy of your children's transcript/grades. Remember, you will need to provide proof of your children's medical shots showing they are all up-to-date as part of the school registration.

FAMILY DISCUSSIONS

If your children are old enough, be sure and have family meetings and one-on-ones with your children. Find out what they are thinking and feeling. Perhaps even seek out counseling because it is an emotional time for everyone and lots of new things are taking place. Children lose friends and confidence and can become fearful of the unknown during this time.

CHECKLISTS

The best tip we found was creating a checklist to help us keep everything straight. Here is a sample checklist we created that you can use to get yours started.

1. Generate a contact list for
   a. Bank accounts
   b. Medical clinic and doctor
   c. Drug stores
   d. Dental offices
   e. Eye doctors
   f. Kids' school
   g. Insurance
   h. Gas, electric, and water companies
   i. Cable and television
   j. Garbage

    k. Realtors

    l. Real estate closing company

    m. Old Human Resources

    n. New Human Resources

    o. New boss—cell and email

    p. All relatives

    q. All neighbors and kids' friends

    r. Insurance carrier information for car, home, and so on.

    s. Car rentals

    t. Moving company

    u. Driver of moving van (private cell phone)

    v. Car transport company

    w. Hotels

    x. Airlines

    y. Credit cards (in case you lose one)

2. US Post Office address change

3. Last day of service for utilities on old home

4. Closing on old home: date, time, and documents

5. Copy of kids' medical shots

6. Copies of last year's tax returns

7. Copy of kids' school records

8. Passports

9. Suitcase of personal items (keep with you during the move)

10. Prescriptions

11. Copy of relocation package

12. Maps

13. List of key websites and passwords

This is not an all-encompassing list but should be a good start to help you generate your checklists.

THE INFORMATION VAULT

One action we took during the move was to get a large briefcase that could hold a three-ring binder, notebook, and our laptop. We kept all our contact and relocation information in the briefcase. People would call or things would happen and we would need to instantly access information. Without it we would have been lost. Once all your belongings are packed up and on the truck you will have no access to this information. Without it many things will grind to a halt.

FINAL GOODBYES

A group party for you and your friends and family is good and I encourage some type of party for your children if they are of an age where they understand. This not only lets everyone put some closure on the move, but also make sure that the friends you have made over the years get a chance to say goodbye. It also makes a wonderful opportunity to invite people for future visits and express how welcomed and desired they would be.

You can always stay in touch via emails, phone calls and now, of course, we have such great programs such as SKYPE that allows you to see and talk to one another live on your computer. My wife and I have cultivated many friends from living in different states throughout my career, our children also remain friends to this day with out-of-state friends.

We recently just threw a surprise birthday party for our son and invited his high school and college buddies from other states where we lived; it was amazing how their friendships sustained also. Pictures are a wonderful way of remembering everyone. Finally, don't forget to send thank you notes to those special friends and coworkers who have helped you with all your transition preparations, and for those who have been there to support you.

GETTING SETTLED AT THE OTHER END

Your biggest decision once you arrive at the new location is finding a place to live. Your relocation coordinator should have a real estate agent all set up to help you. The first big decision is finding the best communities to live . Our priorities for selecting communities were school district rating, housing quality, and commute to work. If you have your priorities established your real estate agent can quickly narrow down the number of communities to start looking for a home.

After you have narrowed down your final choices, one excellent thing to do is drive to the community early in morning and try the commute during the morning rush hour, and again during the afternoon rush hour. See if it's doable.

On your house hunting visits, if you have time, visit the area visitors bureau or go online and get information so that you can be prepared and know about the different resources you may need, such as the library, churches, sporting arenas, and favorite stores. This will help your family get a quicker feeling of being connected with family interests. This will also assist in furthering your feelings of getting established in a new area more confidently. Knowing perhaps the same stores are available for shopping, add to comfort of the move.

Also, my wife spent some time making a visit to the health clinic my company offered for healthcare, mainly because we had children of school-age for whom she felt it was important to get established and find out their services right away.

Make the effort to visit the schools with your kids. Take them to school personally if necessary, and encourage the kids to invite classmates to your house so that you can meet the students and keep your children moving in the right direction of making new friends.

Take up the offers from your new coworkers who invite you over for dinners so you can begin to make your spouse and children feel comfortable in their new city. Ask for your spouse to meet you for lunches, if possible; it will help them to get out and visit restaurants and have special time devoted to the marriage.

Work with your spouse and be encouraging if they want to find employment. Help with contacts, preparing resumes, or anything else for your individual situation. Perhaps it is just putting them in contact with someone who would be helpful to know in the area, or a possible volunteer opportunity you heard about at work.

Finally, have family meetings as often as needed to see how each member of the family is doing with the transition. Let everyone share about their day. Some may adjust better than others. Find out who needs help. Family discussions are absolutely necessary.

If necessary, I highly recommend seeking counsel to facilitate more difficult problems. Our daughter found she had a hard time adjusting to her new school and was more comfortable in a private school. Once we listened to her needs and made the change, she was on the "A" honor roll! Our son wanted to play a major sport and with our support, ended up on a varsity team in school; not an easy task when changing mid-year during high school. Listening, working with school counselors and staff, made it possible. Making appropriate changes to meet the needs of your family is very key.

STARTING THE NEW JOB

Starting a new job is going to require building a new network of business contacts and learning all the new company rules and regulations. To help build your network spend a few extra minutes socializing with people and let them know you are new to the company. It is amazing how this little fact can open people up and make quite a world of difference. Another good tip is to ask your boss who you should call on when you have simple questions about the company and its rules and regulations. Ask for a "buddy" who is a person you can go to when you have questions. A good buddy can save you hours of work and frustration. Finally, try socializing with your coworkers after-hours. You will be surprised at the number of people who have also relocated for their job and have gone through this experience. They understand what you are going through and often are willing to socialize after work and share all they have learned about the area.

SUMMARY

There are so many things to think about and prepare for when job relocations are necessary. There is a great deal of careful work and coordination; possibly the largest of tasks you will take on to coordinate. It's time consuming and can be stressful just because of the sheer nature of the multiple items associated with relocation. The first step to planning your move is to understand your relocation package. If you have any uncertainties contact your Human Resources department immediately.

Start your relocation plan by preparing a timeline on tasks to accomplish for your relocation. This timeline will keep you focused and will enable you to include others in the planning. Make use of all the resources available to you when generating your plan. Consider the impact on your spouse's career and be sensitive to your children's feelings. Make use of checklists and assemble your information vault. Use a calendar and develop a timeline for all the key events.

It is going to take time to get settled and feel comfortable in your new home and job. You can reduce the amount of time by planning family events and getting involved in your kids' school activities. Keep in touch with old friends and make new ones. The move to a new job can be positive for both you and your family, but is going to take work, understanding, and time.

Have you identified any career actions you want to take as a result of reading this chapter? If so, please make sure to capture these ideas before you forget by recording them in the notes section at the back of the book.

ASSIGNMENTS AND DISCUSSION TOPICS

1   Why are preparing and making a plan so important?
2   Do you think checklists are important?
3   Who should you contact when you have questions about your relocation package?
4   What does it mean to plus-up your salary to cover any additional taxes?

PART 7

FURTHER EDUCATION AND LIFELONG LEARNING

CHAPTER 45

HOW TO GET AN EDUCATION FROM YOUR COMPANY

From the previous chapters one can see that higher education is an absolute must for career development. However, obtaining this is not a simple undertaking, nor does it guarantee career advancement. You must make absolutely sure that the large amount of time you will be investing in furthering your education will pay off. This chapter identifies how to use and get the most from your company's educational system. Helpful guidelines are discussed to ensure that the type of further education you choose will be appreciated by your company and will result in career advancement. Choosing the correct type of further education for you benefits everyone.

WHY FURTHER EDUCATION IS A WIN–WIN SITUATION

Most companies gladly support more education for their employees and there are several reasons for this. First, an employee who has received additional education is often exposed to new and improved methods. These new methods may allow them to solve problems more quickly and easily. Applying these new methods to the job can result in cost savings for the company or better products. The cost savings can be significant and easily pay back the employer's educational costs. Education usually has a good return on investment for the company. Second, educational expenses can be recouped by the company since they are tax deductible.

A third reason companies support further education is shared knowledge. Oftentimes, a company will send one employee to a class. When the employee returns they must share with their fellow employees the new knowledge they have gained. In turn, the company can effectively educate many employees through the training of only one person.

Another reason companies support employees returning for further education is the contacts the student can make at the classes (Figure 45-1). Often classes offered at technical symposiums are attended by people with similar backgrounds looking for solutions to their problems. Your classmates may even be customers or competitors. By attending technical symposiums and making contacts, an employee may learn what the competition is doing or encounter future customers.

Finally, most large multimillion-dollar engineering contracts require a company to have university professors on staff acting in an advisory role. These consulting professors provide the high level technical expertise often required to successfully complete the project. Therefore, sending an employee to take courses and establish one or more contacts while at the university is extremely important to the company.

For all these reasons companies gladly offer further education or tuition reimbursement plans for their employees. Both the company and the employee will benefit.

▶ *Career Tip.* Further education is a win–win situation. You win and the company wins!

FIGURE 45-1 Further education is a win–win situation.

Your first challenge once you have decided to return for further education is to find out how your company's tuition reimbursement plan works. The best starting point in answering this question is your supervisor. Contact your supervisor and discuss your ideas with them on returning for further education. Watch for their reaction. If they think it is a good idea, you now have one of the most important people in the company supporting you. If they object to you returning, find out why. Maybe their department has no budget for you. In this case, don't panic; try to get them to budget for your education in the next planning cycle. They may also say no because the timing is bad and they need you at work. If this is the case, discuss with them when the timing would be better and start your plans there. Make sure they know you will not let them down while you are returning for your education.

Your supervisor may object for still another reason—insecurity. They may feel threatened by your returning to school. If this is the case, go slowly. Your supervisor will probably not tell you this directly, but continue to make up a thousand and one excuses why you cannot take the classes. If you sense this is the case, proceed cautiously. This is a major roadblock. Give your supervisor time to adjust to the idea. Maybe this is what they need. You might also point out that by giving you the opportunity to pursue further education, they are really developing your career and that is part of the supervisor's job. Developing you as a junior employee is part of a supervisor's job and they will look good doing so. It is considered a feather in their cap. If none of this works you should consider a department move. It would be very difficult without support from your supervisor while you are attending school.

As difficult as it may sound, you may have to change supervisors or even departments. If you should change supervisors, search out a supervisor who will support you with encouragement and funding. In any case, just remember:

▶ *Career Tip.* Do not take no for an answer about further education. It's your career!

Often your supervisor may not be up on all the latest rules and regulations regarding your company's tuition reimbursement program. Your next contact should be Human Resources or Personnel department. The Human Resources department is usually responsible for administering the employee tuition reimbursement program. Meet with Human Resources as soon as you can and discuss the details of your company's plan. Learn everything you can about the program. Here is a sample list of the things you might inquire about.

1. Which employees qualify for the program? Do I qualify?
2. Is there a limit on tuition reimbursement costs? Do they cover tuition, books, travel, etc?

3. Am I required to pay for classes in advance? What conditions must I meet to qualify for reimbursement?
4. What courses and majors apply for tuition reimbursement?
5. Are there restrictions on the universities or schools I can attend?

▶ *Career Tip.* While you are at the Human Resources department make sure you get the names of coworkers who are also using, or have used, the program. Contact these people immediately and meet with them.

The company cafeteria during lunch is a very good place to do this. Meet with as many people as possible and discuss their experiences. Any helpful direction or shortcuts they have learned could save you hours later on. By contacting other people who have used the program you may find someone who is returning for the same courses you plan on taking. Their help could be valuable in setting up your program, or helping you with classes they have already completed. These contacts within the company allow you to draw upon them for support during tough times.

Normally, the company tuition reimbursement program requires the completion of several forms. These forms will require information about the courses, expenses, and explanations of how the courses relate to your work. (Courses must be related to your work in order for the company to claim the expense as tax deductible.) Forms can scare people away from ever getting started. Don't let them intimidate you. In fact, if you handle the situation correctly, filling out the forms and getting approval can actually help advance your career.

The tuition reimbursement forms will usually require several levels of management to sign and approve. The reason for this is to make sure that management is aware of the tuition costs that they, or you, will be accruing and be able to budget for them. Don't ask anyone to get your forms approved for you. Do it yourself by scheduling time for a one-on-one meeting with each manager who must sign. This gives you an excellent opportunity to meet your managers who will most likely be the same people that approve your raises and promotions.

▶ *Career Tip.* One-on-one meetings with your manager to discuss your further education are a great career move.

A one-on-one meeting with the managers who must approve your forms is an excellent way to get visibility. Go to these meetings well-prepared and ready to discuss your further education: which courses you are planning to take and the benefits you see coming from the courses. Ask them if they have taken any similar courses or you might also inquire who the manager recommends you should contact when you need help.

If you are required to obtain the signature and approval of upper management, use the time wisely. Be sure to point out all the help your supervisor has been, how the class will benefit the company, how you plan on sharing with others in the company, and what you will learn. Remember, after each person has signed and given their approval, shake their hand and thank them. They, too, will be investing a significant amount of money in your education, and they deserve a "thank you" for it.

DIFFERENT TYPES OF CONTINUING EDUCATION (PROS AND CONS)

Choosing the best type of continuing education for you will depend upon your long-term career goals. There are various forms and types of education that you can pursue. Advanced degree programs leading to a master's or PhD are offered through most major universities in the United States. The starting point is contacting your local university and requesting their extension or night school programs. Programs usually exist for advanced degree programs in both engineering and business.

Another avenue available to receive graduate level education is through remote courses. Most universities use net meeting or video feeds to transmit their courses throughout the local area and even across the country.

▶ *Career Tip.* Check with your company to see if they have a remote hookup available to receive classes. This is an excellent way to take graduate courses.

Further noncredited education is available through seminars offered by experts throughout the country at various times during the year. The best sources for finding out what types of seminars are being offered are the company library and trade journals. These seminars are usually intense week-long seminars taught by experts from universities or industry. Their purpose is to provide the student with the opportunity to quickly "come up to speed" on the latest technological advances.

Another noncredited method of obtaining further education is by attending symposiums offered by the various engineering or business societies. By joining an engineering society, you will automatically be sent information about upcoming symposiums and training seminars. These symposiums provide excellent reviews of fundamentals as well as knowledge on the breakthroughs. They offer the opportunity for you to meet with people of a similar background and exchange ideas. You can find out what other people across the nation are doing to solve problems. They also provide a very good means for new job contacts. During these symposiums, many companies post their job openings. You can quickly compare what other people are making and which companies are growing and expanding. This is an excellent means to find out what other career opportunities are available for you.

Finally, some companies offer in-house training or courses given by experts from within the company. This training usually consists of short courses taught at lunch time or after-hours. Contact your supervisor or Human Resources department to find out if any internal courses are planned. Instructors of in-house courses make excellent contacts to have within the company. There is no better way to meet the company's experts than in a non-threatening one-on-one after-hours course. After taking the course, these in-house experts become valuable resources for you to call upon in times of trouble. They also become another vote of confidence for you. When engineering teams are being staffed, management often calls upon their experts to make recommendations for staffing the new, advanced engineering projects. If the expert has already met you in their class and knows your ambition and capabilities, it is highly likely that they will vote in your favor.

▶ *Career Tip.* If you are a company-recognized expert give a class in your area of expertise!

The best courses for in-house training are those that teach you how to use your company's resources to get the job done quicker, easier, and hopefully, with higher quality. Good examples of these courses are those in computer training. They may include courses in such things as training on spreadsheets, email, computer-aided design (CAD), computer-aided manufacturing (CAM), and computer-aided engineering (CAE).

Another excellent idea is to participate in company "Brown Bags." Brown bag meetings refer to lunch time meetings where people gather to discuss and learn, and exchange ideas while eating their lunch. Oftentimes, everyone brings their brown bag lunch hence, where the name came from. These informal meetings can serve a multitude of purposes. The Brown Bags can set up a lecture series with guest speakers. Usually, a department will organize brown bag meetings to have guests who speak about subjects important to the group. They can also be used to help train people in a new process or tool. I organized brown bag meetings so I could learn about other parts of the company. I had guest speakers from other groups and divisions come and present what they were doing. It was a great educational experience as well as an excellent means to network. The skills I learned in arranging the brown bags carried over into my own work meetings and helped me significantly enhance my ability to organize and conduct meetings.

In each area of further education you choose, each type has advantages and disadvantages of which you must be aware. First, all further education is not a guarantee of career advancement. It only becomes a guarantee of advancement when it is combined with excellent job performance. Obtaining further education and performing poorly on the job is of no benefit.

▶ *Career Tip.* Further education is a basic building block to career advancement. Take advantage of the opportunity!

Second, further education qualifies you for higher level job assignments. If the job assignments are not available in the company at the moment you complete your degree, you may be stuck in the same job as before. Or you may have to change departments or even companies to receive the advancement you deserve.

Third, you may appear to the group you are working in as the same old person. They may give you the attitude of "No Big Deal." This can be very discouraging, so you need to recognize it and deal with it appropriately. Finally, coworkers may be jealous as you finally complete the degree and management is giving you more attention. Diplomacy and tact is required on your part when this starts happening.

Returning to night school at a university for a credited degree has its advantages and disadvantages. Your first advantage is better career opportunities at your present position. If in the future you change jobs, the degree has the advantage of being recognized by other companies. The degree stays with you each time you change employment. The disadvantage is that you often have to travel great distances and attend classes at night. Returning for a degree or certificate from a university is also a multiyear commitment, which can be difficult. Most people lose interest or have other commitments (i.e., family and children) that interfere.

Remote or online courses offer you a chance to attend classes from various universities around the country that you might not normally get the chance to attend. In addition, webinars, remote, satellite, and Internet courses may have a limited selection and may not lead to a degree. However, they can still be of benefit to you. Again, evaluate your specific education goals.

Taking after-hours courses offered by the company allows you to meet the technical experts in the company on a one-on-one basis. In addition, you may meet other coworkers in the company who are probably working on problems similar to yours. This is a great way to network in the company. After-hours company-sponsored courses are not usually accredited or recognized by universities. Therefore, they do not lead to a degree and their value quickly diminishes once you leave the company.

For successful career development you will probably use all of the above-mentioned forms of further education at some time in your career. The key is to balance the type of education you choose with your career objectives. If you aspire to become CEO of the company, then attending night school at an accredited university for an advanced degree is what you must do. If you wish only to update your technical background, courses offered at symposiums are your best bet. Remember, further education doesn't hurt and it always serves to broaden a person's capability.

▶ *Career Tip.* The broader your background, and the more knowledgeable you are, the better the chance for career advancement and job security.

After exploring the opportunities for further education, you may find people who are still hesitant. People often use the excuse of "I just don't know." In response to this, I remind people of the days when they were attending college, struggling to make it through one class after another. In particular, I remind them of the high costs, no medical benefits, no vacation pay, little help from others, and the uncertainty about getting a job when they finish. I ask them if I could show them a way to go to college and, at the same time, receive a full salary, paid tuition, full medical benefits, and a paid vacation would they be interested? Not only that, but if they had difficulties with a class, resources exist in the company to help and fellow workers who have taken the class may be able to assist you. And when they completed the program they would have a greater chance of a raise or promotion, would they be interested? I also mention that I had supervisors pay for my books, mileage, and parking fees since the class directly related to my work at the time. These are all benefits a company tuition plan offers. Visit with your department manager and find out what is available at your present company.

LEVERAGING FURTHER EDUCATION
FOR CAREER ADVANCEMENT

Many engineers return to school for an advanced degree in the hopes that when they receive the degree they will also get promoted by their company. The truth of the matter is, most often the engineer is not immediately promoted and it may take years before it happens. Engineers often become discouraged and leave the company in hopes of better opportunities.

I have participated in several committees that were assigned to study this problem and here is what we found. The engineers returned for an advanced degree but rarely showed any evidence of their growth at work. They continued to perform their work day after day, and the manager never saw any real improvement or benefit from the schooling. The people around the engineer never saw any changes, so everyone considered that it's no big deal. The problem was basically, the engineers never took the time to show people all they were studying and the benefits to their group.

If you decide to return to school, you must be proactive in sharing what you are learning with your manager and others around you. You must market yourself as growing and acquiring knowledge. DO NOT wait until you complete the degree before you start this activity; then it is too late. Sharing what you have learned should be something you do after each course.

Since it takes almost a year to get promoted, you should start marketing yourself heavily about one year before you complete the degree. Also, you should be having career discussions with your supervisor about what assignments you will be handling once you have completed your degree. Your rationale for this is that, with your advanced degree, you will be much more capable of taking on more challenging and leading assignments. Naturally all this is setting the stage for your eventual promotion. In my experience, engineers who have done this advance marketing of themselves usually ended up getting promoted before or at the time they completed their program and received the degree.

▶ *Career Tip.* Having career discussions with your manager while working toward an advanced degree is a good career move. This will ensure at a promotion will follow your completion of the degree program.

SUMMARY

In summary, overcoming the educational barriers in your company is a must for career advancement. The best way to hurdle the educational barriers is to use your company's tuition reimbursement. Continuing education is a win situation for you and a win situation for your company. The starting place to return for further education is your supervisor or Human Resources department. You must map out an educational plan that best fits your career objectives. Further education can be obtained through night schools at your local university or community college, symposiums offered by engineering societies, or company-sponsored courses. Each type of continuing education has its advantages as well as its disadvantages.

Have you identified any career actions you want to take as a result of reading this chapter? If so, please make sure to capture these ideas before you forget by recording them in the notes section at the back of the book.

ASSIGNMENTS AND DISCUSSION TOPICS

1   What departments in your company have the funding for further education?
2   Identify others in the company who have returned for further education and meet with them.
3   What type of further education is best for you?
4   Which universities are located close by and what courses do they offer?
5   Are online courses a viable option to taking courses at your company?
6   Does your company offer any internal after-hours courses?

CHAPTER 46

GETTING A MASTER'S DEGREE
SO WORTH IT!

Before we start this chapter let's communicate a very clear and concise answer to the question, Getting a Master's Degree, Is It Worth It?

Yes!

Now that you know the answer to that question let's expand on this answer by exploring the benefits and options available to you.

THE BENEFITS TO GETTING A MASTER'S DEGREE

Let's get immediately to the bottom line regarding a master's degree and how it relates to salary. The National Association of Colleges and Employers data shows a person holding a master's degree makes about 10 to 15% more than the same engineers with only a bachelor's degree [1]. For a newly graduated engineer with a master's degree this amounts to about $7,000 to $9,000 more a year based on engineering salary surveys conducted by the Institute of Electrical and Electronic Engineers (IEEE), *Machine Design Magazine*, and others [2–5]. If you are interested in how your salary compares to others in your field check the websites www.payscale.com or www.hitechsalary.com. Doing an Internet search on "engineering salaries" is another great source.

Let's next look at what a 10% to 15% difference in salary means over the lifetime of an engineer. If you assume the engineer follows the average growth during a career of 40 years, the difference in earning power for obtaining a master's degree can amount up to earning $2 to $3 million more over a career. Since a master's degree is often a ticket to rapid advancement and more opportunity for a leadership position that pays more the figure of $2 to $3 million is actually a low estimate. What could you do with a couple extra million dollars?

The Engineer's Career Guide. By John A. Hoschette
Copyright © 2010 John Wiley & Sons, Inc.

Second, many companies in industry consider a master's degree as minimum qualification to do meaningful technical work. Do you want to be on the leading edge of development and design? Or do you want to be delegated the more mundane and less challenging engineering tasks? The choice is yours.

TYPES OF MASTER'S DEGREES

Not all master's degrees are the same. As shown in Figure 46-1, a person returning for a master's degree has several options. To generalize, there are three basic categories an engineer can pursue when returning for a master's degree. These include:

1. Master's degree in business.
2. Master's degree in technology management.
3. Master's degree in engineering.

A master's in business is usually obtained through the school of business in most universities. This type of master's degree has nothing to do with engineering but is pure business in content. The common types of degrees in this area are a master's in finance, business administration, accounting, or economics. The master's in finance is primarily focused on running the financial aspects of the corporation and generally people who obtain this degree work in the banking/investment industry. A second type of master's degree in the school of business is a master's in business administration. With this type of degree you are taught the administrative aspects of a company manager. These include such things as department budgeting, personnel administration, and project management. Pursuing these degrees means leaving engineering and going into management dealing with money and personnel issues within the company. Most often you become a leader for non-engineering teams.

At the opposite end of the spectrum is returning for a master's degree in engineering. If you pursue this course, you are looking at developing your technical expertise and will remain an engineer. These types of master's degrees follow the general engineering disciplines like electrical, mechanical, chemical, or computer science. These programs are run out of the school of engineering similar to your bachelor's degree.

In the middle category is a new type of degree many universities are offering to help engineers with their careers. These degrees are referred to as master's in Technology Management, Technology Development, Technology Innovations, or Production. The focus of this new category of master's degrees is on educating a person with an engineering degree how to become a successful manager in a high-technology company. I have

Types of Master's Degrees

Master's in Business
• Master of Finance
• Master of Business Administration
• Master of Accounting
• Master of Economics

Pursuing these degrees means leaving engineering and going into management dealing with strictly money and management issues of the company.

Most often you become a leader for non-engineering teams.

Master's in Technology
• Master of Technology Management
• Master of Technology Development
• Master of Production
• Master of Technology Innovation

Pursuing these degrees means leaving engineering and going into management dealing with engineering and technology issues of the company.

Most often you become an engineering department manager responsible for leading and directing the technical team with finance and technical responsibility.

Master's in Engineering
• Master of Chemical Engineering
• Master of Software Engineering
• Master of Mechanical Engineering
• Master of Electrical Engineering

Pursuing these degrees means remaining in engineering and becoming a technical expert in your field.

Most often you become a technical staff engineer, leading the design of products. You are responsible for the technical performance of the product and do not deal with the financial or management issues.

FIGURE 46-1   Comparison of types of master's degrees.

had the opportunity to teach programs at Tufts University in Massachusetts, Santa Clara University and Stanford in California, and the University of Minnesota [6–9].

These degree programs are tightly aligned and often run through the School of Engineering versus the School of Business. The instructors are usually former engineers who have become managers of engineering teams.

The students receive training in engineering, business administration, and managing people. They are well-structured classes that prepare the students to become the future leaders of high-technology corporations.

SELECTING A MASTER'S DEGREE PROGRAM AND STARTING

The first step is to determine your long-range career path. Are you going to remain technical or switch into more of a manager role? To help you with this decision talk to engineers who are managers, like your supervisor. Next, spend some time talking to engineers who decided to remain technical. Compare the different types of tasks both do on a daily basis. Which do you find yourself more aligning with—manager or technical staff? This will be a good indication as to which is the best choice for you. Next, network with people who have recently returned for their master's degree. Ask what universities they attended and why? How did they balance life, work, and school? What made them choose the path they selected?

The second step to getting your master's degree is to contact the local universities in your area and educate yourself on the types of programs they offer. What you have to do to get into the program, how much does it cost, and what types of degrees do they offer? Meet with a university counselor and discuss your goals and plans. Use the university guidance to structure a plan leading to the degree you want.

In the same time frame you are talking to the university, check with your supervisor and the Human Resources department to determine the policies and regulations that need to be followed. What qualifications do you need to meet to get the company to pay your tuition?

Once you have completed your due diligence, you must put the wheels in motion and enroll in a program.

Most university master's degree programs start in the fall and run through a 2-year program. Therefore, the optimum time to start the process for returning for your master's degree is in the first quarter of the year. It will take 4 to 6 months to fill out all the forms and get approval for acceptance. Do not wait until the last minute to apply. If you wait until late summer to start, more than likely you will not be accepted in time and consequently have to wait until the following year to start.

▶ *Career Tip.* Most master's degree programs run on a 2-year cycle that starts in the Fall. You need to start the process of applying in early spring in order to be ready for the fall start. The sooner you get started on earning your master's degree, the quicker you finish.

You can do this. It is not as hard as one would think it would be. It will require you to make some adjustments in your life to fit everything in, but it is all doable.

▶ *Career Tip.* The only person holding you back is you!

When you finally start to attend classes and socialize with other people like yourself who are returning to school, you find it gets easier and the enthusiasm is contagious. You will be socializing with people who are going through the same trials and tribulations as you are. Use your class time to socialize and network with others. Your classmates can be excellent sources of information on how to handle, and effectively deal with, problems you may be encountering.

AVOID THE FATAL ERRORS

One of the fatal mistakes that you can make when returning for your master's degree is not to plan for the impact it will have on your family or significant other. Returning for a master's degree is going to take you away from your family for a significant amount of time in order to attend classes and complete homework assignments. This means other family members will have to make up for your absence—especially your spouse in terms of daily and weekly chores. If you do not have a plan in advance on how you are going to return to school, and get your spouse's buy-in, you are headed for trouble.

The people who I have witnessed getting in trouble are the people who thought returning for a master's degree was an excuse not to have to contribute to the daily family life. The engineer returning to school assumed they would work normal 8 to 10 hour days, attend classes after-hours, and then spend all the other weeknights and weekends completing assignments. They failed to realize, or just assumed, the spouse would do double duty since they were no longer around to help the family out. They decided to check out of the marriage for two years. Most spouses do not like to do double duty for the two years it takes to get the degree.

▶ *Career Tip.* Obtain your family members' support when returning for a master's degree. Determine your plan and actions to ensure a good life balance.

My experience when teaching in master's degree programs was that a few students would fail to involve the family in the decision and the net result was, by the time the students had earned their degree, the spouse was fed up and filed for divorce. Please don't let this happen to you. Sit down early on and involve your spouse, children, or significant other, in your plan to return to school. Make sure they understand why you are returning and the benefits and challenges that will be involved. If you get their input before you start, you will be setting yourself up for success. However, you must be the one who comes up with a plan to maintain your contributions to the family since you will be away and studying so much. Including the people in your life who will be affected will make them more supportive.

GREAT TIPS FOR MAKING IT EASIER

Here are tips that should make it easier for you:

1. Find at least two people in your company who have successfully completed the program you have selected. Network with them and discuss the classes, assignments, and problems they encountered. Human Resources is a great place to find out who these people are in the company.
2. Get copies of assignments from previous students to see what was required. The scope of the project, the final output, and grading criteria. Use these as templates and starting points for your assignments.
3. Have your contacts at work who have been through the program, review your assignments or projects to get feedback on how to improve.
4. The first night of class pass around an attendance list and obtain everyone's cell phone number, work number, and email address.
5. When assigned a team project work virtually after-hours. Have a meeting spot or conference call-in number. Discuss team member roles and responsibilities. Keep great meeting notes and distribute by email. Spend time at your first meeting just getting to know each other. Go around and have each person discuss their background and why they are returning for their degree.
6. When special assignments are due, have the team members show up extra early to run through material or presentation material.
7. Ask to see if the instructors will come to your company and make a presentation. If the instructors are willing to conduct a lunch hour brown bag or after-hours presentation, line up your boss and other workers who might be interested in the material.
8. Here are some great tips for dealing with the family:
   a. Combine your study time with the kids' study time. Sit around the kitchen table doing your homework together. My kids' grades actually improved after they saw dad studying so much.
   b. Make study time fun by giving rewards when the kids or you complete it. Finally, reward your spouse and children every time you complete a major milestone or class. One night my kids actually reminded me that my project was due and I needed to complete it so we could all go out afterward as a reward. I told them we would go roller skating as a reward and they were really looking forward to this.

    c. Take the kids to the university with you at night and study there. Let them see where you are going and what you are doing. I even found a study hall next to a gym. So when we all completed our homework, we got to play basketball as a family.

    d. Spend a date night with your spouse alone where the two of you can reconnect.

    e. Reward yourself. Determine what healthy and morally uplifting thing you can do to reward yourself. It does not have to be big and the rewards should be all along the way. Not just at the end when you complete the degree.

    f. Finally, plan something big when you receive your degree. Maybe a family vacation or weekend get-away for you and your spouse.

SUMMARY

Hopefully by this point in the chapter you are thoroughly convinced, with all things considered, getting your master's degree is well worth the effort. With the proper planning and guidance you can do this.

If completed successfully, it will pay large dividends for the remainder of your career. There are multiple choices when selecting what type of master's degree you are going to pursue. Make sure you understand the differences and select the degree you are most comfortable with. Plan for the impact it is going to have on your family and be proactive in managing problems when they arise. Do not make the fatal mistake of obtaining a master's degree and losing your family. You do not have to do this entirely on your own. Many others have successfully obtained their master's degree. Seek these people and coworkers out and call upon them for help.

Have you identified any career actions you want to take as a result of reading this chapter? If so, please make sure to capture these ideas before you forget by recording them in the notes section at the back of the book.

ASSIGNMENTS AND DISCUSSION TOPICS

1 Conduct a web search of engineering master's degree programs in your local area. Identify the potential programs you would be interested in.

2 Select one of the programs and explore what the program requires.

3 Contact people at your work who already successfully completed the program and get their guidance about the program.

4 Develop a plan of tasks leading up to enrollment and starting.

5 Discuss your ideas with your spouse or significant other.

REFERENCES

1. Konc, Andra, National Association of Colleges and Employers, www.naceweb. org/info_public/salaries.htm.

2. Patel-Predo, Prachi, "The Data-US Engineering Salaries," *IEEE Spectrum Magazine*, August 2008, pg. 64.

3. Burt, Victoria, "Dream Jobs—Annual Salary Survey," *Machine Design Magazine*, April 24, 2008, pg. 58.

4. Mcgee, Marianne, "Pay Crunch," *Information Week Magazine*, April 26, 2008, pp. 28–37.

5. Bokorney, Judy, "Salaries Still Rising," *Evaluation Engineering Magazine*, April 2007.

6. Tufts University. Gordon Institute (TGI) is a nationally recognized center within the School of Engineering that focuses on innovation, entrepreneurship, and engineering leadership. TGI has recently been honored by the National Academy of Engineering for Innovation in Engineering and Technology Education. For further information go to gordon.tufts.edu.

7. Santa Clara University. The goal of the Santa Clara University Engineering Management and Leadership program is to support the development of technical project managers. To this end, we require that half of the Engineering Management degree units be devoted to a technical stem, drawn from one or more of the other engineering departments. The remaining units are in management-leadership-related studies. For further information go to www.scu.edu/engineering/emgt/grad/progms.cfm.

8. Stanford. The Management of Science and Engineering (MS&E). Department provides education and research opportunities associated with the development of knowledge, tools, and methods required to make decisions and to shape policies, to configure organizational structures, to design engineering systems, and to solve problems associated with the information-intensive technology-based economy. For further information go to www.stanford.edu/dept/MSandE/about/index.html.

9. University of Minnesota, Center for the Development of Technology Leaders (CDTL). For high-tech companies, business success is all about mastering the gray zone—that area of the company where business, engineering, science, and technologies converge. CDTL programs and activities help high-tech professionals, managers, and leaders, grow their businesses by showing them ways to explore and maximize growth in the gray zone. For more information go to www.cdtl.umn.edu.

CHAPTER 47

LIFELONG LEARNING
GROWING OR DECAYING

There is no choice in the matter of lifelong learning if you want to remain a viable contributing engineer throughout your career. Either you endorse it, live by it and continually grow, or you quickly decay and become obsolete. Lifelong learning is a critical career skill every engineer should possess. Change is the only thing constant in life and lifelong learning is how you stay on top of all this change.

The reason lifelong learning is such a critical skill is the extremely rapid pace of technology development and information exchange in our global economy. Globalization and international markets are rapidly driving engineers to seek out new tools, more efficient methods, and lower-cost better performing technologies, with the ultimate goal of bringing better products to market in a shorter time. This can only be accomplished by continual growth and improvement fueled by constant learning year after year [1].

Regardless of what stage of your career you are in, whether you just graduated, in the middle of your career, or nearing retirement, you must constantly pursue a learning plan that keeps you updated with all the latest advances in engineering [2].

ENGINEERING KNOWLEDGE HALF-LIFE: HOW LONG BEFORE YOUR SKILLS ARE OBSOLETE

The engineering knowledge half-life is defined as the time it takes for half of everything an engineer knows to become obsolete or forgotten. Studies suggest, the average half-life for engineering knowledge is now in the 5 year range [3]. Specific estimates were even made for various types of engineering backgrounds that indicated a software engineer's half-life was only 2.5 years, and mechanical engineer's half-life was 7.5 years [4]. Based on this estimate of

The Engineer's Career Guide. By John A. Hoschette
Copyright © 2010 John Wiley & Sons, Inc.

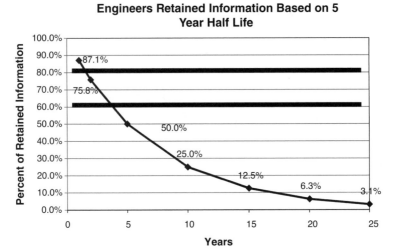

FIGURE 47-1   Retained knowledge for 5 year half life.

half-life for software engineering, what a student learns his sophomore year is 50% obsolete by the time he graduates. Figure 47-1 shows the percentage of knowledge retention for an engineer in a field with a 5 year half-life. The graph identifies the engineer has lost 50% of their knowledge base in 5 years and is down to only 29% left in 10 years. If you assume the person does not update their technical knowledge during their career, by 29 years, they will only have about 3% of the knowledge they once had. These are very alarming numbers and will hopefully convince you of the necessity for lifelong learning to survive.

Another observation made by career specialists is that having just one career in your lifetime is a thing of the past. They are now predicting the average person will have 2 to 3 different careers in their lifetime. If you stop and think about it, starting a new career is going to require a significant amount of new training or learning. Combining the rapid technical obsolescence with changing careers several times, clearly underscores the need for lifelong learning.

WHAT IS LIFELONG LEARNING AND WHY IS IT GOOD FOR MY CAREER?

Professor Beth Todd of the Mechanical Engineering Department at the University of Alabama has defined lifelong learning to have two major components. The first component is the ability to learn on your own. Hopefully as part of your formal education at the university, you learned how to perform independent research and learn things on your own. Simply put, it is the ability to learn on your own and use this talent when faced with problems in your career.

Todd identifies the second major component for successful lifelong learning is the realization that continuing education throughout your career is absolutely necessary. Old skills must be refreshed and updated with new and improved methods, as well as totally new skills that must be added to the engineer's repertoire.

▶ *Career Tip.* Lifelong learning is a necessary and key ingredient in the formula for career advancement.

Georgia Stelluto in her article "Leadership: A Matter of Choice in Lifelong Learning" points out that if you see yourself as a leader, you will choose to learn at a deeper level than most people around you [5]. Lifelong learning will be a cornerstone to your career helping to mold you and those, you lead.

▶ *Career Tip.* Great leaders make lifelong learning a cornerstone for building their career.

SIMPLE THINGS TO HELP FACILITATE LIFELONG LEARNING

Many great ideas exist for helping you to cultivate your lifelong learning skills [6–8]. Here are some simple actions you can take to help develop a habit of lifelong learning.

Always have a book to read or audio CD/IPOD to listen to. Visit your local bookstore, university bookstore, or look online for books to read when you have a few spare moments. Carry it with you or put it in a place where you can easily access it. Lunch time, late at night, or even audio books you can listen to on the commute to work are ways to catch up on the latest advances.

Put what you learn into practice. Nothing motivates you to learn more than trying something new and achieving great results. So if you learn a new technique, make a conscious effort to try it out.

Learn with a group or friend. Join others in workshops and group learning events. It is a great way to socialize and have fun learning at the same time. Joining an engineering society is a great way to get exposed to new methods.

Request a course catalog. Go online and request a course catalog from your local university to discover what new courses are being offered.

Teach a class. Put together the material and teach a new class at your workplace or local university. Research the material and develop the curriculum. The class does not have to be technical in nature, but can even be a hobby class.

SUMMARY

Lifelong learning provides you with continuous rewards throughout your career and is also good insurance in bad economic times. Lifelong learning requires a mindset of continuous self-improvement. It is good for you as well as your company and a means for both of you to remain prosperous in an ever-changing world.

Have you identified any career actions you want to take as a result of reading this chapter? If so, please make sure to capture these ideas before you forget by recording them in the notes section at the back of the book.

REFERENCES

1. Meredith, John, "Even Busy Engineers Can Take Advantage of Lifelong Learning," *Electronic Design Magazine*, April 12, 2007.
2. Johnson, Vern, "Lifelong Learning Is Necessary for Career Success, According to Survey," *IEEE USA Today's Engineer*, August 2003.
3. Smerdon, Ernst, *Lifelong Learning for Engineers, Riding the Whirlwind*, National Academy of Engineering, Workshop on Careers, Volume 26. 1996.
4. American Society of Mechanical Engineers (ASME), http://www.asme.org/Education/Courses/GMET/Need_Lifelong_Learning.cfm.
5. Stelluto, Georgia, "Leadership: A Matter of Choice in Lifelong Learning," *IEEE USA Today's Engineer*, April 2005.
6. Young, Scott, "15 Steps to Cultivate Lifelong Learning," http://www.lifehack.org/articles/lifestyle/15-steps-to-cultivate-lifelong-learning.html.
7. Field, John, and Leicester, Mal, *Lifelong Learning: Education Across the Lifespan*, Routledge, 2000
8. Sutherland, Peter, and Crowther, Jim, *Lifelong Learning: Concepts and Contexts*, Routledge, 2006.

PART 8

FINANCIAL PLANNING

CHAPTER 48

BASIC FINANCIAL MANAGEMENT 101

Understanding the basics of personal finance is key in ensuring you have a sound financial strategy in addition to your technical career strategy. Superior technical skills complemented with excellent personal financial planning lead to a solid, successful, and rewarding career.

The purpose of this chapter is to review with you the fundamentals of personal financial management. If you have never completed a financial plan or visited a financial advisor, this chapter will bring you quickly up to speed on the key actions you should be considering when formularizing your personal financial plans. For those of you who already have a financial plan and are working with a financial advisor, excellent. You should consider this chapter as a double check to make sure you have all the basics covered.

What makes people great financial planners? They think and act differently. Thomas Stanley in his book, *The Millionaire Mind*, identifies the following characteristics that makes millionaires different [1].

1. No high lifestyle consumption
2. Well-educated
3. Fastidious investors (saving 20% of yearly income)
4. Planners and budgeters

If these characteristics do not sound like you at all, then it is time to seriously consider changing your lifestyle and come up with a financial plan. If these characteristics sound like you then keep going you are on the right track.

Emerald Publications, in their financial planning workbook, identify the following six basic steps to financial success and security [2].

1. Protecting what you have
2. Take control of your cash flow
3. Invest wisely
4. Manage your taxes
5. Save for retirement
6. Estate planning

PROTECTING WHAT YOU HAVE

The first area of personal financial planning is protecting what you have in the event that life takes a bad turn. Protection is having the proper insurance coverage. The following are several types of insurance that financial advisors recommend you purchase basic coverage to protect your most valuable asset—you.

Medical Insurance. Number one on the list of insurance policies you should have is medical coverage. You absolutely need this. The average daily hospital stay is 5 days at a cost of $22,596 according to Pacific Business News [3]. At this price you could easily use all of your savings in no time with a major illness due to an extended stay in the hospital without medical insurance. Sign up for coverage at your company and review the coverage annually. Medical plans annually change their coverage and fees; you should be aware of these changes and adjust your policy as required. Knowing and following the health insurance provider policies can save you a significant amount of money when comes to using emergency services, hospital choices, and selection of specialists. Correspondingly, not knowing and failing to follow these policies can cost you hundreds of dollars of extra fees. Take the time to read and understand the medical coverage provided by your company's health plan to know exactly what is covered and what is not. Medical coverage is often complemented by dental and vision coverage. These insurance benefits are equally important to understand.

In the event that you leave a company for a new job, make sure you have medical coverage during the time you leave and start the new job. If something happens to you during this time, you could be at great financial risk. Normally, the old company's health plan will allow you to extend the period coverage beyond your last working day if you purchase a separate policy.

Short-Term and Long-Term Disability Insurance. Most companies offer some type of short-term policy as part of your normal employee benefits. This short-term coverage usually guarantees a paycheck for 2 to 4 weeks if you become ill and cannot work. Check to make sure you have this basic coverage. If not, explore what your options are if you become ill for an extended period of time.

After this initial 2 to 4 week time period if you are unable to return to work, the company may no longer provide a paycheck. Some companies offer automatic long-term coverage and others do not as part of their compensation package. You may have to purchase long-term disability separately. Long-term disability coverage becomes effective after short-term disability ends. Long-term disability insurance coverage only replaces a percentage of your normal pay. The typical replacement coverage is usually in the range of 50% to 60% of your normal pay. If you are forced into a long-term disability situation, you could quickly be in trouble if you do not have any reserve savings.

Life Insurance. Next on the list is life insurance. Life insurance pays a lump sum to the beneficiary you name on the policy. If you are single, you should purchase enough life insurance to cover your funeral and estate liquidation costs. If you have a family, you should have as a minimum, enough coverage provides living expenses for your family for at least 1 year if not more. There are several types of life insurance you can purchase. Working with an insurance agent or financial advisor you can help you determine what type is best for you. The best buy on life insurance is usually through your company benefits plan. However, I recommend supplemental life insurance to your company plan, at a young age, in the event that you become unemployed.

Auto and Home Insurance. Next on the list are automobile and home insurance. Most states now require by law, automobile insurance, so there are no options here. You must purchase this on your own. However, the amount of coverage can be adjustable. As most insurance agents will point out, liability is the biggest risk. If you cause an accident and a semi-trailer with valuable cargo is destroyed, you are responsible for everything your insurance policy does not cover. A semi-trailer with cargo can easily run $1 million to $2 million dollars. Therefore, it is good to know your liability coverage limits when selecting your policy. Most companies do not offer automobile insurance as a benefit option. You will have to select your own. Make sure you comparison shop before selecting.

Home/Condo/Apartment/Renters insurance is another must-have policy. You work hard and use your earnings to purchase housing and many personal items. A simple fire in your building/home can wipe out everything you own and leave you in financial ruin. Protect your hard-earned assets with insurance coverage. An insurance agent can quickly assess your needs in this area.

There are many other types of insurance you can purchase and selecting these will depend on your needs and risk. Working with a financial planner can help you determine what other types of insurance you may need for your particular situation.

TAKE CONTROL OF YOUR CASH FLOW

To gain control of your cash flow, you will need to develop a monthly and yearly budget to follow. An example of a spreadsheet showing the monthly bills for an entire year, and the projected cash flow associated for each month, is shown in Figure 48-1. The far left column identifies the bills and income. The columns to the right are the months of the year.

Monthly Bills	J	F	M	A	M	J	J	A	S	O	N	D	Total
Pay Myself First (15%)	$675	$675	$675	$675	$675	$675	$675	$675	$675	$675	$675	$675	$8,100
Automobile Gas	$65	$65	$65	$65	$65	$65	$65	$65	$65	$65	$65	$65	$780
House Pymt (P&I, Taxes)	$1,800	$1,800	$1,800	$1,800	$1,800	$1,800	$1,800	$1,800	$1,800	$1,800	$1,800	$1,800	$21,600
Energy (Electric & Gas)	$250	$250	$250	$120	$120	$120	$250	$250	$120	$120	$250	$250	$2,350
Grocery	$350	$350	$350	$350	$350	$350	$350	$350	$350	$350	$350	$350	$4,200
Entertainment	$75	$75	$75	$75	$75	$75	$75	$75	$75	$75	$75	$75	$900
Cell Phone	$49	$49	$49	$49	$49	$49	$49	$49	$49	$49	$49	$49	$588
Savings for One Time Bills	$500	$500	$500	$500	$500	$500	$500	$500	$500	$500	$500	$500	$6,000
TV/Computer	$120	$120	$120	$120	$120	$120	$120	$120	$120	$120	$120	$120	$1,440
Credit Card	$120	$120	$120	$120	$120	$120	$120	$120	$120	$120	$120	$120	$1,440
One Time Bills													
Car Insurance		$500								$500			$1,000
House insurance			$500						$500				$1,000
Clothing Allowance			$300						$300				$600
Car Maintenance					$400					$400			$800
Water	$50			$50			$50			$50			$200
Tax Preparation				$500									$500
Vacation							$1,000						$1,000
Year End Holidays												$700	$700
Total Monthly Bills (Monthly + One Time)	$4,054	$4,504	$4,804	$4,424	$4,274	$3,874	$5,054	$4,004	$4,674	$4,824	$4,004	$4,004	$52,498
Income (Take Home)	$4,500	$4,500	$4,500	$4,500	$4,500	$4,500	$4,500	$4,500	$4,500	$4,500	$4,500	$4,500	$54,000
Net Monthly Cash Flow Income - Bills	$446	-$4	-$304	$76	$226	$626	-$554	$496	-$174	-$324	$496	$496	$1,502

Net Yearly Cash Flow

FIGURE 48-1 Example cash flow worksheet for personal finances.

In the top portion of the spreadsheet, the bills that are paid monthly are identified. These include such bills as rent or mortgage payments, energy bills, groceries, and others normally paid on a monthly basis.

The next section down contains a list of the major bills that occur only periodically during the year. These include such bills as house insurance, car insurance, clothing, and others.

Directly below these two sections is the "Total Monthly Bills" row. This is simply the sum of the monthly and one-time bills for each month.

The next section is the Income (Take Home) pay for each month. This section lists the monthly take-home pay you receive after taxes.

The bottom row or line of the spreadsheet is the Net Cash Flow for the month. This is simply the income less the bills you pay out. You should note for this example that there are months highlighted in bold borders where the cash flow is negative. This means that in these months, you have more bills than income, and will need to withdraw from savings in order to meet your bills.

Now let's analyze the spreadsheet. First, the Net Yearly Cash Flow for the year is positive. This is good since it indicates you are living within your means and should have some money left over at the end of the year. This extra money can be used as a special reserve for emergencies and unplanned expenditures. If you set up a cash flow spreadsheet like this and the yearly net cash flow is negative, this means you are living beyond your means and need to immediately cut back and get your spending in control.

▶ *Career Tip.* Complete a cash flow analysis and make sure you are living within your financial means.

Financial advisors all agree that the first payment you should make is to yourself. They recommend you live on 80% to 85% of your take-home pay and invest the remaining 15% to 20% of your monthly income for retirement and other long-term needs. This will ensure a sound financial position for you in the years ahead.

▶ *Career Tip.* Pay yourself first.

There is good motivation to pay yourself first as George McClure, resource member of the IEEE-USA Career and Workforce Policy Committee points out. "If it is not there, you don't have a temptation to spend it." Another way people are getting into financial difficulty is by using their home equity like an ATM. Spending the equity of your home to pay monthly bills is not sound financial management. Build up your savings reserve by paying yourself first and living on the rest.

In addition to savings, financial advisors and mortgage institutions recommend that you do not spend more than 28 percent of net income on mortgage payments or rent. If you have other debt, advisors recommend keeping your total credit card, mortgage, and other monthly loan payments less than 33% of your take-home pay [4]. A new development today is the cutting back of salary by some corporations—ten percent is not unusual. By keeping your debt down and planning a reserve into your budget, you should be alright if you find yourself suddenly facing a salary reduction.

Now let's look at how you cover the months where there is a negative cash flow. To cover these months you need a reserve or savings account to draw upon. To make sure you have enough in reserve, you need to compute the amount to save each month. Shown in Figure 48-2 is a table listing the large one-time bills, the amount of the bill, the months it is prorated over, and finally the amount you need to save each month to ensure there will be enough to pay these bills. For the example shown, the table identifies that one needs to put away $483.33 each month in order to pay all these bills. If you look at the cash flow spreadsheet, there is a monthly savings bill or deposit of $500. This deposit each month will build up a reserve to ensure during the negative cash flow months you have sufficient funds in your savings account to cover the extra bills when they come due.

The next major step to managing your cash is to pay down and eliminate all your credit card debt. If you have high credit card debt, it makes better financial sense to pay this down before investing. Watch your credit card interest rates; they can be as high as 15% to 20% annual percentage rate (APR). Although it may be tempting to purchase items beyond your financial means through credit cards, don't be fooled by what appears as low monthly payments. You will pay dearly in the long run. Use credit cards only for emergencies and pay off the balance each month. On credit card interest, there is now the interlocking experience—when a payment is late on one card the others may hike up the APR even though payments there were timely.

Major One-Time Bills	Yearly Total	Due - Months	Saving Per Month Required
Car Insurance	$1,000	12	$83.33
House Insurance	$1,000	12	$83.33
Clothing Allowance	$600	12	$50.00
Car Maintenance	$800	12	$66.67
Water	$200	12	$16.67
Tax Preparation	$500	12	$41.67
Vacation	$1,000	12	$83.33
Year End Holidays	$700	12	$58.33
		Total	$483.33

FIGURE 48-2 Budgeting for one-time major bills.

INVEST WISELY

One of the best investments you can make is profit-sharing with your own company. Many companies offer company stock at reduced prices that you can purchase. Some companies offer a matching plan, where they will match each dollar you invest in a 401(k) plan with a percentage. For instance, they may match you 50% to 100% for each dollar you invest up to a maximum limit. This is like getting 50% or better return on your investment. However, be careful about putting all your investment funding into these company match programs since the company stock can go down also.

Another good investment is participating in your company-sponsored retirement plan. Most 401(k)s offer some company match. A common company match is 50 cents on the dollar up to 6%, similar to getting a 50% return on your money. Most companies offer a wide variety of mutual funds to invest this money in. Target Date Retirement Funds and Asset Allocation Funds are a wise place to invest if these are offered in your plan. They are managed according to your age and risk tolerance. For example, a 2035 Target Date Fund would be suitable for an individual who plans to retire at or near year 2035. The fund is a blend of stocks, bonds, and cash and will automatically get more conservative as you get closer to retirement.

Oftentimes, individuals who choose their own funds have a lower rate of return than those who invest in the Target Date Funds or Asset Allocation Funds. Sometimes, companies will only offer company stock as the only option for the match. This is fine, but be sure to diversify this stock over time if you're able to. Oftentimes, there is a vesting schedule on the company stock and it's wise to sell a portion, when available, to be sure you have a diversified portfolio within your risk tolerance.

The key is having a balanced portfolio of investments. Investing is not a passive process or one-time event. It is a continuous process of evaluating your alternatives and selecting the best approach for yourself. Most experts recommend a diversified investment portfolio. This is a combination of investments in different areas to reduce your risks.

Prior to visiting a financial advisor will you need to collect and assemble your financial information. This will include but is not limited to:

1. Last year's tax return
2. Cash flow analysis
3. Company benefits statement
4. Credit rating
5. Net worth statement (list of all assets, savings, and long-term debt)
6. Long-term financial goals
7. Estimate of available funding for investment

On credit rating, you can get a free report annually from each of the three credit reporting agencies. Having this information all assembled prior to your first visit will save time and make subsequent visits more productive.

Many engineering societies have advisors to assist you in your decisions. If you belong to an engineering society, please check into what they offer. George McClure (g.mcclure@ieee.org), a resource member of the IEEE-USA Career and Workforce Policy Committee focuses on benefits and health. He is interested in new problems that are arising in these areas and interested in your experiences.

MANAGING YOUR TAXES

Online Internet tax filing has come a long way in helping you fill out your tax forms. However, just knowing how to fill out tax forms is not good tax management. Good tax management goes way beyond just filling out your tax forms and occurs a long time before you file taxes. Meet with your financial advisor, and a tax expert, regarding how to manage and reduce your tax burdens. The financial advisor can provide you with information on how to best structure your investments to reduce your taxes. Many investments offer reduced tax incentives that may be better for you. Reducing your tax burden on investments can significantly improve their yield and your wealth over the long run.

Visiting with a tax expert and discussing your financial situation will/ could result in discovering new tax deductions you did not know existed. This is especially true if you are running a business out of your home, have real estate investments, foreign investments, or special needs, such as medical or family dependencies. The state and federal tax laws change yearly, so this should be a yearly activity. The best time to visit with a tax expert is the start of the New Year. Many tax firms will host free seminars during tax preparation times (February to April) which you can attend and ask questions. Attending a tax seminar could save you hundreds or even thousands of dollars, making them well worth your time.

SAVE FOR RETIREMENT

Planning for retirement is a lifelong activity. The younger you start the better off you will be. Financial advisors recommend that you have a sound balanced portfolio and modify your investments as you age.

Starting early is the best investment policy an individual can make. Let's take the example of Joe versus John. Joe starts investing in his 401(k) at age 25 and contributes $10,000 a year for 10 years, then stops contributing. If he earned a 7% rate of return on his money, at age 60, his account would be worth approximately $630,000. Now John doesn't start saving until age 35. He saves the same amount of $10,000 per year, but does this for 25 years, until he is age 60. At the same 7% rate of return, John's account would be worth approximately $630,000. Joe contributed only $100,000 while John contributed $250,000 and their accounts where worth the same at age 60. If Joe would've kept saving that same amount of $10,000 a year until age 60, his account would've been worth over $1.3 million.

Company stock purchase plans are generally a good investment. They usually allow you to purchase company stock at a discount. Company stock can be volatile and risky, so it's important to diversify the portfolio over a period of time by selling off shares of the stock. Most companies have a vesting schedule on the stock where you must hold it for a specific amount of time. It is prudent to monitor exercise dates and vesting schedules. Don't have all of your eggs in one basket by being too heavily weighted in your company stock in your overall investment portfolio. If you're able, reinvest this company stock over time in a diversified portfolio of mutual funds.

The Rule of 72 states that if you divide a given interest rate by 72 it will tell you the amount of years it takes for your money to double. For example, if you earned a 7.2% rate of return, it would take your money just over 10 years to double. This is a nice rule of thumb when projecting retirement accounts at various rates of return. This demonstrates that saving for retirement should be done sooner than later. Complex financial plans are not the answer to a successful retirement. The most important decision you can make is making the commitment to save a comfortable, affordable amount each month in a well-diversified portfolio and stay with it.

According to many financial advisors, people wait until they turn 50 before they start planning retirement. This can be too late. Also, your company retirement benefits and Social Security benefits are continuously changing; therefore, you will need to review these on a yearly basis. One important fact to remember is that your retirement benefits are never guaranteed. For example, many corporations are now cutting out defined benefit pension plans for the option of matching your 401(k) investments with company stock. Some firms recently have even suspended their 401(k) matches due to economic conditions. All this places much more responsibility on the individual to save over a career lifetime and manage their own investments.

Due to the constantly changing state and federal laws, and changing tax structures, I cannot stress enough the importance of meeting with a financial planner to determine the best methods to save for retirement for you.

A balanced approach to retirement savings is recommended by most financial planners. This balanced approach involves Social Security, 401(k) plans, portfolio income, company pension (if available), savings, and even part-time work. Currently, the final age that you are able to retire is dependent on many factors. Social Security, for example, has multiple ages you can retire and multiple payment plans that will affect the amount of your benefit. Your health and medical benefits are other key factors to consider as well as the standard of living that you wish to maintain. To describe this situation in engineering terms, this is a non-determinate set of equations that have hundreds of possible solutions. Your task is to determine the two or three best retirement solutions for your situation.

ESTATE PLANNING

The final area of personal financial planning is in the area of estate planning. This area is for people who have reached retirement age and would like to protect their estate for their beneficiaries. There are basically four types of estate distribution techniques. These are:

1. Wills
2. Jointly-held property
3. Contracts
4. Trusts

All these methods offer different benefits for estate distribution to beneficiaries. Wills provide a direct means to distribute your estate in the exact manner you wish. However, they expose your beneficiaries to large tax liabilities often eating up a significant portion of the estate. Jointly-held property is a very simple method of transferring estate to surviving partners with little or no tax liability. Contracts are another method, but need the counsel of lawyers to properly structure the agreements. Finally, trusts are a method of locking up the estate assets to ensure they are properly distributed, but involve a complex web of tax rules and regulations that should be only done with the guidance of an attorney.

SUMMARY

Understanding the basics of personal finance is key in ensuring that you have a sound financial strategy in addition to your technical career strategy. The basic steps to financial success and security are: protecting what you have, take control of your cash flow, investing wisely, managing your taxes, saving for retirement, and finally, estate planning. Superior technical skills complemented with excellent personal financial planning lead to a solid, successful, and rewarding career. For those of you who already have a financial plan and are working with a financial advisor—excellent. If you have never completed a financial plan or visited a financial advisor, your best career move is to visit a financial advisor and develop your financial plan immediately.

Have you identified any career actions you want to take as a result of reading this chapter? If so, please make sure to capture these ideas before you forget by recording them in the notes section at the back of the book.

ASSIGNMENTS AND DISCUSSION TOPICS

1 What is a net worth statement?
2 How often should you adjust your investment portfolio?
3 How is your credit score determined?

REFERENCES

1. Stanley, Thomas J., *The Millionaire Mind*, Andrews McMeel Publishing, 2000.
2. Cohen, Ivan, *Focus on Financial Management*, Imperial College Press, 2005.
3. Pacific Business News, Oct 24, 2008, http://www.bizjournals.com/pacific/stories/2008/10/27/story2.html?q=cost%20of%20average%20hospital%20stay.
4. Florea, Linda, C., "To Live Frugally, Change Your Mind, Then Your Life," *Orlando Sentinel*, January 25, 2009.

PART 9

FAST TRACK ADVANCEMENT AND CAREER ACCELERATORS

CHAPTER 49

GETTING ON THE FAST TRACK FOR ADVANCEMENT

If you want to move ahead quickly, you have to become a member of the fast-track crowd. The fast-track crowd is acutely aware that working long hours is simply not enough [1–3]. It is just the minimum requirement for keeping a job and not the means to fast-track advancement. The avenues to fast-track advancement lie in another direction; a direction that requires skills in selecting your supervisor, your assignments, and progress reporting. Let's identify some fast-track actions.

SELECT YOUR SUPERVISOR

The quickest way to advance is to attach yourself to a quickly rising supervisor and ride along. What you want to happen is every time your supervisor is promoted, you are also promoted. For this to happen you must select and work for a supervisor who is clearly on the way up.

Fast-track supervisors do not wear name tags identifying themselves nor do they come along that often. You have to be able to recognize them and seek them out. You can identify fast-track supervisors by their great visibility with upper management. You can also identify them by the fact that their groups usually receive most of the special assignments from the vice presidents. Another characteristic trait is that their groups usually receive more awards and recognition than most groups.

Once you have identified a fast-track supervisor, you need to transfer to his or her group. Once in the group you stand a better than average chance for fast-track advancement. Fast-track supervisors generally tend to rate their group members very high. This gives upper management the impression that the group has superior performance that is a result of the fast-track supervisor's outstanding efforts. The fast-track supervisor realizes the value of overrating his or her people rather than underrating them.

The Engineer's Career Guide. By John A. Hoschette
Copyright © 2010 John Wiley & Sons, Inc.

Another characteristic of fast-track supervisors is their tendency to know how to get upper-management visibility for their employees. A fast-track supervisor will always have employees nominated for awards. Therefore, you stand a much better chance of getting an award for your accomplishments when you work for a fast-track supervisor.

Once you are working for a fast-track supervisor you need to make yourself his or her right arm. You need to be alert and always ready to put in the extra effort and special demands. You must give it your all and be loyal and work as much as possible. Your primary objective is to make your supervisor look good, so they will be promoted [4,5].

Of course, simply executing all their commands will not automatically result in promotions. You must demonstrate that you have the skills and knowledge to perform your job successfully. Your ability to get the job done and exhibit excellent work are basic requirements for successful career advancement.

Let's look at the negative aspects of settling for just any supervisor. Suppose you settle for a supervisor who has been in the same position for the last 10 years and has shown no growth. This supervisor's career may have peaked and he or she has probably been passed over for promotion. They are as high in the company as they can go. Working for this supervisor could actually represent a roadblock to your career. If he is not moving up, you will more than likely also not be moving up either.

Another case in point is settling for an average performing supervisor who does not make waves. Supervisors who receive only average ratings tend to give only average ratings. Supervisors who want to only do their job and go home tend to stay away from upper management. If he is not getting upper-management visibility, then most likely you won't either. These are only some of the examples as to why it is so important for you to select your supervisor and not settle for just anyone. A fast tracker quickly realizes the value in selecting his supervisor. If at all possible you want to choose your supervisor.

DEVELOP EXCELLENT COMMUNICATION AND PRESENTATION SKILLS

If you communicate in an average manner and your presentations are just average you can only expect to advance at an average rate. To get on the fast track you must have excellent communication skills and excellent presentation skills. The excellent communication skills are manifested in the written and oral reports you give to your supervisor and upper management.

Your written reports must clearly reflect your superior performance. All reports need to be well-organized and neat in appearance. The results should be reported in a manner that allows the reader to clearly graph the significant conclusions. Graphical and mathematical analyses are a must. If possible, the reports should at least be done using Microsoft Word or PowerPoint charts that include graphics and color.

Generating excellent written reports is not a natural talent. It is something that is acquired through practice and training. To learn impressive ways of writing you can take writing courses. I highly recommend that you attend two types of writing courses. The first is a technical writing course and the second is an advertisement writing course. The technical writing course will show you how to organize technical material. The advertisement writing course will show you how to accentuate the positive and give it the additional flair to make your communication stand out from the rest.

A technical report can be extremely dry and hard to read. Even if your report represents a great scientific breakthrough, it still may get ignored due to poor writing. You need to jazz up your reports with a little advertising flair if you do not want your hard efforts ignored. The advertising course will show you how to do this. Learning how to write good technical reports with a bit of advertising flair is a must for the fast tracker.

Having great oral communication skills is even more important. You absolutely must know how to give a good oral report. Most upper-level managers do not have the time to read long reports. They want a summary of the significant points and they want it now. Oral reporting is also not a natural skill. You must develop this skill through training and practice. To develop this skill take a speech class or join Toastmasters. Toastmasters is an organization that meets on a periodic basis and its members practice giving speeches and oral reports. The organization provides step-by-step training and guidance on how to improve your presentations. Joining Toastmasters is an excellent way to develop your oral reporting talents and even meet upper-level managers who often attend the group's meetings.

Practice is absolutely necessary to perfect your oral reports. One way to practice is by giving oral reports at home and use a video recorder to tape yourself; play back to critique yourself. You will be surprised at all the mistakes you make. By videotaping yourself you can quickly see what you have to work on. It's always better to make the mistakes at home than in front of your supervisor, coworkers, or at meetings.

Some people ask me if this is really necessary and do they need to practice? To this question I always respond with a large and emphatic yes! Athletes practice for hours; actors and actresses will rehearse their lines for hours; lawyers will practice their courtroom presentation; salesmen always practice their pitches; politicians practice their speeches; clergy will practice their sermons. Why? To give the best performance they can. They want you to see them at their best. This is something natural that nearly all professions do. Practicing is something that most other professions consider imperative. It would appear to be good commonsense to practice any presentations you plan on giving. Why is it that engineers think they do not need to practice presentations?

Finally, you must be able to make great presentations. You must acquire the skills to speak in front of a group at meetings. For this you need to learn how to create presentation material, how to operate projectors, laptops, what makes a good chart, how to handle questions, and how to speak with authority. Great presentations are an absolute must for fast-track development. Remember, your supervisor will want the person with the best speaking and presentation skills to give the report to the vice president. Continue on if you're still interested in becoming a fast tracker.

GO FOR THE SPECIAL PROJECTS

Choose projects that will give you the most visibility. Projects that lock you in a lab with little or no visibility require special effort by the fast tracker to get that visibility. These projects may be technically challenging but they can keep you away from the eyes of management. The fast tracker realizes that selecting your project is very important. If you have a choice of assignments, choose the supervisor's pet project. This is another fast-track move.

THE RIGHT TIMING FOR GETTING ON AND OFF PROJECTS

The fast tracker is said to be acutely aware that there are good times to get on a project and there are good times to get off a project. At the beginning of the project there is usually a lot of visibility by management to ensure the project gets started on the right foot. Upper management is normally involved and monitoring initial progress on the project. Getting on a project at the beginning provides the engineer with extra visibility. The beginning of a project is when most of the promotions are usually handed out.

As the projects progress unforeseen problems may arise. These problems may result in schedule delays and large cost overruns. Most major problems will surface after about a year into the project. If you can see that the project is headed for disaster, the fast-track response will be to get off before the problems are discovered. This is an excellent time to leave a project. The fast tracker can honestly say the problems occurred after she or he left. This allows the fast tracker to be dissociated with any problem projects and always presents a successful image.

When a project hits major problems, management often assembles teams to fix the problems. They sometimes refer to these as "Tiger Teams" or "Get Well Teams." The fast tracker will join a project that has severe problems only as a member of these Tiger or Get Well teams. Team members are usually assigned by upper management to solve the difficult problems. The fast tracker realizes the special teams have extra visibility with upper management due to the pressure to get the project back on track. Often the special teams must report daily progress to upper management. A fast-track engineer who has excellent reporting and presentation skills is now in a great position to move ahead quickly. This is the only good way and good time to get on a problem project.

As the problems are solved and the project comes to an end, the visibility of the project dramatically decreases with little opportunity for career advancement. However, engineers often continue to stay on the project with little or no benefit to their career. The engineer may stay too long because of the large amount of clean-up work needed to close out the project. Or the engineer may stay on the project longer than she or he should because the work was very interesting. Not wanting to leave the project is only natural, since you have spent so much time and effort getting the results. These are some of the reasons why the engineers stay on a project longer than is good for their career. The fast tracker is aware when a project has entered the final stage and has little or no benefit for his career. Fast trackers quickly move on to the next career advancing project.

DRESS AND ACT THE PART

Study the dressing habits and mannerisms of your management. Determine the dress code and dress accordingly. Watch for mannerisms displayed by your supervisor. The fact of the matter is that supervisors tend to promote people who look like them and act like them. If it worked for them, then why not copy the successful formula?

SEEK OUT OPPORTUNITIES TO EXCEL

Opportunities to excel present themselves every day. Most people refer to these opportunities as big, unsolvable problems. Consequently, they avoid them as much as they can. By avoiding them they lose the opportunity to shine. The fast tracker looks for these opportunities to excel and selectively chooses to work on them. Naturally, choosing the opportunities to excel that your supervisor considers the most important is key to rapid advancement. The fast tracker knows that the more desperate the supervisor is about the problem, the better the opportunity.

What if the opportunities to excel are on unsolvable problems? Don't worry, this is even better. The supervisor does not expect the employee to solve every problem but just his willingness to try is what will separate him or her from the rest.

MAKE YOUR OWN LUCK

The fast tracker realizes that most projects are not going to be highly successful and result in a promotion. As a matter of fact, the fast tracker realizes just the opposite–that most projects will experience severe problems and setbacks. She accepts this and plans for it. A fast tracker will always have backup or contingency plans. If you have backup plans that you can fall back on and keep the project moving ahead, it lends great strength to the appearance of a successful project.

▶ *Career Tip.* Making good backup plans is like planning for luck.

Stick to your plans and put in the effort to see things through to the end. Don't give up at the first sign of trouble. There is an old saying by Thomas Edison that best captures the definition of success. "Success is 99% perspiration and 1% inspiration." As a matter of fact, you can actually look at trouble as a good sign. If there are troubles, they will need you all the more. This is also called job security. With trouble comes extra visibility. With extra visibility and great performance comes promotion. Therefore, trouble means job security and can even be the start of your next promotion. This should help you see the positive side of trouble and, hopefully, help you fear it less when it rears its ugly head.

These are only some of the tips that are available to you on fast tracking for engineers. Other sources of fast-track tips can be found in the trade journals, business magazines, and even professional society meeting proceedings [4]. You have to be alert and search for them. The best thing you can do is collect as many articles on the subject as you can. Place them in a binder and at least twice a year review the articles.

HOW TO GET A GIGA-IMPROVEMENT IN YOUR CAREER

People often ask what they can do to quickly accelerate their careers. What actions should they be constantly striving for that will most benefit their careers? To answer this question I have identified the value of various actions you can take to significantly increase your chances for career advancement. The following is a simplified priority action list for career advancement. By following this prioritized list of actions whenever you can, you should, hopefully, realize a Giga improvement in your career.

Priority List for Career Advancement

One picture is worth 1,000 words
One model is worth 1,000 pictures
One successful project demonstration is worth 1,000 models

One email is worth 1,000 handwritten notes
One phone call is worth 1,000 emails
One face-to-face meeting is worth 1,000 phone calls

One phone invitation is worth 1,000 email meeting invitations

One face-to-face invitation is worth 1,000 phone invitations

Complimentary food at a meeting is worth 1,000 face-to-face invitations

Bottom Line. You realize a Giga-improvement in career advancement by having a face-to-face meeting with upper management while providing food, when demonstrating your successful project.

PRIORITIZE AND FOCUS YOUR CAREER ACTIONS

Being a Certified Master Black Belt in Six Sigma has taught me to prioritize actions and focus energy on only those actions that produce results. You get your best results on "value-added" tasks whereas spending time on "non-value-added" tasks does not produce results. It is easy to apply these principles to your career with the net result of fast tracking your career. If you stop and think for a minute, there are career actions that are truly value-added actions and these should always be top priority and where you focus your actions. These actions directly result in career advancement and absolutely must be done in order for you to accomplish your objectives. Usually, there are only a couple of tasks that qualify for true value-added. Let's look at some truly value-added career actions.

Value-Added Career Actions (Primary Effect)

- Interview for a new job and accept the offer
- Ask for a promotion
- Ask for better assignments
- Make your company and their products extremely successful

Basically, these are the only real tasks that directly result in career advancement. Fast trackers understand this and focus the majority of their energy on these tasks.

Then there are non-value-added career actions that are nice to do but do not directly result in career advancement. Non-value-added tasks usually make you feel good about what you are doing, or sometimes they are necessary to accomplish your goal but do not produce results by themselves. In Six Sigma terminology, they are cost incurring actions necessary to get the product out the door. For instance, packing and shipping the product is an example of a non-value-added task.

Non-Value-Added or Necessary Career Actions (Secondary Effect)

- Having a career discussion with your boss
- Writing resumes, cover letters, and putting a portfolio together

- Conducting a job search, reading job ads
- Networking for jobs
- Returning for more education

Can you see the difference between non-value and value-added career actions? Fast trackers realize they need to focus most of their energy and time doing only value-added tasks. They realize that the non-value-added tasks are necessary, so they minimize their effort and time on these tasks.

Finally there are wasteful or counterproductive career tasks that should be avoided at all costs. These include:

- Complaining to your coworkers about your job
- Arguing with people
- Settling for the same old ways and never improving
- Reading career articles and never taking an action
- Blaming everyone else for your lack of advancement

If you want to become a fast tracker you need to seriously consider where you put your efforts. Are they value-added career actions or non-value-added actions?

SUMMARY

If you want to move ahead quickly, you have to become a member of the fast-track crowd. The fast-track crowd is acutely aware that working long hours is simply not enough. It is just the minimum requirement for keeping a job and not the means to fast-track advancement. Select and work for a manager who is a mover and shaker and who is quickly rising, and then help them get promoted.

To get on the fast track you must have excellent communication skills and excellent presentation skills. The excellent communication skills are manifested in the written and oral reports you give to your supervisor and upper management. Go for the special projects and look at problems as opportunities for career advancement. Consider the project you are on and the impact on your career. When is the best time to come on the project and when is the best time to leave? Dress the part and act like a leader. Finally, prioritize your career actions and focus your energies on those that produce result and are truly valued-added.

Have you identified any career actions you want to take as a result of reading this chapter? If so, please make sure to capture these ideas before you forget by recording them in the notes section at the back of the book.

REFERENCES

1. Sweeny, P. Dwyer, "Getting on the Fast Track," *Machine Design Magazine*, Nov 10, 1988, 49.
2. Careers Fast Track, http://www.careersfasttrack.com.au/.
3. Connelly, Brendon, "Fast Track Yourself," Sept 2005, http://www.slackermanager .com/2005/09/fast_track_your.html.
4. Ontario Society of Professional Engineers, "Fast Track Career Series for Engineers," http://www.ospe.on.ca/pd/index.cfm?task=viewprogram&pdprogramid=141.
5. McGarvey, R., "Working with the Boss," *USAir Magazine*, September 1992.
6. Ringer, R., *Looking Out for #1*, Fawcett Crest, 1977.

PART 10

BE A CONSULTING ENGINEER

CHAPTER 50

WHAT IS A CONSULTING ENGINEER? IS THIS THE RIGHT CHOICE FOR YOU?

Have you ever thought about becoming a consulting engineer? Working for yourself sounds like a dream job. If you are thinking along these lines this section of the book offers many tips and guidelines to help you develop a successful consulting business. First let's look at what consulting engineers do.

WHAT DO CONSULTING ENGINEERS DO?

Consulting engineers typically work for themselves or may be associated with a consulting firm. Most often, consultants own their business and basically offer their time, knowledge, expertise, and experience to companies to solve problems for a fee. Engineering consultants often play a multi-functional role such as technical expert, advisor, specialist, manager, or part-time employee [1]. Companies hire consultants for several reasons. The most common is when a problem exists and the company does not have the technical expertise in-house to solve the problem. Their hope is by hiring a consultant they can quickly solve the problem, save time, money, and effort.

Another reason companies hire consultants is to obtain training in a new technology or technical expertise. This training allows the company's engineering teams to rapidly come up to speed in a very efficient manner. Consults are also hired to give their opinions about designs and products, to help improve the functionality, reduce the cost of manufacturing, and help avoid mistakes or suggest ways to reduce the potential for customer lawsuits.

The Engineer's Career Guide. By John A. Hoschette
Copyright © 2010 John Wiley & Sons, Inc.

545

Another reason to hire a consultant is a short-term surge in the amount of work that cannot be handled by the present engineering staff. The company can hire short-term or part-time to help to complete the work. By hiring engineering consultants it offers the company a quick solution to an immediate short-term problem without having to hire permanent staff who would be without work after the job is completed.

The benefits of hiring a consultant can be enormous. Hiring a consultant with the right experience in any given area can cut days, weeks, and months off a project and result in significant cost savings, which easily pays back the consultants salary many times over. It can be the difference between success and failure for a company when developing and launching new products.

WHAT ARE THE PROS AND CONS TO CONSULTING?

There are many attractive benefits to becoming a consultant engineer [2]. Top of the list of benefits is that you are most often in business for yourself and are your own boss. You will get all the credit for a job well done and can earn the entire income. You are also involved in a variety of projects, learning new things, and always challenged. You can select your jobs and when you have built-up your business clients seek you out, this provides good job security. You are also more immune to the sudden layoffs experienced when working in a corporation.

You have tax advantages since you are in your own business. Your office can be in your home which means no dress code, no drive to work, and flexible hours. These are all very attractive benefits to be a consulting engineer and why so many engineers decide to elect to do so.

There are also cons about becoming a consulting engineer that one should be aware of before launching into a consulting business. You are in your own business and if it is just yourself, you will do everything. In addition to getting paid for your technical work, you will have to do nonengineering tasks as well. You will have to take care of the business side of things which involves marketing yourself, financial aspects, dealing with clients, and handling the contractual items. Being in your own business you can expect to be working more hours than if you were employed for a corporation; however if you love the work, this is not a problem. If you have the right skills these tasks will come easy for you and being a consultant is probably right for you. Let's next look at what skills you will need to be a successful consult engineer.

DO YOU HAVE THE RIGHT SKILLS TO BE SUCCESSFUL?

There are some very important skills you must possess or acquire if you are going to become a successful consultant. They involve your technical skills, business skills, and general work ethic.

Unique and Market Demand Skills. The first is having a unique and highly desirable engineering skill to offer clients [3]. You may possess unique skills but is there a market demand for them? Without market demand you will never get any contracts. It is the combination of these two, unique skills and market demand, that will result in paying jobs.

One way to find out if your skills are unique and marketable is to conduct some research on the potential competition. Do an Internet search to find out how many consultant engineers are presently offering services in your area of expertise. If the market is flooded, what is going to make your services stand out from others? Could you network with any of the presently practicing consulting engineers to see how good the market is? Are they turning down jobs because they are too busy? If this is the case, you have identified a potential business. If the consultants are hurting for work, you may want to consider how this will impact your chances for being successful in a consulting career.

If your research shows that you possess a unique skill and there is a market demand you are on a solid ground for starting a consulting business.

Self-Motivated and Willing to Work Hard. Being an consultant is a significant change from the corporate 8 to 5 world and will require a person to be self-motivating and willing to work hard long hours to get the business started [4]. You will not have the support structure like in a corporation to help you through the difficult times. Also building a consulting business is not easy and will be both physically and mentally demanding. It will require consistent hard work that often-times calls for going above and beyond the norm. If this sounds like what you normally do then you have excellent qualities for becoming a consulting.

Business and Social Skills. Your consulting business will consist of spending a large amount of time talking to clients during face-to-face meetings, over the phone, and via email. You will have to sell your services, manage projects, and conduct sales and technical meetings. All these activities require a person to be highly social and have excellent people skills.

The most successful consultants have a natural ability to walk into potential consulting opportunities and quickly make clients want to hire them. When problems arise on a job, they can communicate how to solve the problems putting their clients at ease. This requires the consultant to have excellent communication and listening skills coupled with the ability to project confidence. Your ability to handle all the personality types of clients and deal with people problems will be critical to your success.

In addition, you will need to have the normal business skills of bidding proposals, signing contracts, billing clients, and paying your bills. You will be a multi-functional, multi-tasking consultant where the actual technical work is only a portion of the work. If handling all this intrigues and motivates you, then consulting is for you.

Work Independently with Limited Resources and Networking Skills. As a consultant you will have to possess the ability to work independently for long periods of time in your home office with little social contact. In addition, you will only have the tools of your home office and must be able to complete the work on these limited resources. When you need to draw on other resources, as required, you must have a network of others you can call upon. This network of support people is built up over time using your networking skills. Being a superior networker is a key element in being a successful consultant.

Excellent Writing, Speaking, and Presentation Skills. Often consultants are selected by upper-level managers and executives based on their written proposals, interviews, and presentations. Your ability to convince these managers that you are the best person for the job will depend on your written, speaking, and presentation skills. Your proposals will be reviewed to determine the technical validity of the offering. Does the consultant adequately document in the proposal and reports everything necessary to show the best solution to the problem. The consultant may be asked to come into the company to be interviewed and present their case on why they should be hired. If this is the case, the consultant will need excellent verbal and presentation skills. Therefore, successful consultants possess excellent writing, speaking, and presentation skills.

Analytical and Problem Solving Skills. The engineering problems faced by consultants are normally the most difficult problems. If the problems were easy to fix the in-house engineer team would have solved it and there would not be a need to hire a consultant. But because the problems are more difficult, the consulting engineer must possess excellent analytical and problem solving skills. They approach problems with enthusiasm and simply do not pick any solution but seek out the best solution. They find ways to bust through, dig under, climb over, or go around barriers to solve problems.

If you possess all the above-mentioned skills then you have what is needed to become a successful consulting engineer. Can you still be a successful consulting engineer if you are missing these skills or some need improvement? Absolutely, you can still have a successful consulting career but you will have to get additional training. Even successful consulting engineers make it a practice to update their training at least once a year in these skill areas. Where can you find training? One good source for affordable training is with an engineering society. IEEE has an excellent consultants network that supports independent consultants and offers many affordable training classes (http://www.ieeeusa.org/business/default.asp)

PROFESSIONAL ENGINEER (PE) REGISTRATION

Some companies have a prerequisite to granting a contract—the engineer must be a registered Professional Engineer (PE) for legal purposes. What is a professional engineer? A Professional Engineer is an engineer who is licensed to practice engineering in a particular state after meeting all the requirements of the law [5]. Like other professions such as medicine (Medical Examiners Board), law (Bar Exam), accounting (CPA Exam) engineering is a profession regulated by certain laws. All states have registration laws governing the practice of engineering. Most states prohibit people who are not registered PEs from advertising, indicating to the public they are an engineer, or practicing as an engineer. The PE license is especially useful when doing work for a city, county, or state organization. Obtaining a PE license is not easy but having one will add to your credibility and may be a necessary license for you to acquire.

What is required to become a PE? The requirements differ slightly from state to state, but include getting a bachelor's degree in engineering, passing the Fundamentals of Engineering Exam, and completing 4 years of experience in your chosen field. Many engineering societies offer training courses for engineers to help prepare for the exams. The National Society of Professional Engineers (NSPE) (www.nspe.org) and IEEE (www.ieee.org) have excellent material and help when going for your PE.

IT'S ABOUT TECHNICAL APPROACH, EDUCATION, CREDENTIALS, AND EXPERIENCE

Simply deciding to become a consulting engineer because you want to and having good skills is not enough. It is also about having the best technical approach, education, credentials, and experience to get the job done. Having hired many consultants myself, I can testify the first thing companies are looking for is an excellent proven technical approach. Is the consultant proposing a tried and true solution to the problem or are they proposing to simply study the problem? Most companies want solutions and not studies on how to solve the problem. Studying the problem is okay if the effort is an analysis of the problem, but totally misses the mark if the company expects a quick fix solution to the problem.

▶ *Career Tip.* Companies do not hire consultants to do something they never did before; they hire consultants for what they have done before and for what works.

If the technical solution is good, then the company wants to know about the consultant's education, credentials, and previous experience. A company will be looking at the education of the consultant. Does the consultant possess the necessary technical degree or multiple degrees? Do they possess a Master's or PhD degree? What credentials does the consultant have? Are they a recognized national expert who has published papers in the area? From what universities did they graduate? Next the company will look at the experience base of the consultant. What other projects was the consultant hired for and by what companies? Does the consultant have any references that can be contacted? If you can show the client you offer a complete package of the best technical approach, education, credentials, and experience to get the job done you are going to be a successful consulting engineer.

SUMMARY

Consulting engineers typically work for themselves or may be associated with a consulting firm. Consultants own their business and have one product to offer themselves. Basically, they collect a fee for their time, knowledge, expertise, and experience for companies to solve problems. Engineering consultants often play a multi-functional role such as technical expert, advisor, specialist, manager, or part-time employee.

There are some very important skills you must possess or acquire if you are going to become a successful consultant. They involve your technical skills, business skills, and general work ethic. These include:

Unique and market demanded skills

Self-motivated and willing to work hard

Business and social skills

Work independently, limited resources, and networking skills

Excellent writing, speaking, and presentation skills

Analytical and problem solving

Consulting engineers are awarded contracts on presenting a complete solution to the problem. A solution that solves the problem as quickly as possible, cost effectively, and has a high confidence of working. All backup by a proven track with the proper credentials.

Have you identified any career actions you want to take as a result of reading this chapter? If so, please make sure to capture these ideas before you forget by recording them in the notes section at the back of the book.

ASSIGNMENT AND DISCUSSION TOPICS

1   What do consultant engineers have to offer clients?
2   Name some key characteristics of successful consulting engineers.
3   Why is having business skill so important for consulting engineers?
4   How do you determine if you have a unique skill to offer?

REFERENCES

1. "What is a Consultant," *Imagineeringezine,* http://www.imagineeringezine.com/e-zine/consulting.html, October 4, 2008.

2. "Benefits of Being A Consultant," *Imagineeringezine,* http://www.imagineeringezine.com/e-zine/conpractice.html, October 4, 2008.

3. Greebaum, Thomas L., *The Consultant's Manual,* John Wiley & Sons, 1990.

4. "So You Want to Become A Consulting Engineer," *Imagineeringezine,* http://www.imagineeringezine.com/e-zine/wanabe.html.

5. "What is a P.E. License?," *Imagineeringezine,* October 4, 2008, http://www.imagineeringezine.com/e-zine/whatpe.html.

CHAPTER 51

SETTING UP AND BUILDING YOUR CONSULTING BUSINESS

Starting any new business takes a considerable amount of planning and resources and starting your own consulting business is no different. If you want to be successful there are many things to consider. How to get started? Do you work part-time or full-time? Do you work alone or team with others? Where do you find the jobs and how do you bid on the work? And last but not least, how do you get paid? This chapter provides the guidance you will need to answer these questions and successfully launch your consulting business.

HOW TO GET STARTED

The first decision you will face in the consulting business is whether to do consulting part-time or full-time when you start. The answer to this question will depend upon each individual, their skills, experience level, and ultimate goals.

Starting Part-Time. One option is to start a consulting business by working it part-time at night or on weekends or for a few days during the week [1]. Some people simply cannot afford to quit their present position due to financial reasons. These people build up their consulting business slowly over time until they can successfully leave their job to do consulting full-time. Starting a consulting business part-time allows one to try consulting to make sure this is what they really want to do on a permanent basis while minimizing the financial risks. If the person ultimately decides that consulting is not for them it provides a safety net since they have their existing job to fall back on.

Starting Full-Time. Starting your consulting business on a full-time basis is the other option. You leave your present job and start consulting full-time. This is the sink or swim option where you must quickly develop your

The Engineer's Career Guide. By John A. Hoschette
Copyright © 2010 John Wiley & Sons, Inc.

business for income. One benefit to this approach is the fact that you will be highly motivated to seek out new clients since you will have no other job or income to fall back on. Another benefit of quitting your job and starting full-time is that it allows you to focus all your energies on the business and not be distracted by an existing job or other activities. Working full-time also speaks loudly to potential clients about how serious you are about the business. Some engineers start their consulting venture this way because they have been laid off and cannot find work; this may be the only option available to them.

If you select to go the route of quitting your job and consulting full-time, if possible make sure you have at least one or two clients lined up and under contract before you turn in your resignation notice to your present employer. Starting your consulting business with several paying clients launches your business on a solid foundation.

Teaming with Other Consultants. Another great way to start your consulting career is finding and teaming with other consultants who are already established and successful. These consultants have established clients and know the tricks of the trade as they say. You do not need to team with someone who you are in direct competition with, but a consultant who has complementary talents. For instance, if you are electrical and have a hardware focus you might consider teaming with a software engineer. Or if you are electrical with a focus in motors, you might team with a mechanical engineer. Successful consultants are always looking for other consultants who have complementary skills to team with to enhance the overall chances for winning contracts. They also realize if a client is already hiring consultants for one reason there is probably a good chance they will hire other consultants for other complementary tasks.

Join a Consulting Firm. Before going out on your own, you might consider joining an established and successful consulting firm. You will have the opportunity to observe and learn how to market yourself, do client prospecting, bid jobs, and complete work. Oftentimes when working for a consulting firm, your assignments are short in duration and you rapidly move from job to job. This process of moving from job to job exposes you to new and improved methods, as well as seeing what does not work. Having learned what worked at one company and successfully being able to apply this to other companies is exactly what makes consultants so valuable and sought after.

Teaming with other consultants or joining a consulting firm has many benefits for someone trying to break into the business. By doing either of these you do not have to start from scratch and invent the wheel all over again. You are exposed to the in-house proven methods of successful consulting that you can adopt and utilize when you are ready to operate independently.

Regardless of the method you select for starting your business, the best action you can take in all cases is to join a consulting network. Conduct an online search for "engineering consultants network" to determine potential networks that you can join in your area.

Join a Consulting Network. Most likely all your formal training has been in engineering and you probably have had little formal training in business development and consulting. This lack of training can hurt your chances for successfully starting up your consulting business. Therefore, returning for more business training is also a critical part of successfully starting up your consulting business. Unfortunately, most people do not have the time to return to school and earn a four-year business degree while consulting at the same time.

So how do you acquire these business skills? There is no better way than to join a consultant network. Consultant networks are set up to help engineers with the business side of consulting [2]. These networks offer focused classes on nearly every aspect of the consulting business from how to find clients, bid contacts, do the work, and the most important getting paid.

The engineering consultant networks often meet after business hours and are well attended by other engineering consultants. In addition, many companies will contact engineering consultant networks to find consultants. There is no better place to network than at a consultant's network meeting where you can obtain business training, tips, and leads for potential jobs all in one place. Attending engineering consultant network meetings regularly for about the first 6 months after you start your consulting business is a smart move. Check with your local engineering society to determine when and where the consultant's network meets. IEEE Consultants Network is one example of a highly developed and successful networking society spread throughout the United States [3].

This gets us to the next action recommended to successfully launch your consulting business, which is developing a business plan.

DEVELOPING A BUSINESS PLAN

Writing a business plan requires you to think through all the aspects of your consulting business and formulate a plan for success. The business plan identifies the focused, concise, and predetermined actions you intend to follow when starting and conducting your business. It will help you shape the type of engineering consulting you will be offering and what your practice does and does not do.

There are many excellent and free sources that can help engineering consultants develop a business plan [4–6]. Here is one example of a business plan outline. It starts with the executive summary where in one spot the key points of your entire plan are summarized. This is a statement of your consulting business objectives, mission, and keys to success.

Generalized Outline for Business Plan

Executive summary
Objectives, mission, keys to success
Company summary
Ownership
Start-up plan
Location
Description of services
Technical areas of consulting
Unique capabilities
Tools and support material
Market analysis/Marketing and sales plan
Companies and industry needs
Size of market
Revenue—potential for income
Competition
Marketing budget
Networking and teaming plans
Management summary
Office and legal considerations
Financial—expenses and cash flow
Bidding contracts
Risk summary and prioritize actions

The next section is the company summary where you identify the ownership, start-up plan, and locations you will be operating the business out of. Next is the description of services you will be providing. This is a summary of your technical area of consulting, what unique capabilities you offer, and the tools or support material needed.

Following this is the marketing analysis section where you identify the industry and companies you will be providing consulting. Part of this analysis is determining the size of the market in terms of consulting jobs available and the potential income for you. The other part of this section is analysis of your competition. Who are your competitors, what do they have to offer, and why are you better or different?

The next section will define your marketing and sales plan. Writing this section requires you to think through how you are going to market yourself to potential clients. Are you going to advertise, put up a website, or cold call? Where is your money and time best spent to find jobs and develop a client base?

The next section is how you are going to manage your business' day-to-day activities. Where is your office going to be located, how are you going to handle bidding proposals, working, and what other services might you need—like tax and accounting services? What operating cash and capital are you going to need? What are the expenses and revenue? What is the projected cash flow of your business? And how do you intend to bid contracts, fixed price, time and material or some other method?

All good business plans include a risk management section. In this section, you identify the major risks associated with your consulting business and the actions you plan on taking to controlling and minimizing these risks. It is usually unmanaged or unforeseen risks and problems that ultimately result in you being unsuccessful. Identifying these risks upfront, and having a plan for dealing with them, is vital to being successful. What are some typical risks? Here is a list of some typical risks you should consider when starting your consulting business.

1. Lack of income early in the business.
2. Underestimating your costs or amount of time needed to complete jobs
3. Chasing unqualified leads
4. Identifying opportunities and turning them into actual paying jobs
5. The length of time it takes to actually get under contract (1 to 3 months)
6. The large amount of marketing and networking required
7. Working on a job while looking for your next job

Writing a good business plan is so important because it becomes the roadmap for operating your consulting business. In putting together this plan it will force you to think through your actions and hopefully identify what the optimum actions you can take which lead you to success. You should complete your business plan prior to launching your business.

Once you have a good business plan your activities should focus on finding and bidding jobs. During the normal course of business you will find some of your business plans will not work and therefore adjustments will be needed. For this reason, it is good to update your business plan on a yearly basis.

Defining Your Consulting Business

Defining your consulting business and exactly what services you are going to provide to what markets, as well as what you are not going to provide, is key to a successful business. The first consideration is how specialized the consulting service you are providing. As shown in Figure 51-1, when you have a generalized expertise you are offering it is highly likely that there will be many more clients looking for this, but also more competition from other consultants and lower pay. A good example is consulting in an Information Services (IS) market where you work networking the computers of a company or maintain servers. Or another example might be as a C++ software engineer developing code for companies. In either case, you have a generalized skill

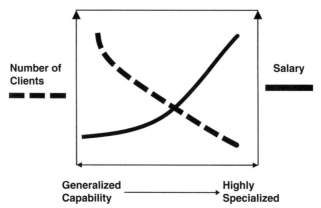

FIGURE 51-1 Generalized versus specialized consulting consideration for your business.

that many companies could utilize. For these types of consulting services usually experience is more important than possessing an advanced research degree.

Contrast this generalized consulting focus with an example of being a highly specialized consultant. For example, let's assume you have a highly specialized expertise in fuel efficiency of commercial aircraft jet engines. In this case there may be only two or three companies interested in hiring you and if you work for one company in the market you will most likely not be working and consulting with others in this industry. However, since you are so specialized, it is highly likely that your pay is going to be significantly more. Most companies which are hiring specialized consulting services want to know they are getting the best out there, so they are going to be looking at credentials and expecting advanced degrees.

When times are good, being a specialized consultant in one industry has big rewards. However, if economic conditions turn bad for the industry you are consulting in, this can be devastating for your business.

Which brings us to the next area of clearly identifying what skills and talents are you going to offer as part of your consulting business. You should generate a list of your top 3 or 4 skills that you want to offer as part of your business. Next do some market research as to what industries these skills are needed in and if jobs exist. If you map these skills as shown in Table 51-1, you can quickly identify which skills and markets you should focus your business on.

For this very simple example you can see that industry 2 has multiple skills you can market and skill number 2 is a good skill to market in all industries. Correspondingly, skill 1 shows no mapping into any of the markets you have identified and, therefore, indicating this may not be a good skill to build your business on.

TABLE 51-1 Skills to Industry Map

Skills	Industry 1	Industry 2	Industry 3	
1				
2	X	X	X	Same skill multiple markets
3		X		
4		X		
		Multiple skills in same market		

CHECKING OUT THE COMPETITION AND LEARNING FROM OTHERS

You are not going to have the time to learn everything from scratch and more than likely there are going to be other engineers offering consulting services similar to you. There is a folklore that says "keep your friends close and your enemies even closer." Well, your competition is not really an enemy but you should know what you are up against and the best way to do this is research their capabilities. How do you best check out what the competition is offering? Pretend you are interested in acquiring consulting services based on what you are offering and start a search to find consulting engineers who offer services in this area. The best place to start is with an on-line search. You should be searching using the keywords that you plan on using for your consulting service. Check out the competitor's web sites. What services do they offer, how do they bid their time, who is or were their clients and, what industries are they working in? How is their web site set up, hosted, and organized? Do they reference papers or books written? What society memberships and credentials do they hold? More than likely your clients will be searching in the same fashion, seeing what you are seeing. So knowing in advance what they are going to find can help you set up your web site. Is the competition also going to be checking you out? Absolutely!

Spending several days online searching web sites, job postings, engineering journal ads, newspaper ads, and engineering consultant networks can provide a wealth of information to help you come up to speed quicker and significantly enhance your chances for success.

SUCCESSFUL MARKETING PLANS

What is your plan for marketing your consulting business? A good marketing plan involves significantly more than just developing a web site and often developing a company web site is only a small portion of the plan. Jay Conrad Levinson in his book, *Guerrilla Marketing: Secrets for Making Big Profits from Your Small Business,* is a must-read for a consulting engineering [7].

In his book Jay identifies some great ways to market your business on limited budgets and gives many practical tips on how to get more for your marketing dollars. Here are some of Jay's recommended methods plus others.

1. Networking and marketing on an eye-to-eye basis
2. Seminars and demonstrations
3. Trade shows
4. Cold calling (phone lists)
5. Company web site
6. Classified advertising (trade journals, society publications)
7. Emailing
8. Direct mailings
9. Brochures

All the experts agree that the most successful method of marketing is by networking and talking face-to-face with potential clients about your business. For this reason the best thing you can be doing is networking with people and marketing your business whether it is face-to-face or over the phone.

Where are the best places to network? Engineering society meetings, trade shows, and conferences where other engineers are attending. Nothing is more impressive than presenting a paper on a difficult problem you solved to a group of fellow engineers, all who are looking for solutions for their problems.

If you belong to engineering societies you may have access to phone and email lists. There are also companies that sell contact lists; however these lists are often expensive and not filtered for your type of clientele and only yield a few leads for thousands of names. Use caution when purchasing contact lists and check how up-to-date the contact information is and how the list can be filtered to your business needs.

Having a company web site is another good marketing tool. The web site should present your qualifications or biography, the focus of your business and any supporting material like customer endorsements, papers you have written or books, you have published. In addition, make sure your contact information is easy to find and, if appropriate, any links to other web sites. Having your company brochure on the web site to view and also download is another good marketing tool. Having your web site hosted where your domain name ends in your company name.com projects a more professional image. There are many web sites that allow you to do this for very minimal costs and have support tools that allow you to easily construct your web site. To find these sites simply search online.

Placing an ad in engineering trade journals and/or society publications is another good way to market your company. These publications go directly to your target audience and are often read by people looking for consulting help.

Sending emails is the least expensive method of marketing if you already have an email contact list. Emails provide you with a means to send a colored company brochure for little cost directly to a potential client. Regular mail is another method of marketing but involves quite a bit more since you will need a contact list, printed hard copies of your marketing brochure, and paying postage. This method should be reserved for potential clients you have qualified as good candidates.

Your company brochure is the heart of your marketing plan. The brochure should summarize all the important aspects of your offering, contact information, and project a professional and polished image. This marketing piece will be used on your web site, sent via email, and left with clients after interviewing, as a reminder of your business. Developing an outstanding brochure is where you want to focus your early efforts when starting up your company.

In addition, you will need marketing tools like business cards and a 30-second commercial about your business. Your business card should have your contact information on the front and a summary of your company's offerings on the backside. Your 30-second commercial is how you tell people what you do and the outstanding points about your business. You use your 30-second commercial when networking and greeting potential clients.

Have you also considered a company name and logo? Your company name should be easy to remember and indicate the type of business you are in. Company names are registered with federal and state governments; make sure you check with these agencies so that you have not selected a name that is already in use by another company.

SETTING UP AN OFFICE

The least expensive way of setting up your office is by using your home. If you have a room where you can shut the door and work without being interrupted, it is ideal. To support your business you should have a good computer/lap top with Internet and email access, a printer with scanner capability, a separate business phone line, and faxing capability. Make sure you have a professional sounding voice mail so that clients can leave a message when you are out of the office. If possible a separate business cell phone is good to have, so clients can reach you no matter where you are. You should have a complete set of software to support both the business and technical aspects of your business. If you have to meet with clients you can use hotel lobbies, libraries, or quiet restaurants.

Once you are established and have sufficient income flowing in, you can move your home office to a professional building where you rent a small office and have support staff answering phone calls, handling your mail, and conference rooms for meetings and presentations.

Regardless of your office location you should set up regular business hours and work during those hours. Avoid using office hours for doing chores or running errands.

Meet with a banker and set up a separate banking account under your company name for depositing checks and paying bills. A separate bank account is one of the items the Internal Revenue Service (IRS) looks for when auditing your taxes. It is good to establish a relationship with a banker in the event that you need to get a loan for your company. Having a pre-established bank account and line of credit will make getting the loan much easier.

You will have to register your company with federal and state agencies and obtain a tax identification number. This can be your Social Security number to start with. Many corporations require you to have a federal tax identification number to consult for them. You will also have to declare the type of business you are running: sole proprietor, partnership, limited liability corporation (LLC), or corporation. The tax laws and legal liabilities for all these are different and this is where a good attorney or tax accountant can advise you.

Once your business is up and running it is recommended you meet with an accountant to help set up the business books and a tax specialist to help you with tax payments. As an independent consultant you must pay taxes on the income you earn. A tax specialist can also advise on what expenses of your business are tax deductible and what receipts you will need to keep.

Another place to check is the Small Business Administration for federal and state governments to see if you qualify for any programs that could support your business. These agencies are especially looking to help small businesses that are women- or minority-owned.

Finally, you should check with your local municipalities or city hall to see if there are restrictions on running a business out of your home. Some municipalities request you fill out a questionnaire regarding your business. They are worried about the environmental hazard aspects of your business and the impact of operating a business out of your home on the surrounding neighbors; especially if you are going to have a large number of clients calling on you regularly.

WHERE TO FIND JOBS

The next challenge you will face in launching your own consulting business is where to find the jobs. The answer to this question will depend on the skills you offer. However, generic to all consulting businesses, there are places you should be searching on a regular basis to find leads. Here are some of these places.

Job Ads. One good source for a consulting job is the job ads. Companies will post job ads on their web site or advertise in newspapers. When job openings are hard to fill and run for multiple weeks or even months most companies will consider hiring a consultant to fill the position. If you have the skills to fill the position, then you are in excellent shape to approach the company about consulting. When you find job ads like these, you should contact the hiring manager directly if possible, and discuss filling the opening as a consultant. Acting in your favor is the fact that you can start immediately and do not have to quit another job. This allows the company to instantly fill a position that has been open for a long period of time.

One of your regularly planned weekly tasks will be to search company web sites for postings and newspapers for positions that fit your background.

Consulting Company. Consulting companies are always looking for engineers they can hire for existing and future jobs. These consulting companies run job ads and also like to collect resumes of many consultants so that they can list these people as part of their firm's talent bank. Therefore, searching the job ads and web sites of consulting companies is another good place to look for jobs. If during your searching you find something related to your business, set up a meeting with the consulting firm and interview.

Many consulting firms have standing agreements and contracts with corporations to supply help. Therefore, these consulting firms are usually the first place corporations go to when considering hiring help.

Corporation Contractor Department. Many corporations have a policy to hire a certain percentage of contract personnel because of the up and down fluctuations in the workload. Corporations like the ability of bringing on temporary contract personnel to help with the workload. When work is completed they simply let them go. This is much simpler than hiring permanent employees and then having to deal with lay-off issues when the work ends. To facilitate this activity many corporations create a special group in their procurement department that does nothing but hire part-time contractors. The procurement group is responsible for putting out notices and bid requests for contract personnel. Meeting with this procurements group and being listed as a qualified contractor for a corporation gets your name on the list of people to contact when work becomes available. By using your marketing skills to penetrate a corporation and making contact with the contractor procurement group. This group is another excellent place to look for jobs.

Government Small Business Agencies. Federal and state government agencies have a preferred small business bidders list they maintain and automatically send bid notices when procuring contractors and consultants. You can have your company added to the list by contacting these agencies. In addition, these lists of small businesses are also sent to major corporations who bid contracts where oftentimes a percentage of the bid has to be awarded to small businesses.

Business Section of Newspaper and Trade Journals. Another good place to search for jobs is in your local newspaper or any trade journals of your industry. Local newspapers and trade journals will carry stories of major contracts won by local companies. Due to this big contract win and sudden surge in business, these companies may be looking for consultants or contractors to help. Monitoring newspaper and trade journal announcements to identify companies with newly awarded large contracts and then searching the company's web site for jobs or contacting the procurement department are also excellent places to look for jobs.

Engineering Society Meetings. When engineers are having troubles at work one of the places they go for help is an engineering society technical meeting in hopes of finding someone who might have knowledge about the problems they face and how to best solve the problems. Therefore, engineering society technical meetings are another excellent place to look for jobs. And volunteering to present at these technical meetings is an excellent marketing plan. Attending several society meetings each month is a good marketing plan.

Trade Show and Conference. Trade shows and conferences are also excellent places to find jobs. Many people attend these events with the desire to see the latest and greatest advances or find solutions to their problems. Having a small booth may be cost prohibitive for a single consultant but sharing the cost of a booth with five or six other consultants may be well within your financial means. By having a booth at a large trade show you can expect to get exposure to thousands of engineers in a single day.

Since you may not know exactly where your next consulting job may come from, you need to be be prepared to discuss your business no matter where you go. Having your business card and company brochures with you at all times, as well as developing a 30-second commercial, are the best marketing tools in being prepared to find work.

One Contact Should Lead to Another and Another. Finally, one last tip to share is that you are going to experience many "NOs" during your search for finds job. Please do not take the NOs personally and when you receive one it should automatically trigger the response, "Do you know of anyone else who would be interested in my capabilities or can you recommend where I look?" It is highly unlikely you will ever run out of leads if you use this technique. Every contact could lead to at least one or two others naturally creating more leads and enhancing your chances of finding the next job.

BIDDING AND SUBMITTING PROPOSALS

There are several common methods of bidding proposals which include fixed price, time and material, or progress payments. A fixed-price bid means you are bidding the cost of your work for one fixed price regardless of the amount and time it takes to accomplish the objectives. Most companies like to award a fixed-price contract because they know the work will get done for a certain price and costs will not escalate if there are problems. If you bid a fix-price contract you should make sure the scope of work is well defined. You can easily achieve the goals and from your prior experience you will know in advance how long it will take [8]. If you are uncertain about completing any aspect of the work you should add what is referred to as "uncertainty cost factor" to your bid. This usually involves adding 20% to 25% of additional cost to your estimate to cover any unforeseen problems in completing the work.

Another option to bid a proposal is time and material. For this type of bid you simply quote an hourly rate you will charge. The company buys your services and pays you for each hour of work you complete regardless of whether or not the objectives are met. Companies prefer not to do this since it provides little incentive for consultants to work quickly. In addition, the company agrees to reimburse you for any material expense you incur.

The next method of bidding a contract is progress payments. For this type of contract the company pays you just on the progress you make. Your pay is directly tied to key accomplishment or milestones planned as part of the work. As each milestone is successfully completed you receive a payment in an amount predetermined for the milestone. On very long contracts this allows you to receive payments during the contract and you do not have to wait until the very end to get any payments. This type of contract may have milestones that are bid on fixed prices tasks, time and material tasks, or a combination of both.

Make sure when you submit a bid that you are fully responsive to the requirements of the bid and you bid in your costs plus a profit which gets us to the next part of determining how much should you bid.

There are many ways to compute what you should bid as your costs. The following is one example and by no means the only way. The purpose of showing this example is to help you understand what to consider when bidding proposals to ensure you recoup your costs plus make a profit.

The hourly rate is obviously one important proposal bidding factor. If you consider yourself as capable of earning $80,000 salary working for a company and you can live comfortably on this and willing to do consulting at this income level, you can estimate a hourly rate using the following method.

$$\$80,000/(52 \text{ weeks} \times 40 \text{ hours per week}) = \$38.46 \text{ per hour.}$$

Now you need to add an overhead factor to cover all your costs associated with having a business and paying for your own benefits, taxes, legal, and accounting fees. Most companies add an overhead rate of approximately 60% to 80%. This means your hour rate to cover your benefits and overhead costs is

$$\$38.46 \times 1.8 = \$69.23 \text{ per hour}$$

However, this does not take into effect the fact that you are not going to get consulting jobs that will cover 100% of your time for the year. If you only get three consulting jobs during a year and only work 75% of the time possible, this means you need to increase your hourly rate to cover the dead time between jobs. This will bump up your hourly estimate to

$$\$69.23/(.75) = \$92.30 \text{ per hour}$$

Finally, you are in the consulting business not to just cover your costs but to make a good profit in doing so. Assuming you wish to make a 30% profit for all your time and effort as a bonus, this increases you hourly estimate to

$$\$92.23 \times 1.3 = \$120 \text{ per hour.}$$

If your desire is not to simply replace your present income of $80,000 but to make significantly more then you are going to have to charge significantly more per hour. If you would like to double your income having your consulting business, then you would have to double the hourly rate to $240 per hour.

Hopefully this simple example highlights to you all the elements you need to consider when determining your hourly rate. What do consultants bid their hourly rate at? This is considered highly proprietary information and most consultants will not share this information with you. Your hourly rate is going to be based on market demand, your skills, and what you are willing to work for which is strictly a personal decision.

My purpose in sharing this with you is to get you to realize that there are many factors to consider when determining your hourly rate and it is not unusual for consulting engineers to bid an hourly rate of $200 to $300 per hour and get it. Very successful consulting engineers are able to bid hourly rates 2× to 4× these rates because clients are willing to pay for their expertise and what they have to offer. With the risk of sounding trite, it is all about supply and demand.

A general rule of thumb for bidding material is to multiply the cost of any material you purchase by 1.5 to cover taxes, shipping cost, and any loan interest you might incur.

Two items to consider when submitting a proposal are the amount of time you spend on the proposal and how much should you tell the company about solving the problems. If you are bidding a proposal all alone, you do not want to spend more than a few hours putting the bid together unless you are able to re-coup your proposal costs as part of the bid. Most companies do not pay consultants to submit a proposal and when you do not win the contract this is lost time and effort.

The next item of concern is putting too much detail into the proposal which may result in giving the client the answer they need to solve the problem without ever hiring you. You are actually hurting your chances for getting hired. You can get away from giving them too much detail by discussing your approach to the problems and how this successfully solved other similar problems rather than detailing out the solution. The detailed solution will be shared once under contract.

STATEMENT OF WORK, SPECIFICATION, AND THE CONTRACT

The three documents used by most corporations to help define the scope of work for bidders are the statement of work, specification, and contract. The statement of work normally defines all the tasks to be bid and a schedule for when the work is to occur. The specification is a document that describes the technical requirements placed on the product or work and how the performance will be either tested or demonstrated. The contract is the document that normally defines the terms and condition for completing the work. In this document instructions are usually provided on how to submit costs, how the work will be funded, and special contractual items that should be considered. It is well worth your time to read and study these documents so you understand what will be required of you prior to signing on the dotted line. If you are unsure about items in the documents get clarification from the client and even seek legal advice if you have to.

During the proposal submittal and contract negotiations, items often change and agreements are reached that change conditions of the contract or might affect your bid. If this is the case, make sure you get it in writing and have updated and signed documents. If in the future, during the course of your work, you run into trouble, the only legal foundation you will have to support your side of the story are these signed documents.

Another item of concern for the consulting engineer is who holds the copyrights or patent rights to work completed during the contract. Most companies, as part of the contract, will require you to sign over all copy and patent rights to the company. You should be prepared for this and in the event you disagree, have an alternate plan you can propose. If you already hold copy and patent rights, you should highlight and document these prior to signing the contract. This will clarify any confusion as to what rights you hold prior to the start of the contract.

COMPLETING THE WORK

Once you have been selected and accepted the work, your efforts turn toward completing the work. If you are working on-site at the client's company, then your progress and ability to report issues will be easier to do than if you are working off-site. If you are working off-site, you should make an effort to meet with the project leader at least once a week where both of you can sit down uninterrupted and you report your progress and issues. Regardless if you are on-site or off-site, you should be submitting a weekly report that summarizes your progress, any issues of contention, and what is being done to resolve them. This written record is extremely valuable for documenting what was accomplished. With your weekly actions documented it is easier to write a final report at the end of the project highlighting all you have accomplished.

GETTING PAID—IT'S NOT ALWAYS GUARANTEED

Most companies will make payments 30–60 days after the invoice is submitted, approved, and processed for payment. You can mail an invoice to a company only to have it sit on the approving manager's desks for weeks only because they were too busy to approve it and submit for payment. Even once it has been approved it still may take 1 to 2 more weeks to process and mail the check. This means, under normal circumstances, a consultant can expect it is going to take 30–60 days to receive payment once they complete the work and submit an invoice for payment. The bottom line for the consultant is the job is not done when they submit the invoice, but must continue to check for payment.

The best thing you can do as you enter the final days of the contract is draw up an invoice and go over it with the approving manager of the company. This action will allow you to get agreement on the terms of payment and the amount prior to officially submitting it. If there are issues, it gives you time to work out the issues and hopefully not slow down the payment. Taking this action may also save considerable time in the approval process. It will also give you a clear indication of whether you are at risk of receiving a payment.

There are indications a client or company may not be able to make payments or even try negotiating down your cost [9].

1. The funding of the project was severely cut prior to completion.
2. The approving manager, company owner, or executive bonuses are tied to the project.
3. The funding of the project was tied to a new successful product launch that did not occur.
4. There is a sudden and unplanned workforce reduction in the company.

The best way to avoid these disaster scenarios is by weekly meetings with the project manager and starting, prior to the end of the contract, to discuss the conditions for the final payment.

SUMMARY

There are many things to consider when thinking about becoming a consulting engineer. First on the list is how to best start a consulting business, part-time or full-time, work independently or team with other consultants. Next is developing a business plan that clearly spells out how you are planning to operate your business and details out the actions you intend to take. It is your business plan for success. One key component to this business plan is clearly defining the skills and services you plan on offering. When the skills you intend to offer match the needs of the paying clients you now have the right combination for a successful consulting business.

Obtaining your first clients will require you to develop a marketing plan and start the search. Be ready at any time to talk about your consulting business and target the high probability place to find jobs. Develop and use your business tools, office, business cards, web site, and marketing brochures. Increase your chances for success when starting by minimizing costs and maximizing profits. Determine your hourly consulting rate and how you plan on bidding proposals fixed price, time and material, or progress payments.

Be aware of the legal documents companies utilize when asking for proposals and awarding work: statement of work, specification, and contract. Clarify and get agreement on all conditions of contract before you start and get it in writing.

Once the work starts and you are making progress as planned, documenting results and getting paid becomes important. Getting paid does not always go smoothly and may require special action on your part. Be ready!

There is no better place to learn the skills of the trade than through an engineering consultants' network. Check in your local area for consultant networks you can call upon for assistance.

This chapter is by no means inclusive of every item one should consider for running a successful consulting business, but is meant as an overview of the key items to consider when thinking about becoming a consultant. The intent is to make you aware of actions to take and provide guidelines to follow for successfully launching your consulting business. Owning your own business is wonderful and hopefully you are motivated to continue on with your dream to becoming a successful consulting engineer.

Have you identified any career actions you want to take as a result of reading this chapter? If so, please make sure to capture these ideas before you forget by recording them in the notes section at the back of the book.

ASSIGNMENTS AND DISCUSSION TOPICS

1 What should you consider when thinking about becoming a consulting engineer?
2 What is better: working alone or teaming with another consultant?
3 What benefits does teaming with another consulting engineer offer?
4 How important is it to define your business?
5 What are the elements of a successful marketing plan?
6 Where do you find jobs?
7 How do you determine your consulting fee?
8 What actions should you take while you are completing the work?

REFERENCES

1. Greebaum, Thomas L., *The Consultant's Manual*, John Wiley & Sons, 1990.
2. IEEE Consultants Network, http://www.ieeeusa.org/business/default.asp.
3. IEEE Local Network, http://www.ieeeusa.org/business/localnetwork.asp.

4. WWW.Score.org Business Plans, http://www.score.org/template_ gallery.html? gclid=CPLg8aLF8JgCFQG7GgodnVc81A.

5. Good-to-Go Business Plans, http://www.goodtogobusinessplans.com/.

6. Free Engineering Consulting Business Plan, Morebuisness.com, http://www. morebusiness.com/engineering-consulting-business-plan.

7. Levinson, Jay Conrad, *Guerrilla Marketing: Secrets for Making Big Profits from Your Small Business*, Houghton Mifflin Company, 1998.

8. Sokal, Nathan O., "Fixed-Price Versus Time-and-Materials Consulting Contracts: Be Sure You Know What You're Getting Into," *IEEE USA Today's Engineer*, May 2001, http://www.todaysengineer.org/archives/te_archives/may01/te2.asp.

9. Sokal, Nathan O., "Consulting—How to Make Sure You Get Paid for Your Work," *IEEE USA Today's Engineer*, March 2001, http://www.todaysengineer.org/ archives/te_archives/mar01/te1.asp.

APPENDIX *A*

CAREER PLANNING WORKBOOK

Industry expert
Patents
Technical analysis
Customer interface
Product develop
Advanced technical degree

Technical vs Management

People problems
Profits and cost
Motivating people
Presentations
Marketing
Job reviews
Team building
Business degree

Staff Engineer

Company-wide expert
Technical papers
Technical leader

Manager

E5

Team dynamics
Budget and scheduling
Customer interface
People skills

E4
E3

Small team leader
Broaden technical skills
Work with and direct other people

Individual contributor
Develop technical skills
Follow orders

E2
E1

Figure A-1 Long-range careering planning.

The Engineer's Career Guide. By John A. Hoschette
Copyright © 2010 John Wiley & Sons, Inc.

My long-range goal is to follow the path of: _____

(Technical versus Managerial)

Visualization of Your Ideal Job

1. Job title: _____

2. Job salary: _____

3. Company location: _____

4. Home location: _____

5. Commute time: _____

6. Company size: _____

7. Company products:

_____

_____

8. Size of engineering group: _____

9. Job functions: _____

10. Office size: _____

11. Company benefits: _____

12. Freedom to work on: _____

_____

_____

_____

13. Travel: _____

14. Laboratories look like: _____

15. Personal computer: _____

16. Supervisor who: _____

17. Coworkers who: _____

18. Career advancement paths leading to: _____

19. In three years I'll be doing:

_____

_____

_____

20. In five years I'll be doing:

_____

_____

_____

21. In 10 years I'll be doing:

_____

_____

_____

Visualization of Your Worst Job

Things I absolutely will not put up with on the job:

_____

_____

_____

_____

_____

If technical career path, I need to:

_____

_____

_____

_____

_____

Suggested Actions. Return for more technical training, sign up for classes, talk with senior technical people in the company, publish my work, develop new modeling and design skills, broaden my technical knowledge to include other engineering fields, software, chemical, electrical, mechanical, and computer science.

If managerial career path, I need to:

_____

_____

_____

Suggested Actions. Return for more management training, sign up for classes, talk with senior management people in the company, develop my team leadership skills, develop my people skills and dealing with people problems.

If undecided I need to:

_____

_____

_____

_____

Suggested Actions. Talk to both senior technical and management people about their jobs. Determine which is more appealing, technical or managerial. Take assignments involving both technical and management tasks. List the pros and cons of each.

Identify Your Present, 3, 5, and 10 Year Career Goals

This Year	3 Years	5 Years	10 Years
Goal:	Goal:	Goal:	Goal:
Actions Required to Meet This Years Goal: 1 2 3 4 5 Actions for 3 Year Goal 1 2 3	Actions Required to Meet Goal: 1 2 3 4 5 Actions for 5 Year Goal 1 2 3	Actions Required to Meet Goal: 1 2 3 4 5 Actions for 10 Year Goal 1 2 3	Actions Required to Meet Goal: 1 2 3

Years from Graduation	0–5 Years First Job	5–10 Years Early Career	10–20 years Mid Career	20–30 Years Late Career	30–40 Years Retirement
Age	22–30	30–35	35–45	45–55	55+
Company/ Technical Career Goals	Adjust to work environment Learn company ropes Enhance technical training Career planning	Focus on technical specialty Development team skills Higher levels of responsibility Publish papers Return for MBA Lead product development	Tech vs Business Develop leadership skills Update training Supervisor Return for MBA Publish papers Patents	Continue leadership development Technical update Upper management Mentoring junior people Senior role in company/staff	Leveling of career and responsibility Consulting role Teach classes
Personal and Family Goals	Payoff college debt Have fun Financial planning plan for retirement Elder care giving New car	Marriage Purchase home Start family Elder care giving Plan for retirement	Family vacations Child development Child school activities Elder care giving Plan for retirement	Family vacations Children college Elder care giving Plan for retirement	Very late for planning of retirement Wedding of children Grandkids Less pressure

Figure A-2 Typical career actions based on age, circle your age group, circle the goals you want to accomplish.

Based on my age, the career actions I have identified are:

_____

_____

_____

_____

_____

_____

_____

_____

_____

Great Career Actions

- Join an engineering society
- Host a society meeting at your company
- Visit the university bookstore
- Have a career discussion session with your supervisor or mentor
- Write a technical paper
- Attend a conference
- Attend a seminar
- Apply for a patent
- Write lessons learned memo
- Do a product demo
- Write an article for company paper
- Do a new product demo
- Submit a team for an award
- Volunteer to improve something in your company
- Return for more schooling and/or advance degree
- Learn a software program
- Write a report
- Read new technical journals
- Learn about the job level above yours
- Help organize engineers week
- Update your resume
- Take a time management course
- Get your work done early ahead of schedule
- Develop a new simulation/model

Write down the career goals you have identified:

Description   Date to Accomplish

1. _____

2. _____

3. _____

4. _____

5. _____

6. _____

7. _____

Great Family/Social Actions

- Plan family vacations
- Sunday night family meetings
- Date night with spouse
- Attend weekly services
- Plan birthdays
- Plan anniversaries
- Plan volunteer work
- Attend your kids school functions
- Coach a sports team
- Exercise (stress relief)
- Make it home for dinner more often
- Spend "special time" with your kids, computers, read to them, art projects
- Financial planning
- Visit library together

Write down the personal family/social goals you have identified:

Description   Date to Accomplish

1. _____

_____

2. _____

_____

3. _____

_____

4. _____

_____

5. _____

_____

6. _____

_____

7. _____

_____

Skills Assessment Worksheets: Technical (page 1)

Type of Skill	Strength or Weakness	Past Accomplishment Demonstrating Strength or Weakness	Action Required	Date	Priority Level High, Medium, Low
Technical					
Product design					
Technical knowledge in your field					
Technical knowledge in other fields					
Product build					
Product integration					
Laboratory test					
Laboratory research					
Technical publications					
Computer modeling					
CAD design and modeling					
Analysis and modeling					

Skills Assessment Worksheets: Technical (page 2)

Type of Skill	Strength or Weakness	Past Accomplishment Demonstrating Strength or Weakness	Action Required	Date	Priority Level High, Medium, Low
Technical					
Experimental research					
Patents					
Technical awards					
Programming					
Producibility					
Manufacturing					
Others					

Skills Assessment Worksheets: Leadership and Project Management (page 3)

Type of Skill	Strength or Weakness	Past Accomplishment Demonstrating Strength or Weakness	Action Required	Date	Priority Level High, Medium, Low
Project Management					
Setting year goals					
Budgeting					
Organizing teams					
Developing policies					
Developing procedures					
Cost tracking					
Project planning					
Customer interface					
Team formation					
Salary administration					
Department budgeting					

Skills Assessment Worksheets: Leadership and Project Management (page 4)

Type of Skill	Strength or Weakness	Past Accomplishment Demonstrating Strength or Weakness	Action Required	Date	Priority Level High, Medium, Low
Project Management					
Capital planning					
Presentation skills					
Running meetings					
Handling multiple priorities					
Establishing milestones					
Generating team metrics					
Team communication					
Coaching skills					
Mentoring					
Others					

Skills Assessment Worksheets: Interpersonal Skills (page 5)

Type of Skill	Strength or Weakness	Past Accomplishment Demonstrating Strength or Weakness	Action Required	Date	Priority Level High, Medium, Low
Interpersonal Skills					
Motivating					
Team leadership					
Conflict resolution					
Work relationships					
Meeting skills					
Versatility					
Team dynamics					
Communication style					
Customer relationships					
Social abilities					
Mentoring					

Goals Worksheets

Career/Personal Goal	Actions Required	Date	Priority Level High, Medium, Low
	1 2 3 4 5		
	1 2 3 4		
	1 2 3		

Goals Worksheets

Career/Personal Goal		Actions Required	Date	Priority Level High, Medium, Low
	1			
	2			
	3			
	4			
	5			
	1			
	2			
	3			
	4			
	1			
	2			
	3			

Career and Personal Goals Calendar for _____

Jan__Update Plan_____   Jul_____
_____   _____
_____   _____
_____   _____

Feb_____   Aug_____
_____   _____
_____   _____
_____   _____

Mar_____   Sep_____
_____   _____
_____   _____
_____   _____

Apr_____   Oct_____
_____   _____
_____   _____
_____   _____

May_____   Nov_____
_____   _____
_____   _____
_____   _____

Jun__Update Plan_____   Dec_____
_____   _____
_____   _____
_____   _____

SELF-INVENTORY WORKBOOK

Job Abilities and Skills Inventory

Type of Skill	Strength or Weakness	Past Accomplishment Demonstrating Strength
Technical		
Product design		
Product build		
Laboratory Test		
Laboratory research		
Technical publications		
Computer modeling		
CAD design and modeling		
Analysis and modeling		
Experimental research		
Patents		
Technical awards		
Programming		
Producibility		
Manufacturing		

The Engineer's Career Guide. By John A. Hoschette
Copyright © 2010 John Wiley & Sons, Inc.

Project Management		
Planning		
Budgeting		
Organizing		
Developing policies		
Developing procedures		
Cost tracking		
Schedule planning		
Customer interface		
Team formation		
Salary administration		
Department budgeting		
Capital planning		
Presentation skills		
Interpersonal Skills		
Motivating		
Team leadership		
Conflict resolution		
Work relationships		
Meeting skills		
Versatility		
Team dynamics		
Communication style		
Customer relationships		
Social abilities		
Mentoring		

Visualization of Your Ideal Job

1. Job title:

_____

2. Job salary:

_____

3. Company location:

_____

4. Home location:

_____

5. Commute time:

_____

6. Company size:

_____

7. Company products:

_____

8. Size of engineering group:

_____

9. Job functions:

_____

10. Office size:

_____

11. Company benefits:

_____

12. Freedom to work on:

_____

13. Travel:

_____

14. Laboratories look like:

_____

15. Personal computer:

_____

16. Supervisor who:

_____

17. Coworkers who:

_____

18. Career advancement paths leading to:

_____

19. In 5 years I'll be doing:

_____

Visualization of Your Worst Job

Things I absolutely will not put up with on my next job:

_____

_____

_____

_____

My significant   Abilities and skills
accomplishments are: demonstrated

1. _____   1. _____
   _____   _____
   _____   _____
2. _____   2. _____
   _____   _____
   _____   _____
3. _____   3. _____
   _____   _____
   _____   _____
4. _____   4. _____
   _____   _____
   _____   _____
5. _____   5. _____
   _____   _____
   _____   _____
6. _____   6. _____
   _____   _____
   _____   _____

Figure B-1 Inventory of significant accomplishments.

LIST OF ENGINEERING SOCIETIES

Industry Specific Societies

Air and Waste Management Association www.awma.org

Alabama Society of Professional Engineers www.alspe.com

American Academy of Environmental Engineers www.aaee.net

American Association of Engineering Societies www.aaes.org

American Ceramic Society www.acers.org

American Chemical Society www.acs.org

American Ecological Engineering Society www.aeesociety.org

American Institute for Medical and Biological Engineering www.aimbe.org

American Institute of Aeronautics and Astronautics www.aiaa.org

American Institute of Chemical Engineers www.aiche.org

American Institute of Mining, Metallurgical, and Petroleum Engineers www.aimehq.org

American Institute of Physics www.aip.org

American Nuclear Society www.ans.org

American Society for Engineering Education www.asee.org

American Society for Engineering Management www.asem.com

American Society for Healthcare Engineering www.ashe.org

American Society for Healthcare Engineering www.hospitalconnect.com

American Society for Quality www.asq.org

American Society of Agricultural Engineers www.asae.org

The Engineer's Career Guide. By John A. Hoschette
Copyright © 2010 John Wiley & Sons, Inc.

American Society of Civil Engineers www.asce.org

American Society of Gas Engineers www.asge-national.org

American Society of Heating, Refrigerating, and Air Conditioning Engineers, Inc. www.ashrae.org

American Society of Materials, International www.asminternational.org

American Society of Mechanical Engineers www.asme.org

American Society of Naval Engineers www.navelengineers.org

American Society of Safety Engineers www.asse.org

American Society of Test Engineers www.astetest.org

American Water Resources Association www.awra.org

American Water Works Association www.awwa.org

Architectural Engineering Institute www.aeinstitute.org

Association for Computing Machinery www.acm.org

Association for Facilities Engineering www.afe.org

Association for the Advancement of Medical Instrumentation www.aami.org

Association of Conservation Engineers www.conservation.state.mo.us/enginnering/ace

Association of Energy Engineers www.aeecenter.org

Association of Engineering Geologists www.aegweb.org

Association of Iron and Steel Engineers www.aise.org

Association of Scientist and Professional Engineers www.aspep.net

Audio Engineering Society www.aes.org

Biomedical Engineering Society www.bmes.org

Connecticut Society of Professional Engineers www.ctspe.net

Electrochemical Society www.edaf.com

Environmental Engineering Geophysical Society www.eegs.org

Florida Engineering Society www.fleng.org

Georgia Society of Professional Engineers www.gspe.org

Human Factors and Ergonomics Society www.hfes.org

Illinois Society of Professional Engineers www.illinoiseengineer.com

Illuminating Engineering Society www.iesna.org

Indiana Society of Professional Engineers www.indspe.org

Institute of Biological Engineering www.ibeweb.org

Institute of Electrical and Electronics Engineers www.ieee.org

Institute of Industrial Engineers www.iienet.org

Institute of Noise Control Engineering of the USA www.inceusa.org

Institute of Transportation Engineers www.ite.org

International Council on Systems Engineering www.incose.org

International Federation for Medical and Biological Engineering www.ifmebe.org

International Society for Optical Engineering www.spie.org

International Society for Pharmaceutical Engineering www.ispe.org

International Solar Energy Society www.ises.org

Iowa Engineering Society www.iaengr.org

Junior Engineering Technical Society www.jets.org

Kansas Society of Professional Engineers www.kansasengineer.org

Kentucky Society of Professional Engineers www.kyengcenter.org

Louisiana Engineering Society www.les-state.org

Maryland Society of Professional Engineers www.MOSPE.ORG

Minerals Metals and Materials Society www.tms.org

Minnesota Society of Professional Engineers www.mnspe.org

Missouri Society of Professional Engineers www.mspe.org

National Academy of Forensic Engineers www.nafe.org

National Association of Corrosion Engineers www.nace.org

National Council of Structural Engineers Associations www.dwp.bigplanet.com/enginners

National Institute of Ceramic Engineers www.ceramics.org

National Society of Professional Engineers www.nspe.org

Nevada Society of Professional Engineers www.nspenv.org

North Dakota Society of Professional Engineers www.ndspe.org

Oceanic Engineering Society www.oceanicengineering.org

Ohio Society of Professional Engineers

Oklahoma Society of Professional Engineers www.ospe.org

Optical Society of America www.osa.org

Pennsylvania Society of Professional Engineers www.pittsburgpe.org

Pennsylvania Society of Professional Engineers www.pspe.org

Society for Biological Engineers www.aiche.org/sbe/

Society for Mining, Metallurgy, and Exploration www.smenet.org

Society for the Advancement of Material and Process Engineering www.samepe.org

Society of American Military Engineers www.marcorengasn.org

Society of American Military Engineers www.same.org

Society of Automotive Engineers www.sae.org

Society of Cable Telecommunication Engineers www.scte.org

Society of Chemical Industry www.soci.org

Society of Fire Protection Engineers www.sfpe.org

Society of Flight Test Engineers www.ceas.sungsb.edu

Society of Hispanic Professional Engineers www.shpe.org

Society of Manufacturing Engineers www.sme.org

Society of Motion Picture and Television Engineers www.smpte.org

Society of Naval Architects and Marine Engineers www.sname.org

Society of Petroleum Engineers www.spe.org

Society of Photo-optical Instrumentation Engineers www.worldcat.org

Society of Photographic and Industrial Engineers (SPIE) www.spie.org

Society of Plastics Engineers www.4spe.org

Society of Professional Engineering Employees in Aerospace www.speea.org

Society of Rehabilitation Engineering www.resna.org

Society of Reliability Engineers www.sre.org

Society of Software Engineers sse.se.rit.edu

South Carolina Society of Professional Engineers www.scspe.org

Standards Engineering Society www.aes.org

Tennessee Society of Professional Engineers www.tnspe.org

Texas Society of Professional Engineers www.tspe.org

Utah Society of Professional Engineers www.uspeonline.com

Washington Society of Professional Engineers www.washingtonengineer.org

Water Environment Federation www.wef.org

Woman- and Minority-Focused Societies

American Indian Science and Engineering Society www.aises.org

Association for Women in Computing www.awc-hq.org

Chinese Institute of Engineers USA www.cie-usa.org

National Society of Black Engineers www.nsbe.org

Society of Hispanic Professional Engineers www.shpe.org

Society of Mexican-American Engineers and Scientists www.maes-natl.org

Society of Women Engineers www.societyofwomenengineers.org

Vietnamese Association for Computing, Engineering Technology, and Science www.vacets.org

Supporting Organizations

MESA (Mathematics Engineering Science Achievement) www.mesa.ucop.edu

National Action Council for Minorities in Engineering www.nacme.org

National Association of Pre-College Directors (focuses on STEM education/programs) www.jhuapl.edu./NAPD/contents

National Association of Minority Engineering Program Administrators www.nameepa.org

National Coalition of Underrepresented Racial and Ethnic Advocacy Groups in Engineering and Science www.ncourages.org

National Consortium for Graduate Degrees for Minorities in Engineering and Science (GEM) www.wasnd.edu.gem/gemwebapp/gem_00-000. htm

SECME (Southeastern Consortium for Minorities in Engineering) www. secme.orgw

Women in Engineering Programs and Advocates Network www.wepan. org

APPENDIX D

NOTES

INDEX

The Engineer's Career Guide. By John A. Hoschette
Copyright © 2010 John Wiley & Sons, Inc.

W